Dairy Sheep Nutrition

Dairy Sheep Nutrition

Edited by

G. Pulina

Department of Animal Science, University of Sassari, Italy

Technical reader of English translation

R. Bencini

University of Western Australia, Australia

CABI Publishing

CABI is a trading name of CAB International

CABI Head Office
Nosworthy Way
Wallingford
Oxfordshire OX10 8DE
UK

Tel: +44 (0)1491 832111
Fax: +44 (0)1491 833508
Email: cabi@cabi.org
Web site: www.cabi.org

CABI North American Office
38 Chauncy Street
Suite 1002
Boston, MA 02111
USA

Tel: +1 617 395 4056
Fax: +1 617 354 6875
Email: cabi-nao@cabi.org

A catalogue record for this book is available from the British Library, London, UK.

Library of Congress Cataloging-in-Publication Data
Alimentazione degli ovine da latte. English
Dairy sheep nutrition / edited by Giuseppe Pulina; technical proofreader, Roberta Bencini
p. cm
Includes bibliographical references and index.
ISBN 0-85199-681-7
1. Sheep—Nutrition. 2. Sheep—Feeding and feeds. 3. Ewes—Nutrition. 4. Ewes—Feeding and feeds. 5. Sheep milk.
I. Pulina, Giuseppe. II. Bencini, Roberta. III. Title.

SF376.A46 2004
636.3'08'5—dc22

2004006555

ISBN-13: 978-0-85199-681-3
ISBN-10: 0-85199-681-7

First published 2004
Transferred to print on demand 2015

Printed and bound by CPI Group (UK) Ltd, Croydon, CR0 4YY

Contents

Contributors

Giovanni Annicchiarico (g.annicchiarico@tin.it) has a degree in Agricultural Science from the University of Naples. After working on the metabolic profile of farm animals and improving the productive and reproductive efficiency of the Mediterranean buffalo, he became a researcher at the Istituto Sperimentale per la Zootecnica di Roma, sezione di Foggia, Segezia (Experimental Institute for Animal Husbandry of Rome, Foggia Section, Segezia). He is presently working on the nutritional requirements of dairy sheep, their welfare, the quality of their milk, and the chemical, physical and sensorial characteristics of the typical dairy products Canestrato Pugliese and Caciocavallo Podolico.

Marcella Avondo (mavondo@mbox.fagr.unict.it) is Professor of animal husbandry at the Faculty of Agriculture at the University of Catania, Italy. Her main areas of research are: biochemical and management techniques for improving silage quality and production from forages and animal by-products; and feeding behaviour of dairy sheep raised in extensive and semi-extensive farming systems, with particular attention to the factors that influence intake capacity and pasture selection.

Gianni Battacone (battacon@uniss.it) has a degree in Agricultural Science from the University of Sassari, Italy, and a PhD in Animal Husbandry from the University of Perugia, Italy. His research focuses on the relationship between animal feeding and mycotoxin carry-over into animal products, and on the quality of sheep and goat milk. He is a researcher at the Animal Science Department of the University of Sassari.

Roberta Bencini (rbencini@agric.uwa.edu.au) has 12 years of experience in research on sheep milking. Her career in this area started in 1988 when she commenced her PhD on lactation in sheep at the University of Western Australia. Previously, she had graduated in Agriculture at the University of Milano (Italy) and had been a tenured lecturer in Dairy Science and Technology at the Professional Institute for Agriculture of Villa Igea, Lodi, Italy. She successfully directed to completion several research projects on sheep dairy research funded by the Rural Industries Research & Development Corporation (RIRDC, Australia). Currently she is a Senior Lecturer in the School of Animal Biology at the University of Western Australia.

Paolo Brandano (brandano@uniss.it) is Professor of Animal Science of the Faculty of Agriculture of the University of Sassari, Italy. His main areas of research are: characteristics and management of rustic cattle and sheep breeds; techniques of raising ruminants; artificial feeding of lambs and kids; and grazing models for farms. At present he lectures in Animal Husbandry.

Antonello Cannas (cannas@uniss.it), Associate Professor of Animal Science at the Faculty of Agriculture of the University of Sassari, Italy, has a degree in Agricultural

Science from the same University, and a MSc and PhD in Animal Science from Cornell University, Ithaca, NY, USA. He works on several aspects related to feeding and nutrition of small ruminants: feed evaluation techniques and energy and protein requirements, with the focus on the development of prediction equations and models for estimating requirements and animal performances; ruminal turnover and its prediction; urea as a nutritional indicator; and comparative nutrition of small and large ruminants. He teaches 'Sheep and goat production' and 'Feed processing and evaluation' at the Faculty of Agriculture of the University of Sassari.

Aldo Cappio-Borlino (genetica@uniss.it) is Associate Professor of Animal Breeding and Genetics at the Faculty of Agriculture of the University of Sassari. His main scientific interest lies in the mathematical modelling of the regular and continuous component of dynamic processes in the field of Animal Science, such as the temporal evolution of milk production and ruminal degradation of feeds. He proposed an original model for the mathematical description of the lactation curve of dairy sheep, which has been cited in several international scientific papers and has also been used by many researchers. Recently he developed a methodology to analyse time series of milk yield and quality in order to disentangle deterministic and stochastic components.

Mauro Decandia (mdecandia@tiscalinet.it) has a degree in Agricultural Science from the University of Sassari and has been an animal husbandry researcher at the Istituto Zootecnico e Caseario per la Sardegna (IZCS) (Sardinian Institute for Animal Husbandry and Dairy Farming) since 1991. He specialized in the feeding of farm animals at the Institut National Agronomique Paris-Grignon (INAPG) (National Agronomy Institute of Paris-Grignon, France). He has carried out research on the intake and feeding behaviour of grazing sheep and goats. In particular, he has worked on the tannin content of Mediterranean maquis and the feeding behaviour of goats.

Giuseppe Enne (enneg@tin.it) is Professor in the Animal Husbandry Department of the Faculty of Agriculture of the University of Sassari. He works mainly on: techniques for raising farm animals and their reproduction; environmental impact of farm animals; and the quality of the products of farm animals. He is coordinator of numerous national and international research groups. He is a member of committees involved in evaluating European research projects and is President of the Istituto Sperimentale Italiano 'Lazzaro Spallanzani' per la Fisiopatologia della Riproduzione, Milan (The 'Lazzaro Spallanzani' Italian Experimental Research Institute for the Physiopathology of Reproduction, Milan).

Nicola Fois (nfois@tiscalinet.it) has a degree in Agricultural Science from the University of Sassari and has been a researcher at Istituto Zootecnico e Caseario per la Sardegna (IZCS) (Sardinian Institute for Animal Husbandry and Dairy Farming) since 1991. He has carried out research on the nutrition of grazing sheep, estimation of their pasture intake, measurement of their body composition, and their grazing behaviour. At present he is working on the problems involved in raising dairy sheep in the Mediterranean environment, with a focus on organic farming.

Alfio Lanza (malanza@mailbox.fagr.unict.it) is Professor in Animal Husbandry at the Faculty of Agriculture of the University of Catania. He works on feeding and nutrition of farm animals, with the following main interests: the suitability of using agri-industrial products and by-products as feeds; new feeding techniques for sheep and goats; animal physio-climatology; the demography of animal husbandry; optimizing systems for raising sheep and goats; and the possibility of introducing high-quality genetic material into Sicilian cattle and sheep. He was President of the Società Italiana di Patologia e Allevamento degli Ovini e dei Caprini (SIPAOC) (Italian Society for the Pathology and Husbandry of Sheep and Goats) until 1998.

Sebastiano Ligios (sligios@tiscalinet.it) has a degree in Agricultural Science from the University of Sassari and has been a researcher at Istituto Zootecnico e Caseario per la

Sardegna (IZCS) (Sardinian Institute for Animal Husbandry and Dairy Farming) since 1988. He spent 6 months at the Institut National de la Recherche Agronomique (INRA), Clermont-Ferrand, France (National Institute of Agricultural Research, Clermont-Ferrand, France). He has carried out research on the nutrition of grazing sheep, the estimation of pasture intake, the measurement of their body composition, grazing behaviour, and on the effect of the number of hours of daylight on milk production in sheep. In addition he works on agricultural systems and breeding techniques.

Lucio Lutri (llutri@inwind.it) has a degree in Agricultural Science and has been a researcher at the Istituto Sperimentale Zootecnico per la Sicilia, Palermo, Sicily, since 1994. He has been working since 1984 as Technical Assistant in Agriculture for the Regional Agriculture and Forestry Department, focusing on ruminant nutrition in intensive and extensive livestock systems. At the moment he is working mainly on feeding and management of dairy sheep and goats, with a focus on the prediction of nutrient requirements of lactating sheep and growing animals, on animal behaviour and on sheep and goat milk quality.

Nicolò Pietro Paolo Macciotta (macciott@uniss.it) is Researcher in Animal Breeding and Genetics at the Faculty of Agriculture of the University of Sassari. His main scientific interest lies in the application of mathematical modelling in several fields of animal science, with particular reference to animal breeding. He has worked on lactation curve modelling, problems in quantitative trait loci detection in farm animals, optimization of selection schemes for ruminants and applications of bioeconomic models for estimating economic values of milk production traits in sheep.

Giovanni Molle (gmolle@tiscalinet.it) has a degree in Agricultural Science and works as a researcher at the Istituto Zootecnico e Caseario per la Sardegna (IZCS) (Sardinian Institute for Animal Husbandry and Dairy Farming), where he coordinates the animal feeding and nutrition department. Using techniques learned from the Institute of Grassland and Environmental Research of Great Britain and the Agricultural Research Institute of Israel, he has conducted studies on the intake, feeding behaviour, energy costs and body composition of grazing sheep and goats. In addition he has studied the relationship between feeding and reproduction in ruminants, and in particular in dairy sheep.

Massimo Morgante (massimo.morgante@unipd.it) is Professor of Veterinary Medical Pathology in the Faculty of Veterinary Science at the University of Padova (Padua), Italy. His scientific interests are: evaluation of the health of the mammary gland; metabolic control of farm animals, with a focus on dairy sheep, goats and cows; the functioning of the kidney in cattle; the effects of treatment with vitamin E and selenium on mastitis in sheep; the study of the fractional clearances of electrolytes in dairy cows and their relationship with renal imbalances and calcium metabolism; and the haemato-chemical variations in alpaca (*Lama pacos*) during pregnancy and lactation.

Anna Nudda (anudda@uniss.it) has a degree in Agricultural Science from the University of Sassari, Italy, and a PhD in Animal Husbandry from the University of Perugia, Italy. She has worked on the quality of sheep and goat milk and on the physiology of lactation of sheep and goats. At present she is a researcher in the Animal Husbandry Department at the Faculty of Agriculture of the University of Sassari.

Giuseppe Pulina (gpulina@uniss.it), PhD in Animal Husbandry, University of Rome, is a Professor and head of the Animal Science Department in the Faculty of Agriculture of the University of Sassari. He has conducted research on sheep and goat production, with a focus on the quality of the products and on mathematical models for animal husbandry techniques, and on genetic selection and reproduction in dairy sheep and goats. He has been a lecturer in the Faculty of Agriculture at the University of Sassari since 1988 and at present teaches Animal Production in the Agricultural Engineering degree course. He is the national coordinator of the project 'Development of a diet programme for sheep and goats' of the Italian Ministry of Agriculture and Forestry. This book has been compiled from the findings of this project.

Salvatore Pier Giacomo Rassu (pgrassu@uniss.it) has conducted research on raising techniques for sheep, with an interest in reproductive techniques, such as the application of new techniques of artificial insemination, and the anticipation of puberty and of the first lambing in ewe-lambs. He is Associate Professor of Animal Husbandry in the Faculty of Agriculture of the University of Sassari. Since 1994 he has been teaching courses in Animal Production. From 1997 to 2001 he was coordinator of the Operative Unit in Sassari of the RAIZ project of the Italian Ministry of Agriculture and Forestry.

Maria Sitzia (msitzia@tiscalinet.it) has a degree in Agricultural Science from the University of Sassari. She has studied the relationship between the meteorological and physiological variables of grasses and legumes for her PhD. She was also trained at the University of Turin and at the Macaulay Land Use Research Institute of Edinburgh. She has conducted research on the eco-physiology of forage species and since 1993 has been a researcher in Animal Husbandry at the Istituto Zootecnico e Caseario per la Sardegna (IZCS) (Sardinian Institute for Animal Husbandry and Dairy Farming). She has conducted research on the methods of estimating biomass, the evolution of the vegetal structure of grazed pasture and the techniques involved in managing new types of forage. She is currently working on agricultural systems, with a focus on organic farming systems.

Luigi Taibi (l.taibi@isnet.it) has a degree in Agricultural Science and has been a researcher at the Istituto Sperimentale per la Zootecnica di Roma, sezione di Foggia, Segezia (Experimental Institute for Animal Husbandry of Rome, Foggia Section, Segezia) since 1975. He has conducted research on the genetics and feeding of sheep and at present is working on the nutritional and feeding requirements of dairy sheep, their welfare, the quality of their milk, and the chemical, physical and sensorial characteristics of the typical dairy products Canestrato Pugliese and Caciocavallo Podolico.

Timothy T. Treacher has a degree in Agricultural Science from Reading University. After gaining a PhD for studies on the nutrition of lactating ewes, he worked at the Grassland Research Institute, Hurley, UK, for 25 years on a wide range of research on the nutrition and grazing management of sheep. Travel and involvement with FAO, EU and European Association of Animal Production programmes brought him into contact with Mediterranean and North African sheep systems and research. Subsequently he worked at the International Centre for Agricultural Research in Dry Areas (ICARDA) in Syria and taught at the Universities of Córdoba and Pamplona in Spain.

Acronyms

A	Herbage accessibility
AA	Amino acids
ADF	Acid detergent fibre
ADPLS	Apparently digestible protein leaving the stomach
AF	Aflatoxin
AF-B1	Aflatoxin B1
AF-M1	Aflatoxin M1
AFRC	Agricultural Food Research Council
ALP	Alkaline phosphatase
AR(1)	First-order autoregressive process
ARMA	Autoregressive moving average model
AUD	Animal unit per day of grazing
B	Biomass
BCS	Body condition score
BF	Bite frequency
BHBA	β-hydroxybutyrate
BM	Bite mass
bST	Bovine somatotropin
BU	Blood urea
BVD	Bovine viral diarrhoea
BW	Bodyweight
BWC	Bodyweight change
CK	Creatine phosphokinase
CLA	Conjugated linoleic acid
Cn	Cell number
CNCPS	Cornell net carbohydrate protein system
CP	Dietary crude protein
CPD	Complete pelleted diet
CS	Compound symmetry
CSFA	Calcium soaps fatty acids
CSH	Compressed sward height
CSIRO	Commonwealth Scientific Industrial Research Organization
CU	Coefficent of sward utilization
CW	Clean wool daily production
D	Duration of grazing period
DBW	Bodyweight change
DIM	Days in milking
DM	Dry matter
DMA	Dry matter availability
DMI	Dry matter intake
DON	Deoxinivalenol
EB	Energy balance
EC	Energy concentration
ECF	Extracellular fluid
EE	Ether extract
EI	Energy intake
F	Fat
FA	Fatty acid
FCM	Fat-corrected milk
FEP	Faecal endogenous protein
FPCM	Fat–protein-corrected milk
FII	Forage index intake
FIL	Feedback inhibition of lactation
FPT	Failure of passive transfer
FSH	Follicle-stimulating hormone
FY	Fat yield
GE	Gross energy
GF	Greasy fleece annual production
GH	Growth hormone
GI	Gastrointestinal tract
GJ	Gap junctions
GnRH	Gonadotrophin-releasing hormone
GR	Growth rate of the pasture
GSH-Px	Glutathione peroxidase
GT	Grazing time
H	Herbage height
ha	Hectare
HM	Herbage mass
IARC	International Agency for Research on Cancer
IBR	Infectious rhinotracheitis
ICF	Intracellular fluid
Ig	Immunoglobulin

IGF	Insulin-like growth factor
IgG	Immunoglobulin G
I/M	Intramuscular
INRA	Institut National de la Recherche Agronomique
I/P	Intraperitoneal
IR	Intake rate of DM
I/V	Intravenous
LAI	Leaf area index
LCFA	Long-chain fatty acid
LDH	Lactate dehydrogenase
LH	Luteinizing hormone
MCFA	Medium-chain fatty acid
ME	Metabolizable energy
MEI	Metabolizable energy intake
MP	Metabolizable protein
MU	Milk urea
MUN	Milk urea nitrogen
MW	Metabolic weight
MY	Milk yield
NDF	Neutral detergent fibre
NE	Net energy
NEL	Net energy for milk production
NFC	Non-fibre carbohydrates
NMD	Nutritional muscular dystrophy
NP	Net protein
NPN	Non-protein nitrogen
NRC	National Research Council
NSC	Non-structural carbohydrates
OM	Organic matter
P	Milk protein (TN × 6.38)
PA	Plasminogen activator
PAI	Plasminogen activator inhibitor
PCV	Packed cell volume
PDI	Protein digestible dans l'intestin
PDIE	Energy-limiting PDI
PDIN	Nitrogen-limiting PDI
PG	Plasminogen
PI	Potential intake
PL	Placental lactogen
PMN	Polymorphonuclear neutrophils
PMSG	Pregnant mare serum gonadotrophin
PRL	Prolactin
PS	Particle size
PTH	Parathyroid hormone
PUFA	Polyunsaturated fatty acid
RUP	Ruminal undegradable protein
SA	Serum albumin
SCFA	Short-chain fatty acid
SICP	Selective index for crude protein
SR	Stocking rate
SRC	Substitution rate of the concentrate
SSH	Sward surface height
STH	Somatotrophic hormone
t	Tons
T	Grazing time
TD	Test day
TJ	Tight junctions
TMR	Total mixed ration
TN	Total nitrogen
TP	Milk true protein
TSH	Thyroid-stimulating hormone
TUS	Total utilizable substances
UEM	Unité d'encombrement mutton
UEP	Urinary endogenous protein
UFL	Milk forage unit
VFA	Volatile fatty acid
w	Secretory activity per cell

1 Milk Production

Giuseppe Pulina and Anna Nudda
Dipartimento di Scienze Zootecniche, Università di Sassari, Italy

1.1 Sheep Milk

Ovine milk is a complex of substances that occur in solution, in suspension or as an emulsion in water: fat and fat-soluble vitamins as an emulsion; protein and mineral salts bound to casein micelles in suspension; and carbohydrates (lactose), minerals, non-protein nitrogen compounds and hydrosoluble vitamins in solution.

The average composition of milk from ewes, compared to milk of the other three main ruminant species and with human milk, is presented in Table 1.1.

Ovine milk, compared to bovine and caprine milk, is characterized by: (i) higher protein and fat content, which makes ovine milk more suitable for cheese-making because protein is mainly responsible for cheese yield (Casu and Marcialis, 1966; Ricordeau and Mocquot, 1967), and fat influences both cheese yield (Campus, 1991) and the organoleptic characteristics of dairy products; (ii) higher opalescence, mainly due to the reflection of light on opaque particles in suspension (casein micelles, calcium, phosphate and citrate); (iii) greater whiteness, due to the lack of

Table 1.1. Average chemical and physical composition of ovine milk, compared with milk composition of the other three main ruminant species and with human milk.

	Sheep	Goat	Cow	Buffalo	Woman
Water (%)	82.5	87.0	87.5	80.7	87.5
Total solids (%)	17.5	13.0	12.5	19.2	12.5
Fat (%)	6.5	3.5	3.5	8.8	4.4
Ø fat globule (μm)	4.0	3.9	4.4	—	—
TN* (%)	5.5	3.5	3.2	4.4	1.1
Casein (%)	4.5	2.8	2.6	3.8	0.4
Serum protein (%)	1.0	0.7	0.6	1.1	0.7
Lactose (%)	4.8	4.8	4.7	4.4	6.9
Minerals (%)	0.92	0.80	0.72	0.8	0.30
Ca (mg/l)	193	134	119	190	32
Energy (kcal/l)	1050	650	700	1100	690
Density	1.037	1.032	1.032	1.030	1.015
Acidity (°SH)	8.5	8.0	7.1	10.0	—
pH	6.65	6.60	6.50	6.67	6.85
Freezing point (°C)	–0.580	–0.570	–0.524	–0.580	—

*TN: total nitrogen ($N \times 6.38$).

carotene in the fat; (iv) higher resistance to microorganism proliferation in the first hours after milking, due to its specific immunological properties; and (v) higher content of minerals, mainly calcium, which increases its buffer power (Luquet, 1985).

While the lactose content is fairly constant between breeds, milk fat and protein content (expressed as total nitrogen; TN = N × 6.38) vary greatly (Table 1.2). Variations are linked to genetic factors as well as to the environmental (and sometimes experimental) conditions in which the data are measured.

Table 1.2. Fat and protein content of ovine milk in certain dairy breeds (Nudda, 1996: Bencini and Pulina, 1997).

Breed	Fat (%)	TN* (%)
Aragat	5.70	5.49
Awassi	6.70	6.05
Chios	6.60	6.00
Comisana	7.5–10.6	5.9–10.4
Delle Langhe	6.75	5.95
Frisona dell'est	6.64	6.21
Karagouniki	8.70	6.60
Lacaune	7.14	5.81
Leccese	7.93–8.38	5.81–6.30
Manchega	9.07	5.43
Massese	6.79–7.44	5.48–5.96
Merino	8.48	4.85
Sarda	6.69	5.82
Tsigai	7.41	5.45

*TN = total nitrogen (*N* × 6.38).

The milk fat and protein content are 1.5- and 1.2-fold higher, respectively, than the lactose concentration, while the average protein to fat ratio is 0.81, similar to that reported for cows.

In order to reduce the differences in milk, protein and fat yields between animals, a method for standardizing milk production is necessary. Fat-corrected milk (FCM) or fat–protein-corrected milk (FPCM) are two methods of standardizing milk production for comparisons between sheep. Some of the different methods of standardizing proposed use 6.5% fat and 5.8% protein as a basis for conversion (Pulina *et al.*, 1989), but other reasonable fat and protein tests are equally useful:

$$FCM^{(6.5)} = M\,(0.37 + 0.097F)\quad (kcal/kg = 1020)$$

$$FPCM^{(6.5;\,5.8)} = M\,(0.25 + 0.085F + 0.035P)\quad (kcal/kg = 1047)$$

where: M = milk yield (kg); F and P = fat and protein concentration (%).

1.2 Milk Production

In nature, milk has the irreplaceable function of being the only nourishment source for the newborn until the digestive tract is able to digest solid food. The selection process carried out on dairy sheep by humans over thousands of years has resulted in a milk yield markedly higher than that required to feed the newborn lambs and, above all, an increase in lactation period.

1.2.1 The anatomical basis of milk production and lactogenesis

Lactation is a physiological process characterized by the synthesis, secretion, ejection and removal of milk. This physiological function is made possible by captation, or passive and active filtration, of organic and inorganic compounds and water from the blood fluid by the secretory epithelium of the mammary gland. The secretory cells, which are roughly cuboid in shape and markedly polarized in structure, are arranged to form spherical structures (alveoli), with a hollow lumen into which milk is secreted (Fig. 1.1).

The cells of the mammary secretory epithelium are joined together by numerous protein junction complexes (Fig. 1.2). These are composed, from basal to apical order, of: (i) desmosomes (D), whose function is to ensure firm adhesion between adjacent cells; (ii) gap junctions (GJ), which are communicating junctions that form pores between adjacent cells and permit the exchange of molecules with low molecular mass; and (iii) tight junctions (TJ), which encircle the cell and fuse adjacent cell membranes, closing

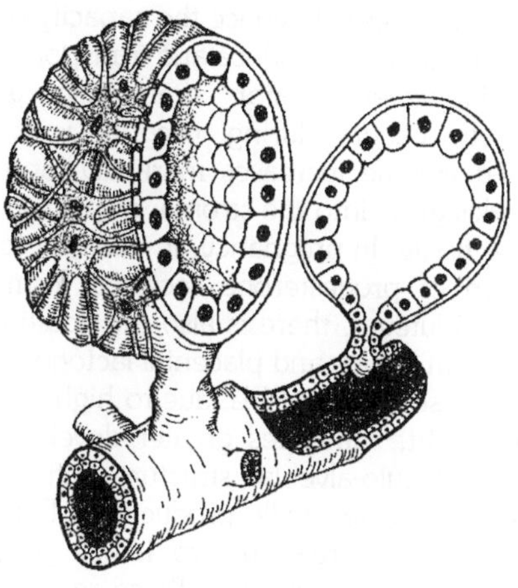

Fig. 1.1. Mammary alveolus (Naitana *et al.*, 1992).

off the intracellular space. The principal function of TJ is to form a barrier between extracellular fluid (ECF), in equilibrium with blood plasma, and the milk contained in the alveoli (Ruckebusch *et al.*, 1991).

The shape and polarization of the secretory cells implies the existence of an extensive system of structural support provided by the cytoskeleton. This is an intricate network of microfilaments which is firmly attached to the inner surface of the plasma membrane and envelopes, and is continuous with the various cytoblood organelles (Mepham, 1987).

The cytoskeleton influences many of the dynamic properties of secretory cells, such as the intracellular movement of molecules toward its apical part. For this reason the cytoskeleton is probably involved in the milk secretion mechanism. Indeed milk precursors (acetate, β-hydroxybutyrate, free fatty acids, glucose, amino acids, and minerals) are adsorbed on the basal membrane of the cells, while the neosynthesized compounds are secreted across the apical membrane into the alveolar lumen.

The mammary epithelium has different permeable characteristics during the different physiological states (Fig. 1.2). During pregnancy, or in the case of diseases such as mastitis or in other conditions (for example, when the intra-alveolar pressure is high due to excessive milk accumulation), the junctions become leaky, permitting the passage of substances between cells from ECF to mammary secretion and from milk to ECF (the paracellular route). During lactation or in conditions in which the integrity of TJ is maintained, the substances pass through the cells (the transcellular route).

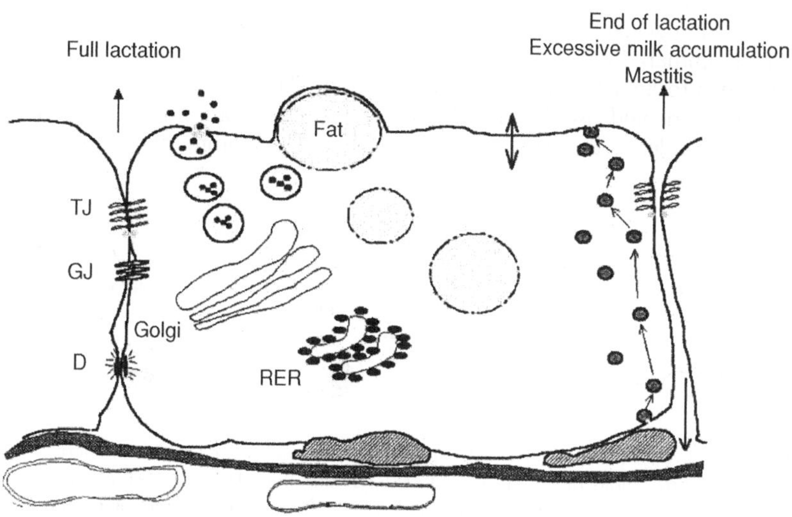

Fig. 1.2. Schematic representation of the mammary epithelium in different physiological states.

1.2.2 Mammary growth and differentiation

The growth and differentiation of the mammary gland is controlled by hormones. Three phases are distinguishable within the mammogenic cycle. These are: (i) the major growth phase during pregnancy, characterized by the formation of the lobulo-alveolar structure; (ii) the synthesis and secretory phases during lactation; and (iii) the phase of involution, characterized by the regression of the lobulo-alvelolar structure. The glands, which first appear in the embryo, have a growth rate prior to puberty similar to that of the rest of the body (isometric growth). After the onset of the ovarian cycle their growth rate is faster than that of the rest of body (allometric growth) (Anderson, 1975; Johnsson and Hart, 1985). Prior to puberty (0–5 months), the mammary gland is characterized by a modest extension of the duct system. With the beginning of puberty (at 5 months), the release of oestrogen, at first alone and then in combination with progesterone when the ovaries mature in their function, causes the glands to undergo dramatic changes characterized by development of the ducts and lobulo-alveolar system. The growth rate of lobulo-alveolar tissue and the duration of the allometric phase are considerably influenced by the nutrition level of the diet (Rattray *et al.*, 1974; Johnsson and Hart, 1985). High energy intake in ewe lambs prior to puberty causes the mammary fat pad to grow excessively and thus to limit the development of ductules (Tolman and McKusick, 2001). Any process that limits the development of these ductules will reduce the capacity of the mature gland to produce and transport milk.

However, major mammary gland growth occurs during pregnancy, when gland volume increases markedly, and the lobules grow into the stroma, displacing adipose tissue. In pregnancy, in addition to the action of progesterone secreted from the corpus luteum, there is also the addition of gonadotrophins and placental lactogen (PL). Higher secretion of PL, due to high placental weight, stimulates greater development of the lobulo-alveolar structure and consequently greater milk production. For this reason, ewes rearing two lambs produce 10–15% more milk than those rearing singles (Rossi *et al.*, 1996). Growth hormone (GH) and prolactin (PRL) (Binart *et al.*, 1999), which are secreted by the pituitary gland, play a relevant role in mammary gland differentiation. Chronologically the development of the lobulo-alveolar structure begins around the third month of pregnancy, continues slowly until the fourth month and speeds up around the fifth month, so that growth is essentially complete in dairy ewes at the time of parturition (Anderson, 1975; Sulochana *et al.*, 1991).

1.2.3 Synthesis of milk

Milk components are principally synthesized by the secretory cells of the mammary gland from precursors absorbed from the circulation. These precursors derive directly or indirectly from nutrients in the diet (Fig. 1.3). The mammary ephithelium also acts as

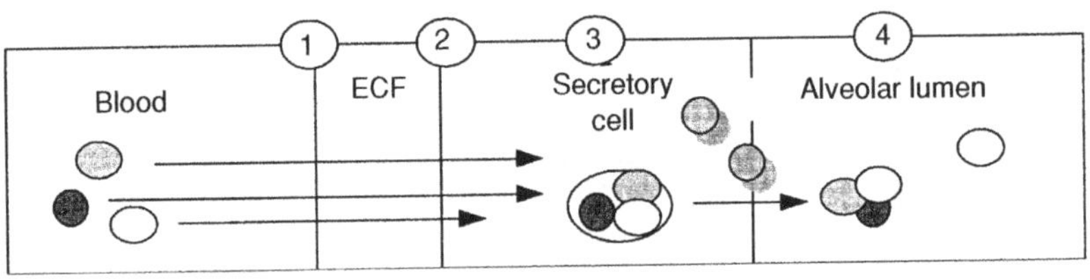

Fig. 1.3. A generalized scheme of milk synthesis and secretion (Mepham, 1987, modified).
1) Passage of precursor molecules from blood to extracellular fluid (ECF); 2) Absorption of precursor molecules from ECF into the secretory cells; 3) Intracellular synthesis of milk components and translocation of these substances through the cell to the apical membrane; 4) Secretion of substances across the apical membrane into the alveolar lumen.

a selective barrier, permitting the transfer of selected substances from the blood to the alveolar lumina without changes in their chemical composition.

Milk is iso-osmotic with blood plasma, but the two are not in chemical equilibrium: milk has more sugar (90-fold), calcium (Ca) (13-fold), phosphorus (P) (tenfold), fat (nine-fold), and potassium (K) (fivefold), but lower protein (0.5-fold) and sodium (Na) (0.15-fold), than blood. Osmotic pressure is chiefly determined by lactose and diffusible ions and there is generally an inverse relationship between lactose concentration and the sum of K and Na concentrations.

Milk fat is composed primarily of triglycerides, which are formed by the linking of glycerol and fatty acids synthesized on the outer surface of the smooth endoplasmic reticulum. Short-chain fatty acids (C6–C10) are built *de novo* within the mammary epithelial cells from acetic acid and β-hydroxybutyrate (BHBA) derived from ruminal fermentation (Peaker, 1977). Milk fatty acids of chain length C18 and above derive from blood lipids, which originate from fat intake in the diet and from mobilization of body fat. Medium-chain fatty acids (C12–C16) either derive from the lipids in the diet or are synthesized *de novo* in the mammary gland (Sauvant and Morand-Fehr, 1978; Mepham, 1987). The mammary gland, and the intestine to a lesser extent, have a desaturase enzyme complex with a high affinity for stearic acid. This is converted to oleic acid. Ovine milk is rich in short- and medium-chain fatty acids, such as caprilic (C8:0) and capric acids (C10:0), but also in conjugated linoleic acid (CLA), which has recently been attributed therapeutic properties (Banni and Martin, 1998).

The milk proteins (with the exception of serum albumin and immunoglobulins) are synthesized in the rough endoplasmic reticulum of the secretory cells from amino acids (AA) extracted from the blood. The AA derive from the post-ruminal digestion of microbial protein and non-degradable ruminal protein, from hepatic transamination reactions and, only when the protein concentration of the ration is scarce, from mobilization of tissue protein. Serum albumin (SA) and immunoglobulins (Ig) are not synthesized in the mammary gland, but pass directly from the blood into the milk. Their levels are very high in colostrum, the first milk secreted immediately after birth, and increase further during mammary gland inflammation and during involution. Non-protein nitrogen (NPN) derives from protein catabolism in the mammary gland and from the transport of analogous compounds from the blood. Urea, which is the highest NPN compound in milk (2–3% of total nitrogen), comes directly from the blood through diffusion, so there is a high correlation between its concentration in blood and milk (Cannas *et al.*, 1998). Poor-quality protein, excessive nitrogen and energy deficiency (which enforces mobilization of body protein stores) in the diet can increase blood urea concentrations through excessive deamination of AA and result in increased urea concentration in the milk.

Lactose is synthesized – by the lactose synthetase enzyme activity in the Golgi apparatus of the epithelial cells – from glucose extracted from the blood. Lactose is a non-permeable disaccharide whose molecule is too large to diffuse out of the Golgi or out of secretory vesicles, so it draws water by osmosis into the Golgi from the cytoplasm. The lactose concentration in the milk is relatively constant (on average about 4.8%) and the volume of milk is directly conditioned by its rate of synthesis (Peaker, 1977).

Minerals and vitamins come exclusively from the blood, and only their concentrations change in the mammary gland. The concentrations of many minerals in milk depend on their intracellular levels. For example Na, K and Cl diffuse very easily, while Ca, Mg, P and citrate diffuse only partly. It is difficult to change blood Ca levels, because they are very tightly regulated. So, because milk Ca comes from the blood, it is difficult to change milk Ca by increasing concentration of Ca in the diet. Chemical elements whose concentration in the milk can be increased by increasing the amount in the diet include I, B, Br, Co, Mn, Mo, Se and Zn.

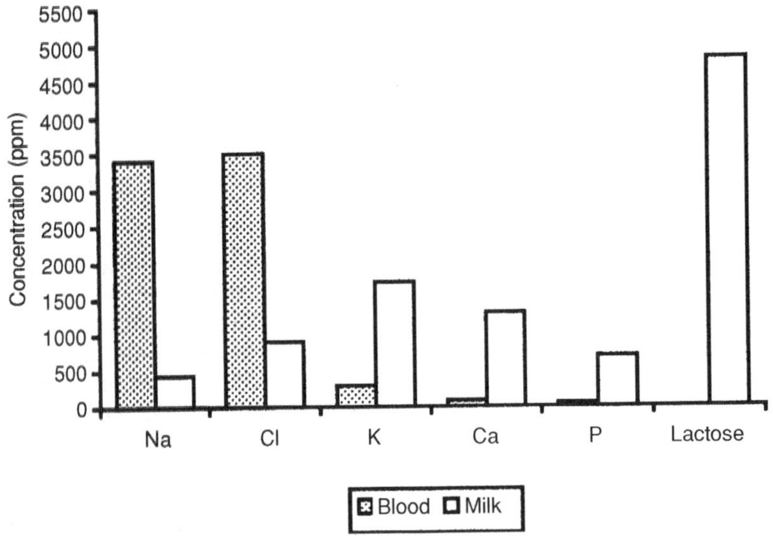

Fig. 1.4. Concentration of selected minerals and lactose (ppm) in blood and milk.

However, milk synthesis depends not only on the supply of nutrients to the mammary gland but also on the regulation of their use through physiological and endocrine controls. Insulin in particular, in interaction with other hormones such as GH and IGF-I, plays an important role in the control of nutrient use for milk synthesis.

1.2.4 Milk secretion

Milk secretion is the transport mechanism of neosynthesized compounds and substances extracted from the blood or interstitial fluid into the alveolar lumen.

Most of the components of the aqueous phase of milk are secreted by exocytosis. Proteins and lactose are packaged into secretory vesicles that move to the plasma membrane of the alveolar cells, where they fuse and release their contents into the alveolar lumen by exocytosis. Because the Golgi vesicles synthesize lactose from precursor galactose and glucose that enter from cytoplasm, and because the Golgi membrane is impermeable to lactose, the sugar is osmotically active and water is also drawn into the terminal Golgi vesicles. Secretory vesicles are thought to be the source of most of the constituents of the aqueous phase of milk, including citrate, nucleotides, Ca and phosphates.

The mechanism for fat secretion is unique to the mammary gland, because lipid droplets become enveloped in the apical membrane and are secreted into the alveolar lumen surrounded by a layer of plasma membrane composed of phospholipids (Fig. 1.2).

Mineral concentration is markedly different between milk and blood (Fig. 1.4). In milk, K, Ca and P concentrations are higher, and Na and Cl concentrations are lower, than in blood because of active regulation mechanisms called 'pumps' (Fig. 1.5a).

Three pumps in particular have been found: (i) the Na–K pump, which regulates the K osmolarity between cytoplasm and milk; (ii) the Ca pump, localized in the plasma membrane, which permits Ca to be transported from the basal membrane to the cytosol; and (iii) a second Ca pump, localized in the Golgi, which exchanges Ca to build casein micelles (Bingham *et al.*, 1993).

In some physiological conditions – such as during pregnancy, mastitis, during epithelial involution or excessive alveolar distension due to extended milking interval – the

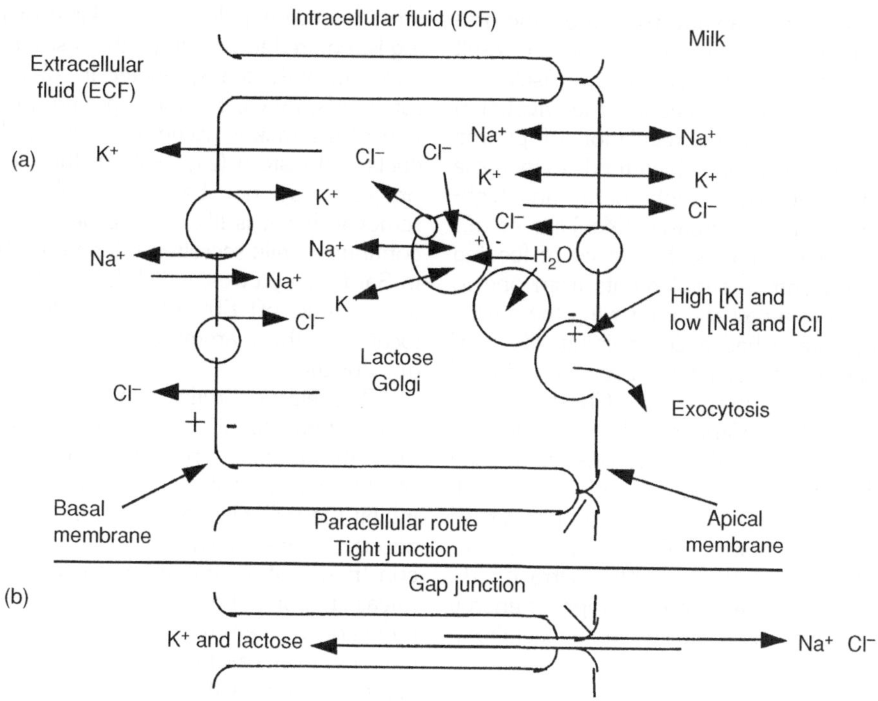

Fig. 1.5. The osmotic balance of the secretory epithelium (Peaker, 1997). (a) Schematic representation of movements of ions, lactose and water between extracellular fluid, intracellular fluid and milk. The directions of the differences in electric potential are indicated by (+) and (–) signs; (b) schematic representation of the paracellular movement of the ions through tight junctions (TJ) between the mammary cells in cases of inflammation of the secretory epithelium or in the final phase of lactation.

TJ become leaky and the paracellular pathway becomes active, thus allowing the passage of substances between epithelial cells, rather than through them (Fig. 1.5b). In such cases, lactose, K, P and citrate move from the lumen into the ECF, and compounds of the interstitial space pass unimpeded into the milk. As a result, Na and Cl concentrations increase in the milk while those of K and lactose decrease.

Because the secretory tissue becomes permeable, the concentration of serum albumin (SA) increases, and serum enzymatic activity (α-1-antitrypsin, *N*-acetyl-β-D-glucosaminidase, glutamic oxalacetic transaminase and γ-glutamil transferase) in milk also increases (Peaker, 1977; Ranucci *et al.*, 1985; Tesei *et al.*, 1988). The leakiness of the TJ affects the cytoskeletal activity, which reduces its dynamic properties in the transfer of neosynthesized compounds towards the apical mebrane of the secretory cells. The lack of secretion of milk components inhibits further synthesis and predisposes the epithelial cells to involution. The involution process involves many factors, in which the plasminogen/plasmin system plays a central role (Turner and Huynh, 1991). Plasmin is the predominant protease in milk and is derived from blood; it occurs in milk together with its precursor, plasminogen (PG), which is mainly associated with casein micelles. These form the substrate for its proteolytic activity. The conversion of plasminogen to plasmin is modulated by plasminogen activators (PA), and, indirectly, by plasminogen activator inhibitors (PAI). The plasma insulin-like growth factor (IGF-I), which acts as a mediator of growth hormone (GH), and the nutritional state of the animal also play a role in regulating PA. The presence of plasmin in milk is undesirable, because it can adversely affect the processing characteristics of the milk, and result in degradation of α-casein and β-casein.

Nutrition can influence the involution process through the regulation of plasma IGF-I concentration, which tends to increase when the intake of high-energy and high-protein diets is increased (McGuire *et al.*, 1992). Increasing the frequency of feeding with concentrate supplements from one to three times a day and improving the quality of the forage had a determining effect on increasing the plasma IGF-I concentration of ewes in late pregnancy (Chestnutt and Wylie, 1995). Variations in somatomedin response due to the nutritional state may partly explain the positive effects on milk yield and milk coagulation properties observed by some authors (Pulina *et al.*, 1993; Cannas *et al.*, 1995; Serra *et al.*, 1995) in dairy ewes whose supply of energy and protein was enhanced in late lactation.

1.2.5 Milk ejection

Milk ejection is the removal of milk from the alveoli through the contraction of the myoepithelial cells surrounding the alveoli and the small ducts when a neuroendocrine milk-ejection reflex occurs. The neuroendocrine mechanism is activated by sensory receptors in the teat at suckling or milking, and by conditioned stimuli of events preceding suckling or milking. The nerve signals reach the posterior lobe of the pituitary gland (neurohypophysis) via the spinal cord and the oxytocin is released into the blood from this gland. On reaching the mammary gland, oxytocin causes the myoepithelial cells surrounding the alveoli to contract, and the stored milk is forced down the galactophore and gradually reaches the cistern. In this way, the secreted milk, in the interval between suckling or milking, is distributed in different proportions between the alveoli and the small-diameter ducts (alveolar milk), and the large-diameter ducts and cistern (cisternal milk).

The increase in intramammary pressure following ejection is not great enough to overcome the resistance of the teat sphincter, so active expulsion by suckling or milking is required to overcome the resistance to milk flow of the teat sphincter. In some dairy ruminants, in which a much greater proportion of the milk is accumulated in the large ducts and cistern (e.g. >50%), the ejection process may not be so crucial to milk removal, but it is likely to be necessary for continuous milk secretion (Mepham, 1987). In Sarda dairy ewes, which have a high cistern volume, 60–80% of the secreted milk is located in the mammary cistern (Nudda *et al.*, 2000).

The ejection reflex can be induced by conditioned stimuli such as the sight and noise of the milking parlour or the presence of the milkers. Stress, pain or fear can inhibit the milk ejection reflex. In these situations, adrenalin, which is a potent vasoconstrictor of mammary vessels, may be released from the adrenal medulla. The reduction of mammary blood flow prevents oxytocin from reaching the mammary gland via the blood, with a consequent lack of ejection.

1.2.6 Removal of milk

Milk stored in the mammary gland can be removed by suckling or milking. Normally, the milk is first removed by the suckling of the lambs in the first few weeks after lambing in traditional sheep milking systems. After a few weeks, the milk that exceeds the nutritional requirements of the lambs is removed by milking, and after weaning all the milk is removed by milking only. Just prior to suckling or milking, the secreted milk is distributed in the mammary gland in two parts: 'alveolar milk' stored in the lumina of the alveoli, smaller ducts and ductules, and 'cistern milk' stored in the large ducts and in the cistern. The ratio of alveolar to cistern milk varies depending on the cistern volume, the breed and the physiological state of the animal. For example, the cistern milk fraction increases, and the alveolar milk fraction decreases, in the presence of a corpus luteum, because luteal oxytocin has a strong galactopoietic action, and therefore permits contact of the milk with the secretory cells for a shorter period of time (Labussière *et al.*, 1989).

Milk removal is essential for the maintenance of lactation. Accumulation of milk in the cistern and duct system, for extended milking intervals, can reduce the rate of milk secretion. For a long time this reduction of milk secretion rate was attributed to the increase in intramammary pressure due to accumulation of the milk in the mammary gland, which stretches the secretory epithelium, and so may cause the compression of blood vessels and reduction of mammary blood flow. As mammary blood flow declines, the availability of hormones (e.g. PRL) and nutrients to the gland is reduced.

This theory was ruled out by research carried out in goats (Wilde and Knight, 1990) and cows (Stelwagen and Knight, 1997), which showed that physiological factors other than intramammary pressure are involved. One such factor which plays a major role in regulating milk secretion is the feedback inhibitor of lactation (FIL), a protein found in milk (Wilde *et al.*, 1995). When milk accumulates in the alveoli, this inhibitory protein also accumulates in the milk and causes a feedback inhibition of milk synthesis and secretion. Frequent removal of milk from the mammary gland minimizes the local inhibitory effect of the FIL and increases milk secretion (Wilde *et al.*, 1987; Wilde and Knight, 1989, 1990). This mechanism of autocrine regulation of milk synthesis and secretion has also recently been found in dairy ewes (Bencini *et al.*, 2003).

1.3 The Evolution of Milk Yield and Milk Composition

In spite of continued milk removal, with associated removal of FIL, milk yield declines as lactation progresses until the mammary gland dries up. As lactation progresses, the secretory cells gradually regress from a state of active secretion to a non-secretory state through a process called involution. Because milk yield (MY) is the product of cell number (Cn) and secretory activity per cell (w), gradual involution happens through either a decrease in the number of mammary cells or a decrease in their activity.

Apart from systemic factors, the involution of the mammary gland is also regulated, as previously mentioned, by local mechanisms such as FIL and the plasminogen–plasmin system. These mechanisms are active during the whole lactation and can accelerate with the following events: (i) inflammation of the mammary gland; (ii) extension of the milking interval; (iii) suppression of milk removal; and (iv) reduction of nutrients to the mammary gland. In the latter case there is a positive relationship between nutrient supply and amount of milk synthesized, because milk synthesis is related to substrates available for milk synthesis (Rook and Thomas, 1983). The availability of substrates in ECF is not the only limiting factor in milk synthesis. One of our recent studies on well-fed primiparous ewes showed that the milk yield of animals with an udder half-milked twice a day (when milking one side of the udder was completely suppressed for 1 week) was half that obtained from ewes with a normally milked gland (618 vs. 1221 g/day). This result showed that increasing the quantity of nutrient substances (which were not utilized by the opposite udder half in which milking was suppressed) did not result in a corresponding increase of their uptake for milk synthesis by the regularly milked udder half. This can be explained by the fact that there was enough nutrient available for the potential synthesis activity of

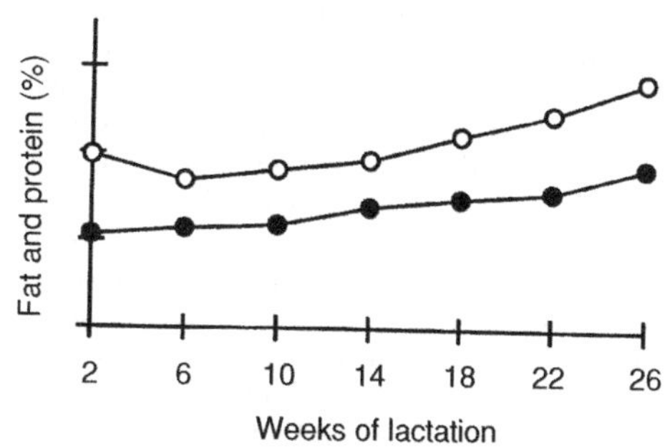

Fig. 1.6. Variation in fat (°) and protein (•) concentrations during lactation (Pulina, 1990).

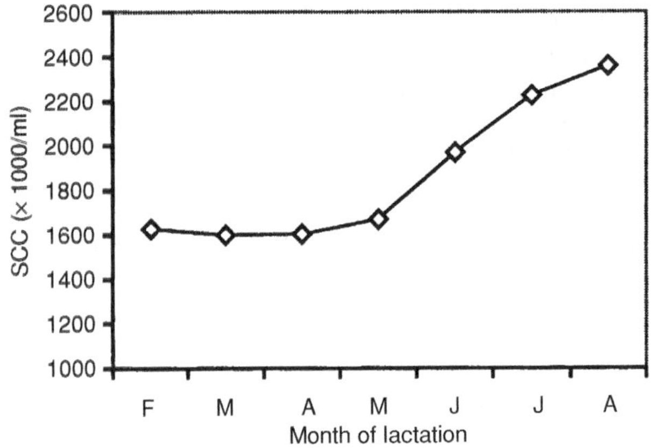

Fig. 1.7. Variation of somatic cell count during lactation (Banni and Martin, 1998).

each cell (w); therefore milk yield was conditioned mainly by cell numbers (Cn) active in the lactation stage during which the experiment was carried out.

As lactation progresses the composition of the milk changes, with increases in the fat and protein concentrations (Fig. 1.6), and in somatic cell count (Fig. 1.7), in particular PMN leucocytes, while lactose concentration, cryoscopy and acidity decrease.

Other changes accompanying the progress of lactation include an increase in the content of Mg and Cl, and a decrease in the content of K. The milk coagulation time and curd firming rate increase while curd firmness decreases.

References

Anderson R.R. (1975) Mammary gland growth in sheep. *J. Anim. Sci.*, 41 (1): 118–123.

ARA (Associazione Regionale Allevatori della Sardegna) (2000) Laboratorio Analisi di Oristano, Italy.

Banni S., Martin J.C. (1998) Conjugated linoleic acid and metabolites. In: J.L. Sebedio and W.W. Christie (eds), *Trans fatty acids in human nutrition*, The Oily Press, Dundee, UK, pp. 261–302.

Bencini R., Pulina G. (1997) The quality of sheep milk: a review. *Aust. J. Exp. Agr.*, 37: 485–504.

Bencini R., Knight T.W., Hartmann P.E. (2003) Secretion of milk and milk components in sheep. *Aust. J. Experimental Agric.*, 43: 529–534.

Binart N., Ormandy C.J., Kelly P.A. (1999) Mammary gland development and prolactin receptors. In: J.A. Mol and R.A. Clegg (eds) *Biology of the mammary gland*, Vol. 480, pp. 85–92.

Bingham E.W., McGranaghan M.B., Wickham E.D., Leung C.T., Farrel H.M. (1993) Properties of [Ca^{2+}+Mg^{2+}]-adenosine triphosphatases in the Golgi Apparatus and microsomes of the lactating mammary gland of cows. *J. Dairy Sci.*, 76: 393–400.

Campus R. (1991) Valutazione della capacità di resa del latte ovino, di razza Sarda, in formaggio Pecorino Romano e Pecorino Sardo. Laurea thesis of Venusti M.G.

Cannas A., Nudda A., Rassu S.P.G., Serra A. (1995) Effects of concentrate fiber content and lambing season on summer milk yield of dairy ewes on stubble grazing. *Proc. XI Congress ASPA*, Italy, pp. 283–284.

Cannas A., Pes A., Mancuso R., Vodret B., Nudda A. (1998) Effect of dietary energy and protein concentration on the concentration of milk urea nitrogen in dairy ewes. *J. Dairy Sci.*, 81: 499–508.

Casu S., Marcialis A. (1966) Contributo alla conoscenza delle relazioni fra composizione del latte e resa in formaggio di tipo pecorino romano. *Sci. Tecn. Latt.-cas.*, 17: 204–213.

Chestnutt D.M.B., Wylie A.R.G. (1995) The effects of frequency of feeding of supplementary concentrates on performance and metabolite and IGF-I status of ewes given silage in late pregnancy. *Anim. Sci.*, 61: 269–276.

Johnsson I.D., Hart I.C. (1985) Pre-pubertal mammogenesis in the sheep 1. The effects of level of nutrition on growth and mammary development in female lambs. *Anim. Prod.*, 41: 323–332.

Labussière J., Marnet P.G., Combaud J.F., De la Chevalière F., Haigron S., De Saint Jan J. (1989) Influence du nombre de corps jaune sur le transfert du lait alveolaire dans la citerne de la brebis. Conséquences sur le niveau de production laitière. *Proc. 4th International Symposium on Machine Milking of Small Ruminants,* Tel-Aviv, Israel, pp. 13–17.

Luquet F.M. (1985) Laits et produits latières. I. Les laits. *Technique et Documentation,* Lavoisier, Paris.

McGuire M.A., Bauman D.E., Miller M.A., Hartnell G.F. (1992) Response of somatomedins (IGF-I and IGF-II) in lactating cows to variations in dietary energy and protein and treatment with recombinant n-methionyl bovine somatotropin. *Nutr. Abstr. Rev.*, 62 (6): 2839.

Mepham T.B. (1987) *Physiology of Lactation.* Open University Press, Milton Keynes, UK.

Naitana S., Nuvole P., Marongiu A. (1992) Lattazione. In: G. Aguggini, V. Beghelli, and L.F. Giulio (eds) *Fisiologia degli animali domestici con elementi di etologia,* UTET, Turin, Italy, pp. 781–808.

Nudda A. (1996) La qualità del latte nella specie ovina. PhD thesis, Università degli Studi di Perugia, Facoltà di Agraria, Italy.

Nudda A., Pulina G., Vallebella R., Bencini R., Enne G. (2000) Ultrasound technique for measuring mammary cistern size of dairy ewes. *J. Dairy Res.*, 67: 101–106.

Peaker M. (ed.) (1977) *Comparative aspects of lactation.* Academic Press, London.

Pulina G. (1990) L'influenza dell'alimentazione sulla qualità del latte ovino. *L'inf. Agr.*, 46 (37): 31–38.

Pulina G., Serra A., Cannas A., Rossi G. (1989) Determinazione del valore energetico di latte di pecore di razza Sarda. *Proc. XLIII Congress S.I.S.Vet.*, Italy, pp. 1867–1870.

Pulina G., Serra A., Macciotta N.P.P., Nudda A. (1993) La produzione continua di latte nella specie ovina in ambiente mediterraneo. *Proc. X Congress ASPA,* Italy, pp. 353–356.

Ranucci S., Fruganti G., Tesei B. (1985) Determinazione di alcune attività enzimatiche in secreti mammari di pecore non macroscopicamente alterati. *Proc. I Congr. S.I.B.C.A.*, Parma, Italy, pp. 148–153.

Rattray P.V., Garret W.N., East N.E., Hinman N. (1974) Growth, development and composition of the ovine conceptus and mammary gland during pregnancy. *J. Anim. Sci.*, 38 (3): 613–626.

Ricordeau G., Mocquòt G. (1967) Influence des variations saisonnières de la composition du lait de chevre sur le rendement en fromage. Conséquences pratiques pour la selection. *Ann. Zootech.*, 16 (2): 165–181.

Rook J.A.F., Thomas P.C. (1983) Milk secretion and its nutritional regulation. In: J.A.F. Rook and P.C. Thomas (eds) *Nutrition physiology of farm animals,* Longman Inc., New York, pp. 314–368.

Rossi G., Cannas A., Macciotta N.P.P., Pulina G., Rassu S. (1996) Base genetica della prolificità e della stagionalità dei cicli riproduttivi nella specie ovina. In: G. Enne and A. Lauria (eds) *La riproduzione in zootecnia.* Quaderni RAIZ n. 1: 101–127.

Ruckebusch Y., Phanéuf L.P., Dunlop R. (1991) In: Decker (ed.) *Physiology of small and large animals.* Philadelphia, Pennsylvania.

Sauvant D., Morand-Fehr P. (1978) Analyse des variatiòns individuélles en nutrition animale: application de l'analyse en composantes principales a l'étude de la secretion lipidìque du lait de chèvre. *Ann. Zootech.*, 27 (1): 75–81.

Serra A., Nudda A., Pulina G., Cannas A. (1995) Effects of concentrate fiber content and lambing season on summer milk quality of dairy ewes on stubble grazing. *Proc. XI Congress ASPA,* Italy, pp. 285–286.

Stelwagen K., Knight C.H. (1997) Effect of unilateral once or twice daily milking of cows on milk yield and udder characteristics in early and late lactation. *J. Dairy Res.*, 64: 487–494.

Sulochana S., Singh Y., Sharma D.N. (1991) Histological studies on the development of mammary gland parenchyma in pregnant sheep. *Dairy Sci. Abstr.*, 53: 1653.

Tesei B., Ranucci S., Fruganti G., Mangili V., Boni M., Avellini G. (1988) Stato sanitario della mammella e variazioni di alcune attività enzimatiche nel secreto mammario di pecore sottoposte a mungitura meccanica. *Proc. VIII Congress SIPAOC,* Italy, pp. 59–70.

Tolman B., McKusick B.C. (2001) The effect of growth rate on mammary gland development in ewe lambs: a review. *Proc. 7th Great Lakes Dairy Sheep Symposium,* Eau Claire, Wisconsin, pp. 143–155.

Turner D.J., Huynh T. (1991) Role of tissue remodelling in mammary epithelial cell proliferation and morphogenesis. *J. Dairy Sci.*, 74: 2801–2807.

Wilde C.J., Henderson A.J., Knight C.H., Blatchford D.R., Faulkner A., Vernon R.G. (1987) Effects of long-term thrice-daily milking on mammary enzyme activity, cell population and milk yield in the goat. *J. Anim. Sci.*, 64: 533–539.

Wilde C.J., Knight C.H. (1989) Metabolic adaptations in mammary gland during the declining phase of lactation. *J. Dairy Sci.*, 72: 1679–1692.
Wilde C.J., Knight C.H. (1990) Milk yield and mammary function in goats during and after once-daily milking. *J. Dairy Res.*, 57: 441–447.
Wilde C.J., Addey C.V., Boddy L.M., Peaker M. (1995) Autocrine regulation of milk secretion by a protein in milk. *Biochem. J.*, 305: 51–58.

2 Mathematical Modelling of Milk Production Patterns in Dairy Sheep

Aldo Cappio-Borlino, Nicolò P.P. Macciotta and Giuseppe Pulina
Dipartimento di Scienze Zootecniche, Università di Sassari, Italy

2.1 Introduction

The mathematical analysis of milk production records over time represents a fundamental step in achieving a deeper understanding of the complex physiological mechanisms that underlie the milk secretion process. Furthermore, the mathematical description of the main features of temporal evolution of milk production is often required for technical purposes, mainly to improve management and breeding choices. Thus, the study of lactation curves is one of the most important applications of mathematical modelling in animal science. This study has been widely developed in dairy cattle, where a standard shape of the lactation curve has been assessed (Fig. 2.1), with first an ascending phase until a maximum (lactation peak), followed by a decreasing phase that goes from the peak to the drying off.

From a theoretical point of view, there are no elements to support the hypothesis of a substantial difference between dairy cattle and sheep in the general shape of the lactation curve. Therefore, most of the models already developed to analyse milk records and describe the lactation pattern of dairy cattle have also been proposed, with a few modifications, for dairy sheep. However, in practice, temporal evolution of milk production in sheep shows several peculiarities that can be ascribed to biological and, mainly, management aspects. In the Mediterranean countries, where the

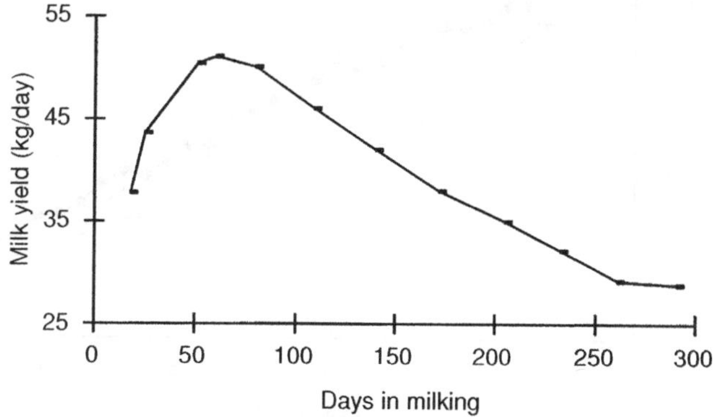

Fig. 2.1. The standard lactation curve of dairy cattle.

largest concentrations of dairy sheep farms exist, reproductive and productive cycles are strictly seasonal, being synchronized with the availability of natural pastures, which represent the most important feeding source; as a consequence, milk production is largely influenced by environmental factors (Barillet, 1985, 1997; Carta *et al.*, 1995; Macciotta *et al.*, 1999). Moreover, in most cases the milk of the first month of lactation is suckled by the lamb, resulting in a limited availability of data for the ascending phase of the lactation. Lactation peak is expected within 3–4 weeks of lambing, i.e. in winter for mature ewes, but it is often smoothed by adverse environmental conditions (low temperatures, scarce feed availability). On the other hand, the favourable climatic conditions that can be found during spring, and especially the large availability of natural pastures, usually result in a 'false' lactation peak in the second half of the lactation (Fig. 2.2), as sometimes also happens in dairy cows with autumn calvings in pasture-based dairy systems (Garcia and Holmes, 2001).

Another important factor that affects the lactation curve in sheep is represented by the type of lambing, i.e. the number of suckling lambs. Several studies have reported a larger milk yield for ewes with multiple births. However, an effect of prolificacy along the whole curve has been detected only in meat and wool sheep (Torres Hernandez and Hohenboken, 1980; Snowder and Glimp, 1990), whereas in dairy breeds the higher production of ewes raising twins is usually evident only in the first part of lactation (Benyoucef and Ayachi, 1991; Gabina *et al.*, 1993; Baro *et al.*, 1994; Gonzalo *et al.*, 1994). Furthermore, this effect can be hidden or even reversed (Pagnacco *et al.*, 1989) if the ewe is undernourished before and after parturition.

Finally, the existence of a large number (20–50% of observed cases) of lactation curves that do not have a lactation peak has been reported for several sheep breeds (Cappio-Borlino *et al.*, 1997). Such a situation has also been observed in dairy cattle (Shanks *et al.*, 1981) and can be ascribed to the low impact of genetic selection in dairy sheep.

2.2 Empirical Models of Lactation Curves

As in other fields of applied biology, mathematical models suggested for studying the lactation curve can be divided into two main classes: empirical models and mechanistic models. This distinction arises from the dif-

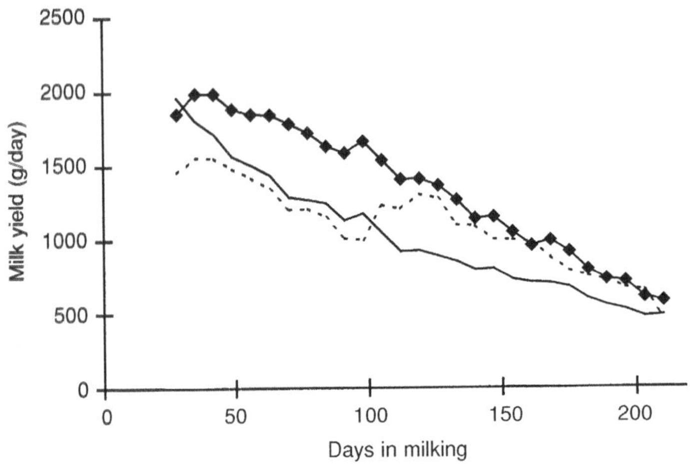

Fig. 2.2. Examples of different shapes of lactation pattern in sheep. -◆-, normal; broken line, with a double peak; continuous line, without peak.

ferent approach used to analyse the main components that can be recognized in biological processes: the deterministic component, usually characterized by a continuous and regular pattern that refers to genetic and environmental factors affecting the phenomenon under study; and the stochastic component, peculiar to each animal and, therefore, unpredictable. Empirical mathematical models are essentially aimed at disentangling these two components without any interest in the mechanisms that underlie the process itself. On the contrary, the mechanistic approach aims to increase the knowledge about a biological phenomenon; thus a mechanistic model is developed as an attempt to translate in mathematical terms an hypothesis about the deep physiological and biochemical processes that regulate the phenomenon of interest.

2.2.1 Mathematical models of lactation curves

Milk production experimental data consist of Test Day (TD) measures recorded at different time intervals from lambing (approximately every 30 days). In dairy sheep, data recording usually starts at the end of the suckling period, i.e. after the first month of lactation, and considering a lactation length of 200 days, TD records available per animal are on average 5–6.

The most simple mathematical method to analyse milk TD records is the fitting of a time function, continuous and differentiable in the whole interval of time that corresponds to the lactation length. As previously stated, several mathematical functions that were originally proposed to model the lactation curve of dairy cattle have also been applied to dairy sheep. The most widely used equation is the Wood's incomplete gamma function (Wood, 1967), which is defined by three parameters:

$$y(t) = a\, t^{b} exp(-\, ct) \qquad [1]$$

where y(t) is the average daily production at time (t) and a, b and c are parameters with positive values which describe the shape of the lactation curve. An advantage of the Wood model lies in the fact that a technical meaning can be assigned to its three parameters: a is an estimate of the initial milk yield, which offers a kind of benchmark of the total yield; b is a measurement of the rate of milk yield increase until the peak; and c is a measurement of the rate of decreasing milk yield after the peak. Table 2.1 shows the values of the parameters for the Wood function for milk yield estimated in some breeds of dairy sheep.

A practical advantage of the Wood function is represented by the opportunity to linearize it by a simple natural logarithm transform:

Table 2.1. Parameter estimates of Wood function for milk yield (g/day) in some dairy sheep breeds.

	Parameter		
Breed	a	b	c
Sarda			
(Pulina and Nudda, 2001)	934	0.181	0.041
Comisana			
(Portolano *et al.*, 1996)	1146	0.197	0.011
Valle del Belice			
(Portolano *et al.*, 1999)	2599	0.034	0.0045
Massese			
(Franci *et al.*, 1999)	1116	0.290	0.0147
US breeds			
(Chang *et al.*, 2001)	758	0.185	0.0108
Awassi			
(Pollot and Gootwine, 2000)	1690	0.320	0.0106

$$\log y(t) = \log a + b \log t - ct \qquad [2]$$

Several other functions have been proposed to model sheep lactation curves (Table 2.2). Some of them fit the data better than the model [1], but with a larger number of parameters whose technical meaning is difficult to define. An example is represented by the fitting of a third-order polynomial to the lactation curve characterized by a false double peak; although this function gives the best fit to experimental data, no relationships between its parameters and technical aspects of milk production pattern can be assessed. However, these models can be used to estimate the production loss due to the absence of the physiological peak by comparing their predicted patterns with the theoretical pattern obtained with a parametric function, such as the linearized Wood function; with this approach, a reduction of about 20% in the total lactation yield has been estimated in Sarda dairy ewes with lactation pattern characterized by a double peak (Cappio-Borlino *et al.*, 1989).

The knowledge of the average lactation curve of homogeneous groups of animals allows for the calculation of missing data, for the extension of incomplete lactations and for the estimation of total milk yield per lactation, which is still the main breeding goal in selection programmes for dairy ewes. Recently, a Bayesian hierarchical non-linear mixed model has been used to investigate the genetic variation of individual lactation curves in some dairy sheep breeds reared in the USA, and to estimate breeding values for the equation parameters in order to find selection goals to modify the general shape of the lactation curve (Chang *et al.*, 2001).

In theory, all these methods allow for the evaluation of the effects of the main environmental factors on milk production by means of the analysis of variance of the parameters that define the function actually used. However, in the analysis of milk production data developed by continuous functions of time, environmental effects are assumed to average out over the lactation, even if there can be effects peculiar to each TD milk yield that may not average out (Jamrozik and Schaeffer, 1997). Moreover, when temporary environmental effects become predominant, the analytical description of lactation curves loses its discriminant power, and the continuous and regular pattern of yield variability is no longer separable from the environmental effects and from the residual random variation. A typical example is represented by the fitting of the Wood model to lactations that lack the production peak, resulting in estimates that are out of the parameter space.

2.2.2 Estimating lactation curves by mixed linear models

The direct modelling of TD yields by mixed linear models, usually called TD models, represents an alternative that can overcome the limitations of the analytical function methodology. These models allow for the reconstruction of the lactation curve by solving for the least squares estimates of the fixed effects of a days-in-milking (DIM) factor on milk yield, taking into account at the same time all the environmental factors affecting milk production, such as lambing season, season of production and parity (Stanton *et al.*, 1992). In addition, TD mod-

Table 2.2. Examples of mathematical functions used to fit dairy sheep lactation patterns.

Model	Author
$y(t) = t/a + bt + ct^2$	Nelder (1966)
$y(t) = a - bt + ae^{-ct}$	Cobby and Le Du (1978)
$y(t) = a + bt + ce^{-0.05t}$	Wilmink (1987)
$\log y = a - bt(1 + kb) = ct^2 + {}^{d}/_{t}$	Morant and Gnanasakthy (1989)
$y(t) = a + t^{b\exp(-ct)}$	Cappio-Borlino *et al.* (1995)

els allow for: (i) a more accurate estimation of the main environmental effects; (ii) a different number of records from each lactation; (iii) the variation of fixed effects estimates across herds and lactation stages; and (iv) the adjustment for different effects of sampling date and the prediction of daily milk yields from a limited number of recorded TD (Pool and Meuwissen, 1999).

Lactation curves of homogeneous groups of animals can thus be estimated by the least squares means of DIM effect nested within the levels of environmental factors. As an example, the following TD model allows for the construction of lactation curves of sheep from flocks located at different levels of geographical altitude (Macciotta *et al.*, 1999):

$$Y = m + F + DIM(ALT) + L + E \quad [3]$$

where Y is the TD milk yield; F are the fixed factors included in the model (in this case flock, type of lambing, year of production, parity); DIM is the fixed effect of days-in-milking; ALT is the fixed effect of the level of altitude of the location of flocks; L is the random effect associated with each individual lactation; and E is the random residual. Figure 2.3 shows the plot of the least squares estimate of the DIM(ALT) factor against days-in-milking.

In fact, lactation curves are clearly separated for all three levels of altitude, especially in the first phase, due to the specific effect of the DIM factor for each level of altitude. Such interaction can be explained by differences in the climate and, mainly, in the quantitative and qualitative evolution of pastures. None of the lactation curves shows the presence of a lactation peak, because the first record occurs on average on the 40th DIM, i.e. after the occurrence of the peak. The slightly increasing phase in the lactation curve of mountain ewes must be interpreted as an effect of the improvement in the conditions of production as lactation proceeds rather than of the physiological rise to the peak yield; early lactation occurs in early winter, environmental conditions are particularly adverse, mainly due to low temperatures and scarce availability of pastures. Estimates of DIM also include a seasonal effect on production, which cannot be considered separately due to the fact that all ewes lamb in the same period.

TD models have been also used to estimate lactation curves for milk fat and protein contents, which are important production traits, since almost all sheep milk is used for cheese processing. Figures 2.4, 2.5 and 2.6 show the lactation curves of milk yield, and fat and protein content in Valle del Belice dairy sheep for three different parities estimated by a TD model similar to model [3] (Cappio-Borlino *et al.*, 1997).

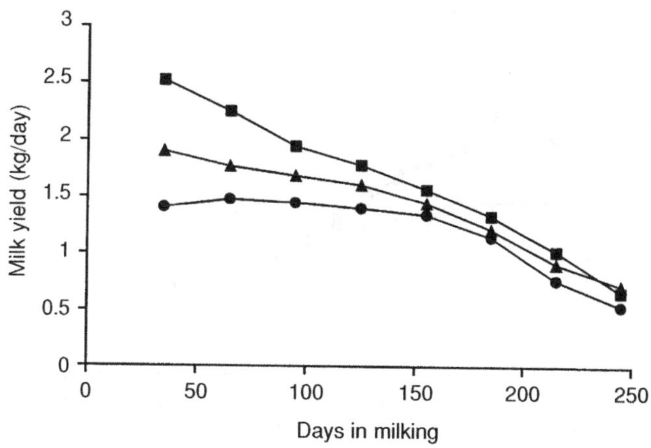

Fig. 2.3. Sarda sheep lactation curves for milk yield separated by altitude of location of the flocks estimated with a Test Day model. ■, plain; ▲, hill; ●, mountain (Macciotta *et al.*, 1999).

Test Day models are able to highlight the higher level of production for ewes at third or greater parity and the greater persistency of lactation in primiparous ewes (Fig. 2.4). This last characteristic, evidenced in most dairy species, is usually explained by the maturation process, which is still in progress during the first lactation and which counteracts the normal decline in milk yield (Stanton *et al.*, 1992).

Curves of fat content do not show substantial differences between different parities (Fig. 2.5), whereas a lower protein percentage in the milk of first and second parity ewes can be observed (Fig. 2.6). This result is usually explained by the reduced rumen functionality, the lower synthesis efficiency of the mammary gland and the preferential utilization of the available amino acids in growing tissues in young animals.

Another interesting aspect evidenced by TD modelling in Valle del Belice sheep is the difference in the rate of increase throughout lactation between fat and protein percentage, the latter being of a lower order. As a consequence, the ratio between fat and protein percentage increased over time, as observed in other dairy sheep breeds.

TD phenotypic models used for lactation curve estimation often include a random factor associated with each individual lactation (L), which allows for the division of the residual variance (σ_P^2) into a component between animals (σ_L^2) and one within animals (σ_E^2). Such a (co)variance structure is known as Compound Symmetry (CS) and it is the most frequently adopted in phenotypic analysis of repeated measures with 'split-plot in time' designs (Stanton *et al.*, 1992; Cappio-Borlino *et al.*, 1997; Carta *et al.*, 2001). The average correlation value among TD measurements within lactation, usually called repeatability, can then be calculated as the ratio (σ_L^2:σ_P^2). Average estimates of repeatability values for milk production traits in sheep are around 0.50, 0.35 and 0.45 for milk yield, fat and protein percentage, respectively.

However, the real (co)variance structure of TD measures within lactation is usually more complicated, as environmental effects on daily milk yield are markedly variable across lactation stages due to the peculiar management of dairy sheep (Carta *et al.*, 1995; Cappio-Borlino *et al.*, 1997; Macciotta *et al.*, 1999). In particular, measures at different days in milking (DIM) on the same lactation can be correlated simply because they share a common contribution from the animal, but measures closer in time are usually more correlated than measures further apart in time. This time-dependent relationship is an expression of the individual variation around the mean lactation curve due to environmental short-term effects such as random variations in dietary quality and intake, weather and minor

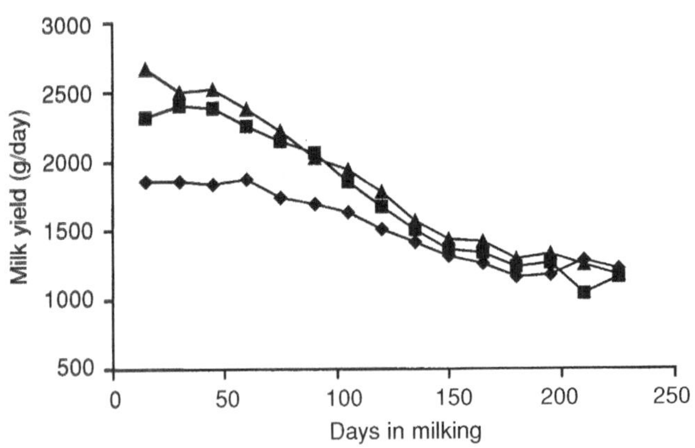

Fig. 2.4. Lactation curves of milk yield for three different parity classes of Valle del Belice sheep estimated with a Test Day model. ◆, 1; ■, 2; ▲, 3 (Cappio-Borlino *et al.*, 1997).

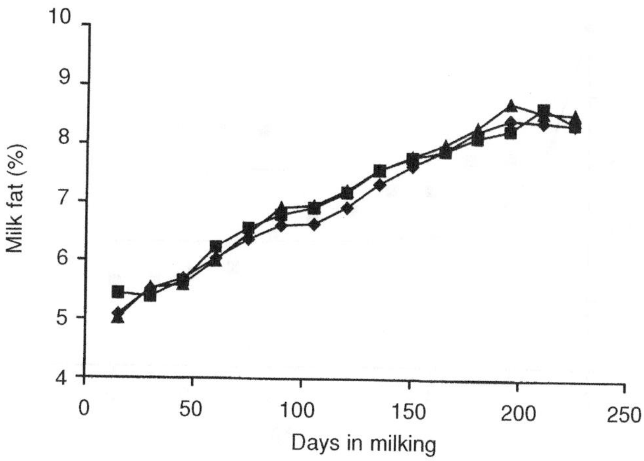

Fig. 2.5. Lactation curves of fat percentage for three different parity classes of Valle del Belice sheep estimated with a Test Day model. ◆, 1; ■, 2; ▲, 3 (Cappio-Borlino *et al.*, 1997).

injuries (Ali and Schaeffer, 1987; Carvalheira *et al.*, 1998), but also to genetic variation. The mixed linear model methodology can be exploited to model directly the (co)variance structure among repeated measures. Suitable (co)variance structures for residuals in mixed models have been suggested for genetic evaluation in cattle (Carvalheira *et al.*, 1998) and for phenotypic analysis in dairy sheep (Giaccone *et al.*, 2000; Carta *et al.*, 2001). In particular, the constant component of the (co)variance can be modelled by a lag-independent parameter (σ_L^2), whereas the time-dependent component of covariation can be modelled by a lag-dependent covariance function with the form of a first-order autoregressive process AR(1) (Diggle, 1988; Littel *et al.*, 1998). Figure 2.7 shows an example of the fitting of an AR(1) process to the actual correlation structure of residuals of TD milk yields.

The definition of a more appropriate (co)variance structure not only affects the goodness of fit of the mathematical model proposed, but it also allows for a more

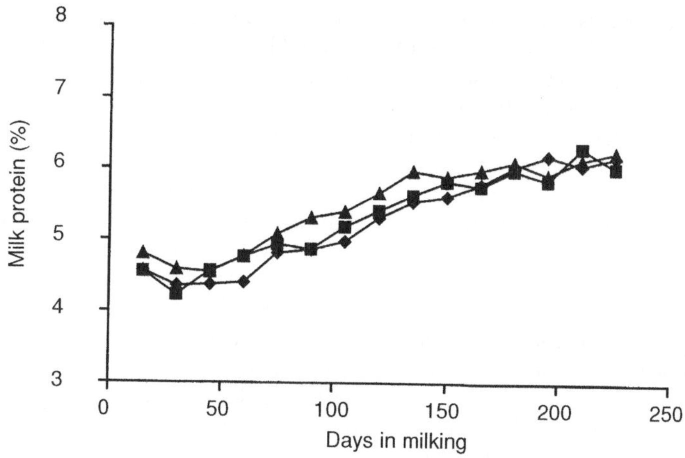

Fig. 2.6. Lactation curves of protein percentage for three different parity classes of Valle del Belice sheep estimated with a Test Day model. ◆, 1; ■, 2; ▲, 3 (Cappio-Borlino *et al.*, 1997).

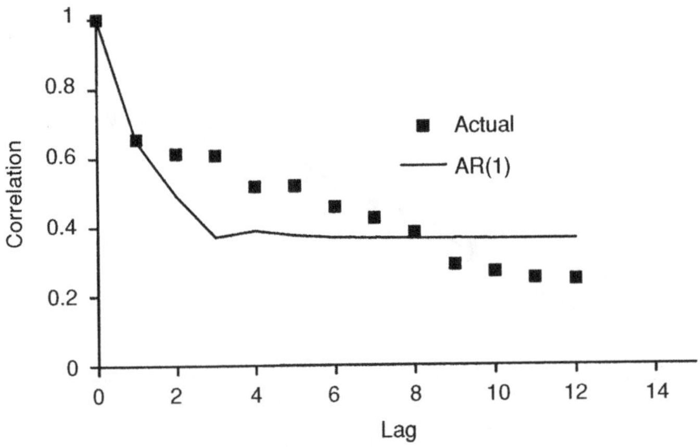

Fig. 2.7. Fit of an autoregressive process of the first order (—) to the correlation pattern (■) among Test Day yields of dairy ewes at different lags between measures.

detailed explanation of some factors affecting the analysed trait. For example, the combined use of the CS and AR(1) structures is often able to detect the nature of random factors causing variation among average lactation curves in dairy ewes. The constant component, estimated by the CS structure, can be due to factors affecting the whole lactation, such as genetic merit of the animal or permanent environmental factors. On the other hand, the existence of an appreciable component of covariation that decreases rapidly with increasing lags can be seen as an indication of important environmental factors that result in short-term effects, whose 'memory' is lost within a few lags. The combination of the two types of covariance components results in a stronger linkage between TD records at different lags for milk yield and protein percentage than for fat percentages. Such different behaviour reflects differences in metabolic processes involved in milk production (Carta *et al.*, 2001).

2.2.3 Modelling of TD covariances within lactation by Time Series analysis models

(Co)variances among milk production data within lactation can be also accounted for by another class of mathematical models, originally developed in the field of econometrics, the Time Series analysis models (Box and Jenkins, 1970; Piccolo, 1990; Hamilton, 1994; Chatfield, 1999). Fitting Time Series models requires a sufficiently large data set. This requirement seems to exclude the use of Time Series methodology for the analysis of data derived from dairy sheep recording schemes (on average 5–6), restricting its use to situations in which all TD within each lactation are available (Deluyker *et al.*, 1990). This constraint, however, can be removed by constructing a series with an index variable that does not have an immediate reference to time. Figure 2.8 shows a succession of data made by milk TD records of 100 Sarda dairy sheep, with seven records each, ordered by lactation and, within lactation, by distance from lambing (Macciotta *et al.*, 2000).

Time Series methodologies possess some diagnostic tools, such as Fast Fourier transform and the autocorrelation function, that are able to recognize the main characteristics of the series under study and to highlight deterministic and random components. The Fourier analysis, also known as spectral analysis, is based on the assumption that any succession of values ordered by an index variable can be obtained as the sum of periodic elementary waves with different angular frequencies. Results of spectral analysis are usually summarized in a periodogram function that plots the average

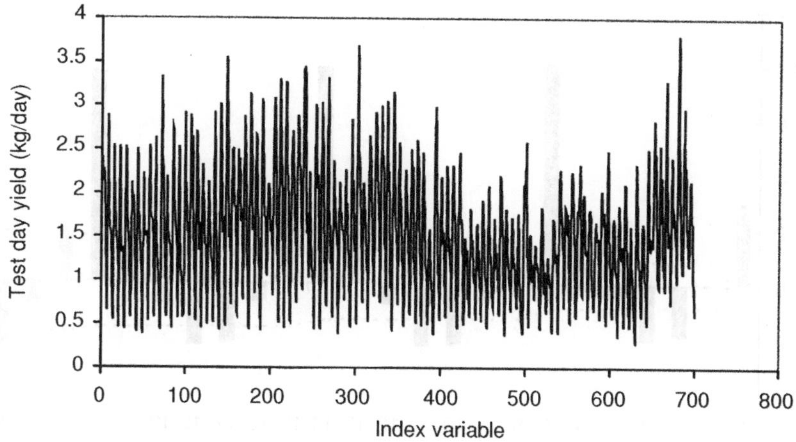

Fig. 2.8. Pattern of 700 TD records of 100 lactations ordered by sheep and by distance from parturition (Macciotta *et al.*, 2000).

squared amplitudes of the elementary waves against the corresponding frequencies. Peaks of amplitude for well-defined frequencies are evidence of deterministic components, whereas a continuous pattern of the periodogram reflects the non-deterministic components. In particular, a periodogram that shows only isolated and discrete peaks underlines a deterministic process, whose future values can be predicted easily from the data; a periodogram nearly parallel to the frequency axis is evidence of processes that are random and unpredictable (i.e. white noise processes); the repeated occurrence of well-defined peaks at equally spaced intervals of frequency highlights the existence of a periodic component. Figure 2.9 shows results of the spectral analysis performed on the series illustrated in Fig. 2.8.

The periodogram shows the occurrence of well-defined peaks at equally spaced intervals of the angular frequencies w = *n**2p/T, where T is the period of seven lags of the index variable. This pattern underlines the existence of a periodic deterministic compo-

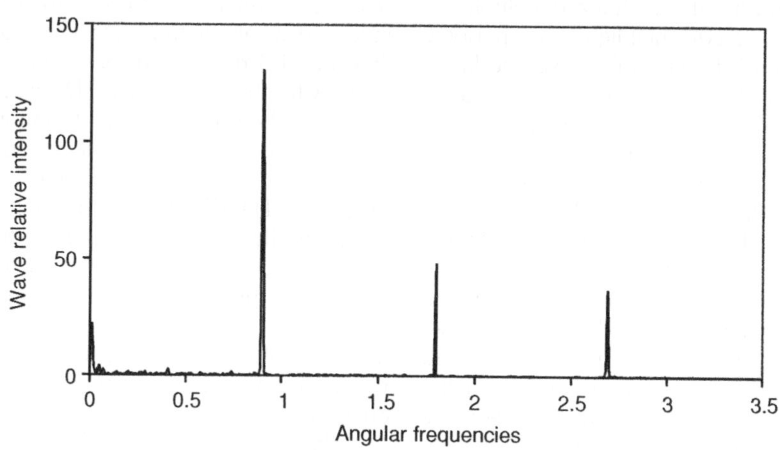

Fig. 2.9. Periodogram of result of Fourier transform of the succession of data reported in Fig. 2.8. The wave intensity is expressed in an arbitrary scale. The angular frequency (w) corresponds to n*2p/T, where T is the lag (Macciotta *et al.*, 2000).

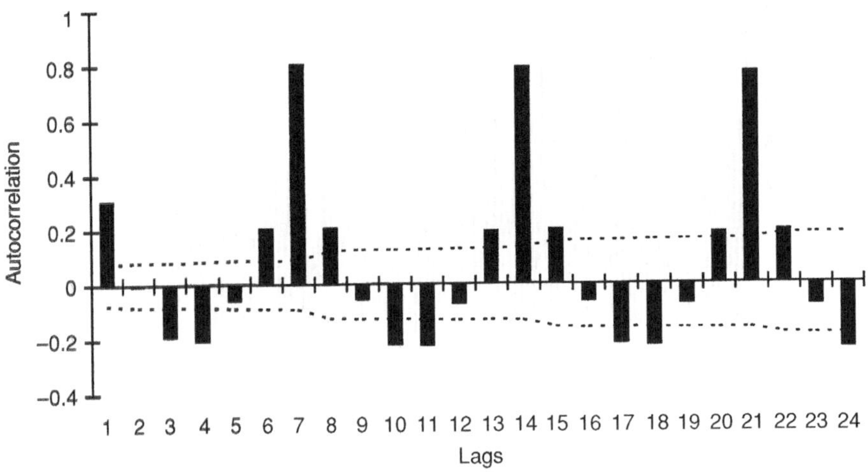

Fig. 2.10. Autocorrelation function of the succession of data reported in Fig. 2.8. Dotted lines are plotted at zero plus and minus twice the standard errors for each coefficient (Macciotta *et al.*, 2000).

nent, which occurs in all lactations and which can be identified with the amount of variability explained by the average lactation curve of the succession. This component, i.e. the lactation curve, can be estimated by using the so-called seasonal periodic decomposition procedure.

The function of autocorrelation ρ(k) at lag k of the TD series is the linear correlation coefficient between TDt and TDk, calculated for k = 0,1,2...

Autocorrelation values, significantly different from zero at lags that are multiples of seven (Fig. 2.10), highlight the periodic component of the variability (average lactation curve), further confirming results of spectral analysis. Moreover, the autocorrelation pattern indicates the existence of an autoregressive component of the first order (AR(1)). On the basis of this diagnostic step, a suitable ARMA (Autoregressive Moving Average Model) model can be suggested. In comparison with models considered in the previous sections, this model is able to take into account both the periodic component of the succession (lactation curve) and the temporal dependence of the (co)variance structure among TD within lactation. Finally, the goodness of fit of the model can be tested by the absence of values significantly different from zero (i.e. that exceed the dotted line) in the autocorrelation function of residuals (Fig. 2.11), thus confirming the white noise of residuals, i.e. their complete randomness.

ARMA models are particularly useful for forecasting missing data, due to their ability to upgrade predictions by using the errors made in the previous estimates. Table 2.3 reports the goodness of ARMA predictions in three different parities of Sarda sheep for two situations of incomplete lactations: only the first two TD records out of seven are available, with the last actual TD recorded at about the 65th DIM; only the first four TD records out of seven are available, with the last actual TD recorded at about the 125th DIM. Total lactation yields are calculated by using all seven actual TD records (AY), two actual and five predicted TD (FY2) and four actual and three predicted TD (FY4). Statistics are: Pearson correlations between actual and predicted productions; the prediction bias; and the ratio between the standard deviation of differences between actual and predicted and the mean value of actual productions.

Values of Pearson correlations between actual and predicted yields are remarkably high, even if predictions are obviously better when four out of seven TD are known

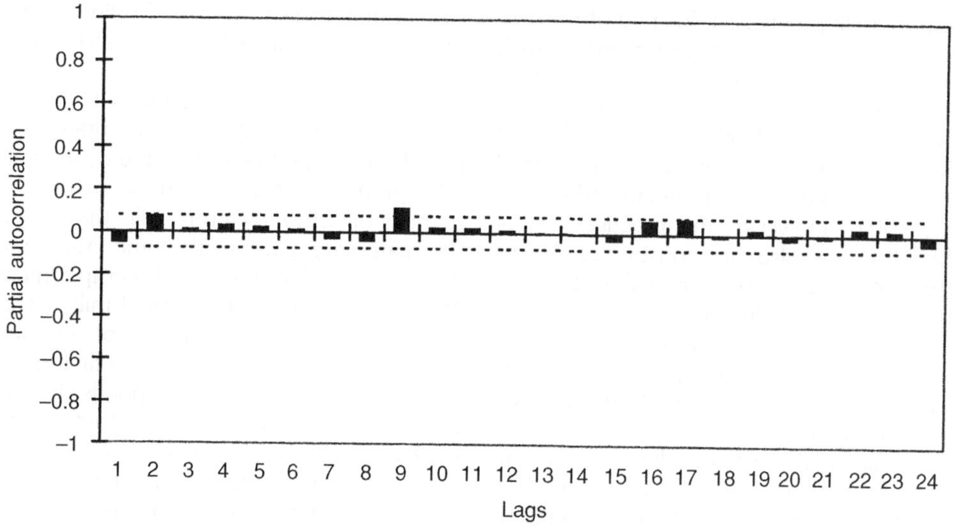

Fig. 2.11. Autocorrelation function of residuals of ARMA modelling fitted to the succession of data reported in Fig. 2.8. Dotted lines are plotted at zero plus and minus twice the standard errors for each coefficient (Macciotta *et al.*, 2000).

(about 0.92 and 0.98, respectively). The magnitude of the prediction bias is negligible and, more importantly, of different signs in the different classes, indicating the absence of any tendency of the model to over- or underestimate TDY. Finally, the last index, obtained as 100 times the ratio of the standard deviation of the difference between actual and predicted productions, and usually proposed as a measure of the goodness of predictions, is about 8.5 (except for third parity ewes) for FY2, whereas it is about 4.5 for FY4. These results are comparable with those obtained in dairy cattle, thus confirming the good forecasting power of ARMA modelling.

2.3 Mechanistic Models of Lactation Curves

The construction of a mechanistic model starts from current knowledge about the process under study. In the case of lactation, this is a very hard task to perform due to both the large number of variables involved in the process and the difficulty of their determination.

Neal and Thornley (1983) proposed a multi-compartmental model of the mammary gland able to explain the standard lactation pattern, on the basis of the most important physiological processes and metabolic pathways that control the mecha-

Table 2.3. Statistics of the predictions for the three parity stages in Sarda ewes (see text for explanations).

Parameter	65 DIM			125 DIM		
	3rd	4th	5th	3rd	4th	5th
Correlation (AY[1], FY[2])	0.91	0.93	0.94	0.98	0.98	0.98
Mean (AY − FY)	−2.73	−2.22	2.52	−1.33	2.29	1.42
$\frac{\sigma (AY - FY)}{\overline{AY}} \times 100$	10.08	8.38	8.40	4.68	4.02	4.30

[1] Actual yield; [2] forecasted yield.

nism of milk production. In spite of remarkable advances in the understanding of milk production physiology (Davis *et al.*, 1999), Neal and Thornley's work remains a relevant attempt at in-depth dynamic analysis of the main processes of mammary gland physiology in domestic ruminants (Pulina *et al.*, 2001). However, the practical use of this model is strongly constrained by its great theoretical complexity and by the large number of parameters involved.

A lactation curve function based on the biology of the mammary gland during pregnancy and lactation has recently been devised (Dijkstra *et al.*, 1997; Pollot, 2000). The major processes considered are: mammary parenchymal cell proliferation, their differentiation into cells showing secretory activity and the reduction in their number as a result of programmed cell death (apoptosis). A mechanistic model has been developed by using two logistic functions representing secretory cell differentiation and death throughout lactation, respectively, and then fitting them to the lactation curve of the Awassi dairy sheep (Pollot and Gootwine, 2000).

A similar approach, even if largely simplified, has been specifically proposed to test a hypothesis about the mechanism that underlies the dimorphism of the lactation curve observed in several breeds of sheep. In fact, together with the 'regular' lactation curves, characterized by the usual shape, there are 'decayed' lactation curves, showing no peak and a continuous decreasing trend; these are observed in 30–50% of lactations (Sakul and Boylan, 1992; Cappio-Borlino *et al.*, 1995; Carta *et al.*, 1995). Empirical mathematical models applied to dairy sheep are able only to describe the regular shape (Cappio-Borlino *et al.*, 1989, 1995; Groenwald *et al.*, 1995). Attempts to fit these models to decayed curves result in parameter estimates that fall out of the range of biological significance. As an example, the fitting of the incomplete Wood's gamma function [1] to decayed curves resulted in a negative estimate of the parameter b value (Portolano *et al.*, 1999). The mechanistic model of the lactation curves of dairy sheep that is able to explain such a dimorphism (Cappio-Borlino *et al.*, 1997) is based on a functional bicompartmental model of the mammary gland (Ferguson and Boston, 1993). It gives a schematic mathematical representation of the physiological theory that the quantity of milk produced each time depends on the number of active secretory cells that are in the mammary gland at the time (Mepham, 1987). The first functional compartment represents a pool of inactivated milk secretory cells, and the second a pool of activated secretory cells. The flow from the first to the second compartment corresponds to the cell activation process. The output of the model = $q_2(t)$ is the number of secretory cells that are in the mammary gland at time t, and can be considered proportional to the milk production at that time. Figure 2.12 shows an analogic representation of the model obtained by using the simulation software STELLA II (STELLA®5.0, High Performance Systems, Hannover, USA).

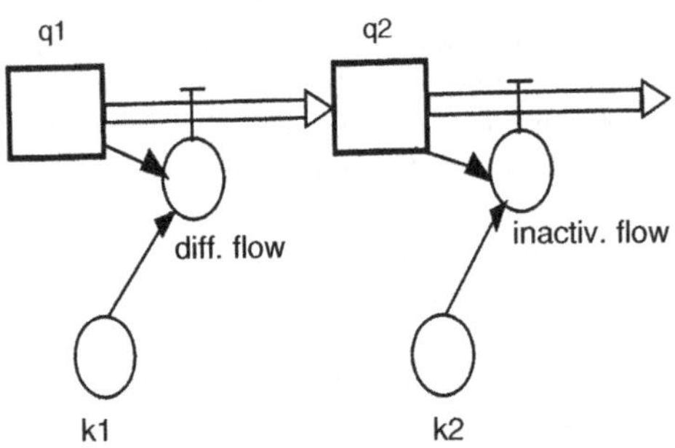

Fig. 2.12. Analog representation of the bicompartmental model obtained by STELLA II software. q1, number of inactivated cells; q2, number of activated cells; k1, cell activation rate; k2, cell inactivation rate (Cappio-Borlino *et al.*, 1997).

The model works according to the following differential equations:

Equation [4a] refers to the number of cells that undergo activation at time t, while [4b] represents the variation in the number of activated cells at time t due to the difference between this activation process and a contemporary process of inactivation;

$$\frac{dq_1}{dt} = -k_1 q_1 \qquad [4a]$$

$$\frac{dq_2}{dt} = k_1 q_1 - k_2 q_2 \qquad [4b]$$

k_1 and k_2 are positive parameters, which measure secretory cell activation rate and inactivation rate, respectively. This system of ordinary linear differential equations is equivalent to a second-order differential equation of the form:

$$\frac{d^2 q_2}{dt^2} + (k_1 + k_2)\frac{dq_2}{dt} + k_1 k_2 q_2 = 0 \qquad [5]$$

The general solution of equation [5] is the function:

$$q_2(t) = \frac{k_1 Q_1}{k_2 - k_1} e^{-k_1 t} + \left(Q_2 = \frac{k_1 Q_1}{k_2 - k_1}\right) e^{-k_2 t} \qquad [6]$$

where Q_1 is the initial number of inactivated cells present in compartment one at the time of lambing and Q_2 is the initial number of activated cells present in compartment two at the same time.

However, in the case of a reduction of the activation process to a single pulse that occurs at time of lambing, immediately after lambing the connection between the two compartments is interrupted, equation [6] reduces to zero and the variation of Q_2 can be described by the differential equation:

$$\frac{dq_2}{dt} = -k_2 q_2 \qquad [7]$$

whose solution is the decreasing exponential function:

$$q^2(t) = Q^2 e^{-k2t} \qquad [8]$$

Solution of equation [5] is in the form of a biexponential function:

$$q2(t) = B1\ e^{-k1t} + B2\ e^{-k2t} \qquad [9]$$

Equation [9], originally suggested by Brody (1945), and afterwards used by Schaeffer *et al.* (1977), is a good empirical model for the regular shape of the lactation curve, given that $B1 > 0$, $B2 < 0$ and $k2 > k1$. On the other hand, equation [8] represents a decreasing curve that could be a good empirical model for the decayed lactation curve. Fitting of equations [9] and [8] to experimental data for regular and decayed shapes is shown in Figs 2.13 and 2.14, respectively.

These results confirm the ability of the bicompartmental model of the mammary gland to explain the dimorphism in the

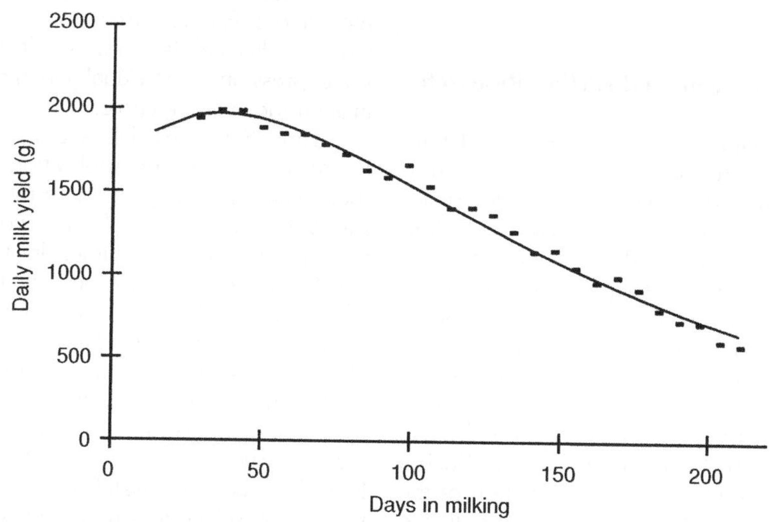

Fig. 2.13. Fitting of the bicompartmental model to regular lactation curves data (Cappio-Borlino *et al.*, 1997).

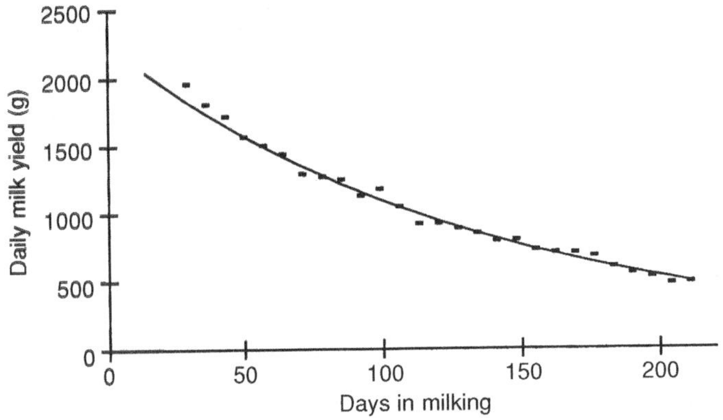

Fig. 2.14. Fitting of the bicompartmental model to decayed lactation curves data (Cappio-Borlino *et al.*, 1997).

shape of lactation curves of dairy sheep in terms of the mechanism of activation and inactivation of the secretory cells.

These mechanistic models could be further developed by including a process of delay of apoptosis, which could explain the higher capacity of dairy sheep, in comparison to dairy cattle, to recover good production levels in the event of temporary adverse environmental conditions. Such a peculiarity of sheep has been recently confirmed in a study carried out on Sarda and Awassi ewes in the final stage of lactation (Nudda *et al.*, 2002).

2.4 Conclusions and Further Research

Fitting a suitable analytical function of time to Test Day records remains the most simple and immediate technique for describing the evolution over time of milk production quantitative traits and for separating the effects of lactation stage from other environmental factors. In fact, it is quite easy to find simple equations, both linear and nonlinear, which are able to fit the average lactation curve of homogeneous groups of animals, but the goodness of fit to individual lactation curves is usually very poor. Such a drawback results in a great limitation of the use of mathematical functions in predicting missing TD records and in extending lactations in progress. However, average lactation curves can be usefully employed to help management decisions, such as the rational planning of feeding for homogeneous groups of ewes.

Test Day models are able to remove environmental variation and to analyse in detail the covariance structure of residuals that often show a high degree of autocorrelation. As in dairy cattle, a possible option for the immediate future of dairy sheep breeding will be the combination of TD models and mathematical functions in repeatability TD models, where average lactation curve functions are included as covariates, and in Random Regression models, in which individual lactation coefficients are the expression of individual genetic variation around the average curve.

Time Series models seem particularly interesting for future applications, due to their ability to recognize deterministic and random components of the evolution of milk production over time. In particular, ARMA models are able to identify the shape of the average lactation curve and to highlight the autoregressive nature of TD residuals within each lactation. Furthermore, ARMA models provide a simple and flexible tool for forecasting missing Test Day records. Such properties of ARMA modelling could be very useful in increasing the impact of selection programmes in dairy sheep populations, where the critical point is represented by the small number of controlled animals. Adaptive signal processing and its application in Time Series prediction will become more important in the future, when dairy animals may have much longer official test intervals, but much more frequent records held by the farm owner. To analyse such a complex wealth of information, more powerful and sophisticated tools of signal processing will be necessary in order to manage, at the same time, several Time Series of different variables and with different recording frequencies. Methods based on cross-correlations among Time Series, artificial neural networks and wavelet decomposition seem particularly appealing for these purposes.

However, for the future, the greatest effort in the mathematical modelling of lactation curves must be directed towards the mechanistic approach. The use of mechanistic models is rather infrequent in several sectors of applied biology, particularly in animal science. This is mainly due to the great difficulty in describing in mathematical terms the complex physiological processes that underlie the phenomenon under study. As far as milk production is concerned, models recently proposed on the basis of the known biology of the mammary gland seem particularly promising, because of their ability to provide a theoretical basis to empirical models of lactation curves, by using a set of differential equations with rather simple analytical solutions.

References

Ali T.E., Schaeffer L.R. (1987) Accounting for covariances among test day milk yields in dairy cows. *Can. J. Anim. Sci.*, 67: 637–644.

Barillet F. (1985) *Amélioration genetique de la composition du lait des brebis. L'exemple de la race Lacaune.* PhD Diss., INRA Paris-Grignon, France.

Barillet F. (1997) Genetics in milk production. In: L. Piper and A. Ruvinsky (eds) *The genetics of sheep.* CAB International, Wallingford, UK, pp. 539–564.

Baro J.A., Carriedo J.A., San Primitivo F. (1994) Genetic parameters of Test Day measures for somatic cell count, milk yield and protein percentage of milk ewes. *J. Dairy Sci.*, 77: 2658–2662.

Benyoucef M.T., Ayachi A. (1991) Mesure de la production laitière de la brebis Hamra durant les phases d'allaitement et de la traite. *Ann. Zootech.*, 40: 1–7.

Box G.E.P., Jenkins G.M. (1970) *Time Series Analysis: forecasting and control.* Holden Day, San Francisco, California.

Brody S. (1945) *Bioenergetics and growth.* Reinholds Publishing Corp., New York.

Cappio-Borlino A., Pulina G., Rossi G. (1989a) La previsione della produzione lattea totale in pecore di razza sarda con l'uso di polinomi. *Produzione Animale,* 35: 35–39.

Cappio-Borlino A., Pulina G., Cannas A., Rossi G. (1989b) La curva di lattazione di pecore di razza Sarda adattata ad una funzione di tipo gamma. *Zoot. e Nutr. Anim.*, 15: 59–63.

Cappio-Borlino A., Pulina G., Rossi G. (1995) A non-linear modification of Wood's equation fitted to lactation curves of Sardinian dairy ewes. *Small Rum. Res.*, 18: 75–79.

Cappio-Borlino A., Portolano B., Todaro M., Macciotta N.P.P., Giaccone P., Pulina G. (1997a) Lactation curves of Valle del Belice dairy ewes for milk, fat and protein estimated with test day models. *J. Dairy Sci.*, 80: 3023–3029.

Cappio-Borlino A., Macciotta N.P.P., Pulina G. (1997b) The shape of Sarda ewe lactation curve analyzed with a compartmental model. *Livest. Prod. Sci.*, 51: 89–96.

Carta A., Sanna S.R., Casu S. (1995) Estimating lactation curves and seasonal effects for milk, fat, and protein in Sarda dairy sheep with a test day model. *Livest. Prod. Sci.*, 44: 37–44.

Carta A., Macciotta N.P.P., Cappio-Borlino A., Sanna S.R. (2001) Modelling phenotypic (co)variances of test day records in the dairy ewe. *Livest. Prod. Sci.*, 69: 9–16.

Carvalheira J.G.V., Blake R.V., Pollak E.J., Quaas R.L., Duran-Castro C.V. (1998) Application of an autoregressive process to estimate genetic parameters and breeding values for daily milk yield in a tropical herd of Lucerna cattle and in United States Holstein herds. *J. Dairy Sci.*, 81: 2738–2751.

Chang Y., Rekaya R., Gianola D., Thomas D.L. (2001) Genetic variation of lactation curves in dairy sheep: a Bayesian analysis of Wood's function. *Livest. Prod. Sci.*, 71: 241–251.

Chatfield, C. (1999) *The analysis of Time Series; an introduction.* Chapman & Hall, London, UK.

Cobby J.M., Le Du Y.L.P. (1978) On fitting curves to lactation data. *Anim. Prod.*, 26: 127–133.
Davis S.R., Farr V.C., Stelwagen K. (1999) Regulation of yield loss and milk composition during once-daily milking: a review. *Livest. Prod. Sci.*, 59: 77–94.
Deluyker H.A., Shumway R.H., Wecker W.E., Azari, A.S., Weaver L.D. (1990) Modelling daily milk yield in Holstein cows using time series analysis. *J. Dairy Sci.*, 73: 539–548.
Diggle P.J. (1988) An approach to the analysis of repeated measurements. *Biometrics*, 44: 959–971.
Dijkstra J., France J., Dhanoa M.S., Maas J.A., Hanigan M.D., Rook A.J., Beever D.E. (1997) A model to describe growth patterns of the mammary gland during pregnancy and lactation. *J. Dairy Sci.*, 80: 2340–2354.
Ferguson J.D., Boston R. (1993) Lactation curve analysis: comparison of gamma function, polynomial and exponential methods toward a mechanistic model of milk production. *J. Dairy Sci.*, 76 (suppl. 1): 268.
Franci O., Pugliese C., Acciaioli A., Parisi G., Lucifero M. (1999) Application of two models to the lactation curve of Massese ewes. *Small Rum. Res.*, 31: 91–96.
Gabina, F., Arrese F., Arranz J., Beltran De Heredia I. (1993) Average milk yields and environmental effects on Latxa sheep. *J. Dairy Sci.*, 76: 1191–1198.
Garcia S.C., Holmes C.W. (2001) Lactation curves of autumn- and spring-calved cows in pasture-based dairy systems. *Liv. Prod. Sci.*, 68: 189–203.
Giaccone P., Di Stasio L., Macciotta N.P.P., Portolano B., Todaro M., Cappio-Borlino A. (2000) Effect of b-lactoglobulin polymorphism on dairy ewe milk-related traits analysed by a repeated measures design. *J. Dairy Res.*, 67: 443–448.
Gonzalo C., Carriedo J.A., Baro A., San Primitivo F. (1994) Factors influencing variation of test day milk yield, somatic cell count, fat and protein in dairy sheep. *J. Dairy Sci.*, 77: 1537–1542.
Groenwald P.C.N., Ferreira A.V., Van der Merwe H.J., Slippers C.J. (1995) A mathematical model for describing and predicting the lactation curve of Merino ewes. *Anim. Sci.*, 61: 95–101.
Hamilton, J. D. (1994) *Time Series Analysis.* Princeton University Press, Princeton, New Jersey.
Jamrozik J., Schaeffer L.R. (1997) Estimates of genetic parameters for a test day model with random regressions for yield traits of first lactation Holsteins. *J. Dairy Sci.*, 80: 762–770.
Littel R.C., Henry P.R., Ammermann C. (1998) Statistical analysis of repeated measures data using SAS procedures. *J. Anim. Sci.*, 76: 1216–1231.
Macciotta N.P.P., Cappio-Borlino A., Pulina G. (1999) Analysis of environmental effects on test day milk yields of Sarda dairy ewes. *J. Dairy Sci.*, 82: 2212–2217.
Macciotta, N.P.P., Cappio-Borlino A., Pulina G. (2000) Time series autoregressive integrated moving average modelling of Test Day milk yields of dairy ewes. *J. Dairy Sci.*, 83: 1094–1103.
Mepham T.B. (1987) *Physiology of lactation.* Open University Press, Milton Keynes, UK.
Morant S.V., Gnanasakthy A. (1989) A new approach to the mathematical formulation of lactation curves. *Anim. Prod.*, 49: 151–162.
Neal H.D., Thornley J.H.M. (1983) The lactation curve in cattle: a mathematical model of the mammary gland. *J. Agric. Sc. Camb.*, 101: 389–400.
Nelder J.A. (1966) Inverse polynomials, a useful group of multifactor response functions. *Biom.*, 22: 1228–1241.
Nudda A., Bencini R., Mijatovic S., Pulina G. (2002) The yield and composition of milk in Sarda, Awassi and Merino sheep milked unilaterally at different frequencies. *J. Dairy Sci.* 85: 2879–2884.
Pagnacco G., Rizzi R., Miglior F., Zanotti Casati M. (1989) Produzione di latte e numero di nati nella pecora delle Langhe: ripetibilità e correlazioni. *Zoot. Nutr. Anim.*, 15: 171–174.
Piccolo, D. (1990) *Introduzione all'analisi delle serie storiche (Introduction to time series analysis).* La Nuova Italia Scientifica Eds., Rome.
Pollot G.E. (2000) A biological approach to lactation curve analysis for milk yield. *J. Dairy Sci.*, 83: 2448–2458.
Pollot G.E., Gootwine E. (2000) Appropriate mathematical models for describing the complete lactation of dairy sheep. *Anim. Sci.*, 71: 197–207.
Pool M.H., Meuwissen T.H.E. (1999) Prediction of daily milk yields from a limited number of test days using test day models. *J. Dairy Sci.*, 82: 1555–1564.
Portolano B., Spatafora F., Bono G., Margotta S., Todaro M., Ortoleva V., Leto G. (1996) Application of the Wood model to lactation curves of Comisana sheep. *Small Rum. Res.*, 24: 7–13.
Portolano B., Todaro M., Giaccone P., Montalbano L., Militi W. (1999) Genetic parameters of the Valle del Belice lactation curves. *Proc. XIII Congress ASPA,* Piacenza, Italy, pp. 230–232.

Pulina G., Nudda A. (2001) La produzione del latte. In: *L'alimentazione degli ovini da latte*, Avenue Media, Bologna, Italy, pp. 9–31.
Pulina G., Cappio-Borlino A., Macciotta N.P.P., Di Mauro C., Nudda A. (2001) Empirical and mechanistic mathematical models of temporal evolution of milk production in ruminants. *Bio. For.*, 94: 331–343.
Sakul, H., Boylan W. (1992) Lactation curves for several US sheep breeds. *Anim. Prod.*, 54: 229–233.
Schaeffer L.R., Minder C.E., Mc Milian I., Burnside E.B. (1977) Non-linear techniques for predicting 305 days lactation production of Holsteins and Jerseys. *J. Dairy Sci.*, 60: 1636–1644.
Shanks R., Berger P.J., Freeman A.E. (1981) Genetic aspects of lactation curves. *J. Dairy Sci.*, 64: 1852–1860.
Snowder G.D., Glimp H.A. (1990) Influence of breed, number of suckling lambs and stage of lactation on ewe milk production and lamb growth under range conditions. *J. Anim. Sci.*, 69: 923–930.
Stanton T.L., Jones L.R., Everett R.W., Kachman S.D. (1992) Estimating milk, fat, and protein lactation curves with a test day model. *J. Dairy Sci.*, 75: 1691–1700.
Torres Hernandez G., Hohenboken W.D. (1980) Biometric properties of lactations in ewes raising single or twin lambs. *Anim. Prod.*, 30: 431–436.
Wilmink J.B.M. (1987) Adjustment of test day milk, fat and protein yield for age, season and stage of lactation. *Liv. Prod. Sci.*, 16: 335–348.
Wood P.D.P. (1967) Algebraic model of the lactation curve in cattle. *Nature*, 216: 164–165.

3 Energy and Protein Requirements

Antonello Cannas
Dipartimento di Scienze Zootecniche, Università di Sassari, Italy

3.1 Introduction

Many studies have been conducted to determine the nutrient requirements of sheep, mainly because this species has often been used as a model for cattle. Indeed, sheep are easier to handle and less expensive to work with. However, because there are only a few countries with a large number of sheep, because sheep are economically less important than cattle and because they are normally managed in extensive, low-input systems, the feeding systems developed for sheep are simpler and more empirical than those designed for cows. Another characteristic of these feeding systems is that they are mostly based on data obtained on meat or wool sheep breeds. Only the French system (INRA, 1987) has dedicated a specific effort toward dairy sheep. However, the rations formulated with this system appear to underestimate feed requirements of lactating ewes, as discussed in Chapter 6 of this book.

All the most important feeding systems have specific sections for estimating feed intake, nutrient requirements and feed value. These aspects are all inextricably connected (Lanari *et al.*, 1983); indeed, the estimation of requirements must take into account the availability of nutrients, whose prediction, in turn, requires an accurate knowledge of animal requirements. However, while many feeding systems use the same feed evaluation approach for all ruminants, requirements are specific for the species considered. A comparison of energy and protein requirement systems for sheep was published in the 1980s by Robinson (1987) and also by Susmel and Cuzzit (1988). However, since then several new or updated feeding systems for sheep have been published.

For this reason, the aims of this chapter are to compare the estimates of energy and protein requirements of the most important sheep systems and to highlight differences in approach, level of aggregation, input required and flexibility. The main focus is on mature ewes, because of their importance in all sheep production systems. In addition, two recently published feeding systems derived from the previous ones will be discussed.

3.2 Main Feeding Systems for Sheep

Several feeding systems for sheep have been published. Those systems that have achieved international recognition are: (i) the Agricultural and Food Research Council (AFRC, 1990, 1992, 1995); (ii) the Commonwealth Scientific and Industrial Research Organisation (CSIRO, 1990); (iii) the Institut National de la Recherche Agronomique (INRA, 1987, 1988, 1989); and (iv) the National Research Council (NRC, 1985). All these systems allow the prediction of feed intake, nutrient requirements and feed value. As stated previously, this chapter focuses only on requirements,

Table 3.1. Efficiencies of conversion of ME to NE for different physiological functions.

	AFRC	CSIRO	INRA	NRC
Maintenance	$k_m = 0.35\ q_m + 0.503$	$k_m = 0.02$ M/D+0.5	k_l	$k_m = 0.729$
Lactation	$k_l = 0.35\ q_m + 0.42$	$k_l = 0.02$ M/D+0.4	$k_l = 0.249 * q_m + 0.463$	ng
Growth	$k_g = 0.78\ q_m + 0.006$[a]	$k_g = 0.042$ M/D+0.006[a] $k_g = 0.063$ M/D–0.308[b]	k_l[c]	k_g[d]
k_g reserves in lactation	0.95 k_l	0.60	k_l	ng
k_g reserves in dry period	as in growth	as in growth	k_l	ng
NE BW loss → NE milk	0.84	0.84	0.80	ng
NE BW loss → NE maint.	0.80	ng	ng	ng

q_m, ME/GE at maintenance feeding level (LN = 1), calculated assuming GE= 4.398 Mcal; M/D, metabolizable energy per kg of DM of the diet (MJoules); ng, not given.

[a] k_g when M/D > 2.271 Mcal/kg DM.

[b] k_g when M/D < 2.271 Mcal/kg DM.

[c] For replacement animals; in the case of meat breeds, a single coefficient is used for all functions, assuming that LN=1.5, $k_{mg} = (k_m * k_g * 1.5)/(k_g + 0.5\ k_m)$, with $k_m = 0.287 * q_m + 0.554$ and $k_g = 0.78 * q_m + 0.006$

[d] $k_g = (1.42 * ME - 0.174 \times ME^2 + 0.0122\ ME^3 - 1.65)/ME$.

which were compared on the basis of the physiological and productive factors known to affect nutrient demand. More details of these simulations will be given in the appropriate sections. The safety margins used by the systems were not considered in this comparison.

Due to the direct connection between requirement estimation and supply of nutrients, dietary metabolizable energy concentration and intake had to be estimated for the calculation of energy and protein requirements. This was done by simulating feed and energy intake using normal physiological and productive conditions. Since feed and nutrient intake are calculated in different ways depending on the system considered, the comparison between feeding systems here described does not allow one to draw conclusions about the accuracy of the requirements predicted.

3.3 Energy Requirements

The NRC and INRA systems use a single energy unit for all functions (e.g. maintenance, milk production, pregnancy): metabolizable energy (ME) for NRC, UFL (forage unit for lactation, which corresponds to 1700 kcal of net energy for lactation) for INRA. More precisely, NRC presents energy requirements in terms of ME and gives a partial description of the net energy (NE) requirements used to estimate ME requirements.

The AFRC and CSIRO systems calculate specific NE requirements for each function. These requirements are then converted to ME using a specific conversion efficiency of ME to NE for each physiological function (Table 3.1). For this reason, the systems could be compared only by using a common unit, ME. Net energy was converted to ME for each system using the efficiencies of that system (Table 3.1). To accomplish this conversion it was necessary to calculate the ratio of ME:gross energy in the diet (called metabolizability or q_m) for the AFRC and INRA systems, and the concentration of ME in the diet (called M/D) for the CSIRO system, both calculated at maintenance feeding level. Requirements were estimated and compared using the same q_m, and the corresponding M/D, for all systems. When required, dietary gross energy was considered equal to 4.398 Mcal/kg DM (CSIRO, 1990).

In all four systems, the calculation of total requirements is based on a factorial approach; therefore the requirement for each important metabolic function is calculated separately and then added to the energy required for other functions.

3.3.1 Maintenance requirements

Energy requirements for maintenance are calculated by using a basic requirement, which is then adjusted for several factors (Table 3.2).

The basic NE maintenance requirements are lowest for AFRC and CSIRO and highest for INRA and NRC. The first two systems are based on the results of the fasting metabolism studies of ARC (1980). The other two systems acknowledge that heat production under fasting conditions involves decreased metabolic rates for energy conservation. For this reason, NRC adopted the extrapolated fasting metabolism values of Rattray *et al.* (1973). INRA based its requirements on values obtained by various techniques (INRA, 1978) (measurements taken after slaughtering, feeding trials, calorimetry studies and blood indicators). Recognizing that calorimetry studies gave lower values than those obtained with the other methods, an average of the reviewed measurements was adopted (INRA, 1978). Sheep maintenance energy requirements per kg of $BW^{0.75}$ with the INRA system are substantially lower than those used for cattle (INRA, 1988), or those proposed for goats (Morand-Fehr *et al.*, 1987), as part of the INRA system. The higher values for goats compared with sheep were justified on the

Table 3.2. Basic requirements and corrections applied to estimate maintenance energy requirements of adult sheep by four different feeding systems.

	AFRC	CSIRO	INRA	NRC
Basic requirements				
NE: kcal/kg $BW^{0.75}$	51.9[b]	60.3–51.9 (NEm)[d]	56.1	56
k_m or k_l[a]	0.643–0.748	0.647–0.758	0.563–0.637	0.729[f]
ME: kcal/kg $BW^{0.75}$	90.5–77.8	93.2–79.6 to 80.2–68.5[d]	95 (99.6–88.1)	76.8[f]
Corrections				
Breed	no	no	no	no
Age	no	1.0–0.84 (0–6 y)	no	no
Grazing activity	yes	yes	yes	no
Cold stress	no	yes	no	no
Temperature	no	yes	no	no
Rain	no	yes	no	no
Cold nights, clear sky	no	yes	no	no
Acclimatization	no	yes	no	no
Hide depth	no	no	no	no
Coat depth	no	yes	no	no
Production	no[c]	0.09 MEI[e]	no	no

[a] The range of k_m (AFRC and CSIRO) or k_l (INRA) is obtained considering q_m ranging from 0.4 to 0.7, for AFRC and INRA and M/D ranging from 7.36 to 12.88 MJ of ME per kg of DM, assuming GE = 4.398 Mcal/kg.
[b] For housed sheep an average activity allowance of 2.29 kcal/kg BW^1 is added.
[c] Total ME requirements (maintenance, production, pregnancy, etc.) are multiplied by a correction factor [1+0.018 (L–1)], where L is multiple of maintenance ME requirements, to account for the decrease in digestibility that occurs at feeding levels higher than 1.
[d] Variations related to the age of the animals; the range reported refers to cows from 1 (highest value) to 6 years old (the lowest); the requirement is highest for newborn lambs (62.1 kcal/kg $BW^{0.75}$ of NE) and minimum for sheep 6 years old or older.
[e] Multiplicative factor; MEI = daily intake of ME.
[f] The requirement of NEm here reported is explicitly given by NRC, even though its requirement tables do not adopt NEm but only ME; k_m and ME values are an estimated average based on NRC tables.

Table 3.3. Energy cost of various physical activities.

AFRC and CSIRO	
Activity	*NE cost per kg of BW*
Standing (vs. lying) (kcal/day)	2.390
Changing body position (2 movements: down and up) (kcal)	0.062
Walking (horizontal) (kcal/km)	0.621
Walking (vertical) (kcal/km)	6.692
AFRC only	
Housed lactating ewes (kcal/day)	2.294
Housed pregnant ewes (kcal/day)	1.300
Outdoor lowland ewes (kcal/day)	2.557
Hill-grazing ewes (kcal/day)	5.712
INRA	
Good-quality and abundant pasture	increase NEm by 20%
Mature and scarce pasture	increase NEm by 30–60%
Extensive, hilly pastures	increase NEm by 70%

basis that while sheep values were generally obtained on animals at maintenance feeding levels, those for goats were obtained on lactating animals (Morand-Fehr *et al.*, 1987). These results are confirmed by the observation, discussed later in more detail, that as feeding level increases, the weight of visceral organs and their metabolic activity increases as well, with considerable effects on maintenance requirements.

Basic ME requirements were more variable than NE requirements, due to the effect of the different efficiencies of conversion of NE to ME adopted, as later discussed (Section 3.3.4) (Table 3.1). The adjustment factors used to correct the basic maintenance requirements vary depending on the system (Table 3.2).

None of these systems correct requirements to account for possible breed differences. Nevertheless, the variability among sheep breeds for morphology, genetic merit and productive attitude is probably as high as for cattle. Differences in maintenance requirements were observed among Italian dairy sheep breeds and even within breeds among groups of sheep of different genetic merit (Pilla *et al.*, 1993).

The effect of age on maintenance requirements is considered only by the CSIRO system, which decreases the maintenance requirements from 60.3 to 51.9 kcal of NE per kg of $BW^{0.75}$ as the animal ages from 1 to 6 years. The AFRC and CSIRO systems estimate the cost of normal physical activity (rumination, normal movements such as walking or standing) by using the same energy requirements for movement originally proposed by ARC (1980), which are valid for all domestic ruminants (Table 3.3). However, while the CSIRO system suggests that an allowance for normal physical activity of non-grazing animals is inherent in the km used to convert NEm to MEm, AFRC does not include this cost in maintenance requirements. For this reason, it adds activity allowances to maintenance requirements, in amounts that vary depending on the bodyweight (BW) of the sheep and mean physical activity. The INRA system does not include an activity allowance in the maintenance requirements (INRA, 1989). The NRC system does not specify whether or not normal activity allowances are included in its basic maintenance requirements.

To account for activity during grazing, AFRC and CSIRO adopted ARC values (ARC, 1980), which are a function of the horizontal and vertical distance walked by the sheep while grazing (Table 3.3). These distances are quite difficult to estimate in practice. Therefore, AFRC also suggests average values for the most common situations. For the same reason, INRA gives estimates of activity requirements, expressed as a percentage of maintenance requirements, for the most common types of pastures (Table 3.3). NRC does not give any requirements for grazing.

Some of the systems adjust maintenance requirements to account for the effects of production. The CSIRO system increases maintenance requirements in proportion to the total daily ME intake (MEI) (Table 3.2), considering that the increase in feed intake associated with milk production or bodyweight gain triggers changes in both size and rate of metabolism of organs and tissues. Similar findings were reported by Ferrell (1988), Ortigues and Doreau (1995) and Ball *et al.* (1998). Graham (1982) (cited by CSIRO (1990)) maintained that because the effects of variations in feeding level on maintenance energy requirements occur slowly, they could not be detected by short-term calorimetric studies. The AFRC system suggests a correction of requirements in lactating animals as well. However, this correction, also proposed for cattle and goats, aims to account for the reduction of digestibility that occurs at high feeding levels. Therefore, AFRC proposes an artificial increase of the requirement to avoid modifications of the feed value of the diets. For this reason, this correction was not considered in the comparison. Maintenance ME requirements were estimated and compared for different BW and milk yields (Table 3.4).

The estimates were made assuming a constant qm for all cases, even though qm is usually higher in high-producing animals, which are fed better diets. The comparison showed that the highest maintenance requirements were those calculated by NRC. However, these values, reported in note a of Table 3.4 and obtained from the NRC (1985) allowance tables, are much higher than those obtained using the maintenance requirements of Table 3.2, which were also given by NRC in the introduction. No clear reasons for this discrepancy could be found. INRA maintenance requirements were, in turn, higher than those of AFRC and CSIRO (Table 3.4). CSIRO maintenance requirements, as said before, increase quite

Table 3.4. Energy requirements[a] for 4-year-old sheep in free stalls, which are 45 DIM and fed diets having a ratio: q_m = ME/GE = 0.6.

FCM[b]	50 kg BW				70 kg BW			
	AFRC	CSIRO		INRA	AFRC	CSIRO		INRA
(kg/day)	Total	Total	Maintenance	Total	Total	Total	Maintenance	Total
Mcal of ME/day[c]								
0	1.53	1.57	1.57	1.79	1.99	2.01	2.01	2.30
1	3.22	3.45	1.74	3.54	3.67	3.90	2.19	4.05
2	4.90	5.33	1.91	5.29	5.36	5.78	2.36	5.80
3	6.59	7.22	2.08	7.04	7.05	7.66	2.53	7.55
Indexes								
0	100	103	103	117	100	101	101	116
1	210	225	114	231	184	196	110	204
2	320	348	125	346	269	290	119	291
3	431	472	136	460	354	385	127	379

[a] Maintenance requirements (Mcal of ME/day) from NRC (1985): 50 kg = 2.00; 70 kg = 2.40. These values, which include the requirements for 10 g/day of BWC, are much higher than those that can be estimated from the requirements reported in Table 3.2.

[b] 6.5% fat-corrected milk yield = milk yield x (0.3688 + 0.0971 x % fat) (Pulina *et al.*, 1989).

[c] Assuming NE of milk equal to 1.030 Mcal/kg, qm equal to 0.6 and M/D equal to 11041 MJ/kg of DM. The resultant values of k_l and milk ME (Mcal/kg) used in this table are: 0.63 and 1.687 for AFRC; 0.621 and 1.712 for CSIRO; 0.607 and 1.751 for INRA, respectively.

substantially as milk production, and therefore MEI, increases (Table 3.4).

3.3.2 Milk production requirements

Both AFRC and CSIRO estimate milk net energy cost of production by using an equation developed from data obtained using wool breeds. This equation includes milk fat content and days in milk as predictors. INRA uses an equation based on milk fat only, which was developed for dairy ewes. NRC gives only a fixed NE for milk, assuming 7.1% milk fat content. Moreover, it does not give information on the milk production and on the efficiency of conversion of NE for lactation to ME used to estimate total ME requirements in lactating sheep (Table 3.5). For all these reasons, NRC milk production requirements were not considered in Table 3.4.

Table 3.5. Equations proposed to estimate milk energy content or milk energy values.

System	Energy (Mcal of NE/kg)
AFRC	0.0328 * BF + 0.0025 * DIM + 2.203
CSIRO	0.0328 * BF + 0.0025 * DIM + 2.203
INRA	(0.00588* FAT + 0.265)*1.7
NRC	1.100 (milk: 7.1% BF, 4.5% CP, 4.8% lactose)

BF: butterfat content, g/kg; FAT: butterfat content, g/litre; DIM: number of days of lactation.

The differences in ME requirements for milk production are due to the different efficiencies used to convert NE to ME (Tables 3.1 and 3.4). Metabolizable energy requirements per kg of milk are highest for INRA, intermediate for CSIRO and lowest for AFRC (note c, Table 3.4).

3.3.3 Maintenance and milk production requirements

Total requirements (sum of maintenance and milk production requirements) were quite similar between CSIRO and INRA and about 10% lower for AFRC (Table 3.4). The large differences found for maintenance requirements were in part diluted by milk production requirements (Table 3.4).

3.3.4 Efficiency of conversion of NE to ME for maintenance and milk production

The efficiencies of conversion of ME to NE vary depending on the type of requirements (Moe *et al.*, 1972; Moe, 1981) (Table 3.1) and on the system considered. AFRC, CSIRO and INRA use the same values for cattle and sheep.

As stated before, AFRC and INRA estimate these efficiencies with equations that use the metabolizability (q_m) of the diet as the independent variable. CSIRO uses M/D, in order to account for the higher efficiency of ME utilization observed for high-quality diets (high q_m and M/D) compared with low-quality ones (ARC, 1980).

AFRC and CSIRO use separate equations for maintenance and milk production (Table 3.1). Both systems adopt the ARC (1980) equation, originally developed by Blaxter and Boyne (1974) (cited by ARC, 1980), for maintenance. More precisely, CSIRO suggests the use of either the Blaxter and Boyne equation or a rewritten form of that equation in which the independent variable is the M/D, assuming an average gross energy of 4.398 Mcal/kg of DM and a maintenance level of nutrition. Both AFRC and CSIRO use the ARC (1980) equation for milk production, although CSIRO proposes also a rewritten form using M/D as the independent variable.

With the goal of using only one NE unit, INRA adopted an equation (Table 3.1), developed by Van Es (1975) (cited by INRA (1978)) for both maintenance and milk production but originally proposed only for milk production. This choice was justified by the similar slopes that k_m and k_l assume when related to q_m.

The NRC model does not clearly state which efficiency was used to convert NE for maintenance and for lactation to ME. In the case of k_m, the value reported in Table 3.2

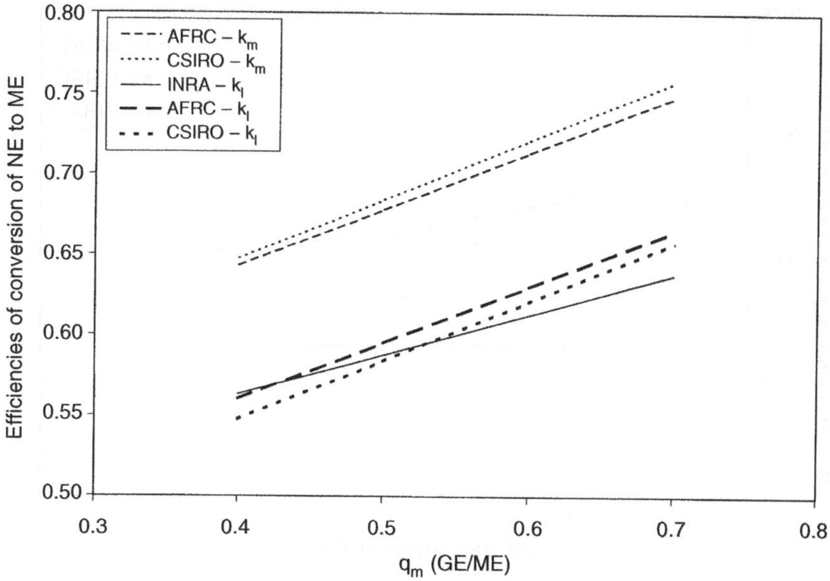

Fig. 3.1. Relationship between dietary metabolizability (q_m = GE/ME) and efficiencies of conversion of NE for maintenance and lactation to ME in AFRC, CSIRO and INRA feeding systems.

was extrapolated from NRC requirement tables. In the case of k_l, NRC reports values between 0.65 and 0.83. It was not possible to extrapolate the actual values used to compute ME requirements because, in the case of lactating ewes, the milk yield used to compute the requirements was not given.

The comparison of the efficiency of utilization of ME for maintenance showed that AFRC and CSIRO estimated higher efficiencies than INRA, which uses the same k_l for maintenance as for lactation (Fig. 3.1). The difference between AFRC and CSIRO on one hand, and INRA on the other hand, ranged between 0.08 and 0.12 units of efficiency. A comparison of the efficiencies used for milk production showed that the differences among systems were much smaller than those found for maintenance (Fig. 3.1).

On the basis of the ratio between maintenance requirements and total requirements, it was possible to calculate weighted efficiencies of these two functions (Fig. 3.2). Their comparison showed that the AFRC and CSIRO systems convert similarly. Their weighted efficiencies decrease, when q_m is kept constant, as milk production increases. Indeed, k_m becomes less important as milk yield increases. The INRA system assumes constant efficiency, at equal q_m, regardless of milk yield. The difference in weighted efficiency of conversion of ME to NE between AFRC and CSIRO on one hand, and INRA on the other hand, ranged between 0.04 and 0.06 units depending on the milk yield considered.

3.3.5 Cold-stress requirements

Cold stress, in general, causes an increase in energy maintenance requirements, but AFRC, INRA and NRC do not use any adjustment for cold. In contrast, CSIRO developed a complex set of equations to account for many environmental factors (temperature, wind, rain, cold nights and clear sky, radiant heat losses), for energy intake, and for animal factors (body weight and surface, wool, acclimatization). The same set of equations, with small modifications, was used to estimate the requirements of cold-stressed cattle (CSIRO, 1990).

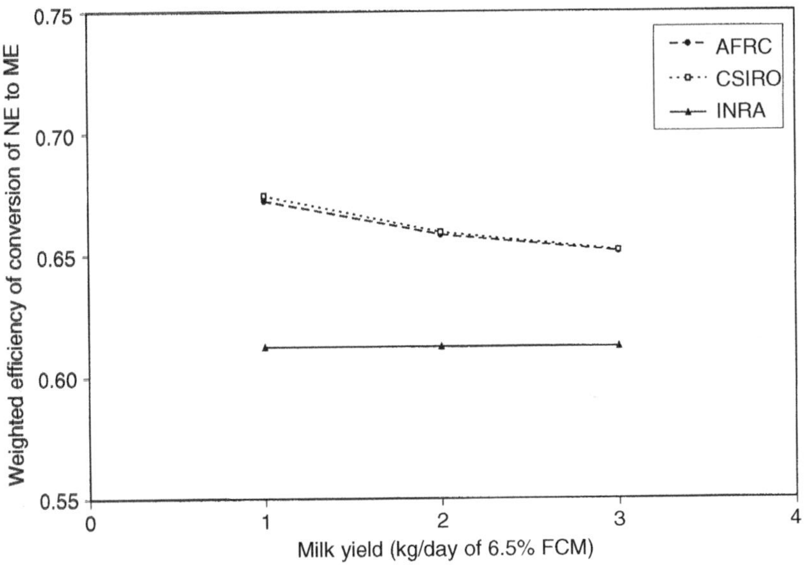

Fig. 3.2. Relationship between milk production and weighted efficiencies of conversion of NE to ME in AFRC, CSIRO, and INRA feeding systems, considering the requirements of 60 kg sheep calculated in Table 3.4. AFRC and CSIRO use different values for maintenance and milk production, while INRA uses the same efficiency (k_l) for both. For AFRC and CSIRO systems the efficiency for maintenance and milk production were weighted based on the relative contribution to total requirements of ME requirements for maintenance and milk production.

The effect of cold stress on maintenance requirements was simulated by considering the effect of wind, rain, temperature, wool thickness and physiological stage on sheep of 50 kg BW (Table 3.6). The results of this simulation showed that lactating animals were less affected by cold stress than were dry animals. This is because the high energy intake necessary for sustaining milk production increases the heat produced by the body and by the rumen, which consequently alleviates the effects of cold stress.

Wool thickness is also very important in reducing the effects of cold stress (Table 3.6), because of its thermoinsulating properties. However, wind or rain can markedly reduce its protective action. In the simulation, the combined effects of all these factors increased maintenance requirements by up to threefold. These effects were much higher than those found in a similar simulation with dairy cows (Cannas, 2000). Indeed, since small animals have a higher body surface per kg of BW than do large animals, they disperse more heat (Blaxter, 1977; CSIRO, 1990). Even though sheep wool is a much better insulator than cattle hair, its additional insulation is less than the effects of body size.

3.3.6 Heat-stress requirements

None of the systems for sheep considers requirements for heat stress. Only the Cornell Net Carbohydrate and Protein System for cattle includes a sub-model to predict the extra energy requirements associated with heat stress (Fox *et al.*, 2004).

3.3.7 Pregnancy requirements

Pregnancy energy requirements are calculated by using different approaches. AFRC and CSIRO use the same approach and provide very similar equations: firstly they calculate the total energy content of the gravid

Table 3.6. Effect of coat depth, wind, rainfall and current mean daily (24 h) temperature on cold-stress requirements of 50 kg adult ewes.

	25 mm coat				50 mm coat			
Wind (km/h)	calm		30		calm		30	
Rainfall (mm/day)	0	30	0	30	0	30	0	30
Adult, dry								
Temp. +5°C	115	134	234	247	100	103	183	195
Temp. 0°C	129	149	267	280	100	114	208	220
Temp. –5°C	144	164	300	313	109	124	233	245
Adult, lactating								
Temp. +5°C	100	100	125	133	100	100	107	114
Temp. 0°C	100	100	137	145	100	100	116	123
Temp. –5°C	100	104	149	157	100	100	125	132

MEI sufficient to satisfy in thermoneutral conditions (15–20°C) the maintenance requirements of dry ewes and the maintenance and milk production requirements of lactating ewes producing 1.5 kg of milk/day with 6.5% fat. Total maintenance requirements are expressed as index, with maintenance requirements in thermoneutral condition equal to 100.

uterus using the ARC (1980) Gompertz model, which uses lamb birthweight and day of pregnancy as variables. As stated by CSIRO, these equations do not take into account the costs associated with mammary gland growth and colostrum secretion, which occur in the last days of pregnancy. Both AFRC and CSIRO use the same equations, with different coefficients, for the pregnancy requirements of both sheep and cattle. A simpler approach is used by the other systems. Taking pregnancy into account, the INRA system adjusts the requirements only in the last 8 weeks of pregnancy. These requirements are adjusted every 2 weeks and are a function of total lamb weight at birth. Similarly, NRC increases energy requirements for the last 6 weeks of pregnancy in amounts that vary depending on the stage of pregnancy and on the expected weight at lambing.

All systems use very low efficiencies to convert ME to NE for pregnancy, with values ranging from 0.133 (AFRC and CSIRO) to 0.17 (NRC). These values have been obtained either by calorimetric studies or by slaughtering trials and varied very little, regardless of the method used (INRA, 1978). The low efficiency of utilization of

Table 3.7. Pregnancy energy and protein requirements for ewes producing a 4.0 kg lamb.

	AFRC	CSIRO	INRA	NRC
Requirements at the 147th day of pregnancy				
NE (Mcal/day)	0.167	0.167	0.176	0.208
ME (Mcal/day)	1.255	1.284	1.304	1.224
NP (g/day)	23	23	23	17
MP (g/day)	27	33	55	26
Total requirements for a 147-day pregnancy				
NE (Mcal)	5.9	5.9	6.0	7.3
ME (Mcal)	44.4	45.4	41.6	43.1
NP (g)	830	830	756	820
MP (g)	977	1186	1800	1242

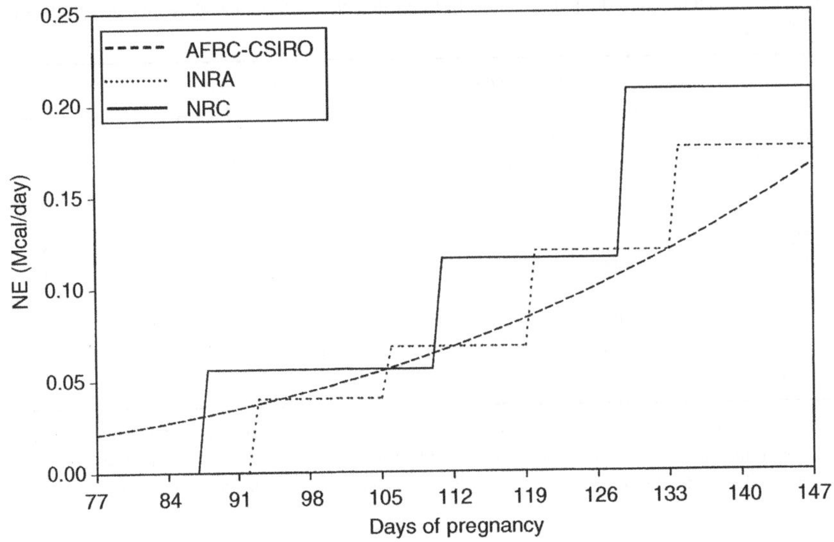

Fig. 3.3. Net energy requirements for pregnancy, assuming lambs weigh 4 kg at birth.

ME is somehow surprising when these values are compared with those used for other physiological functions (Table 3.1). However, this depends on the method of calculation used, because all energy costs of pregnancy (growth and maintenance of the uterus, maintenance of the fetus, increased maternal metabolism) are expressed as a function of conceptus growth.

Pregnancy requirements were compared assuming that ewes deliver a 4 kg lamb at birth (Table 3.7 and Figs 3.3 and 3.4). The NE and ME energy requirements of pregnancy were very similar among systems for both the last day of pregnancy and a complete pregnancy (Table 3.7), despite the fact that the shape of the energy requirement curve (curvilinear or stepwise)

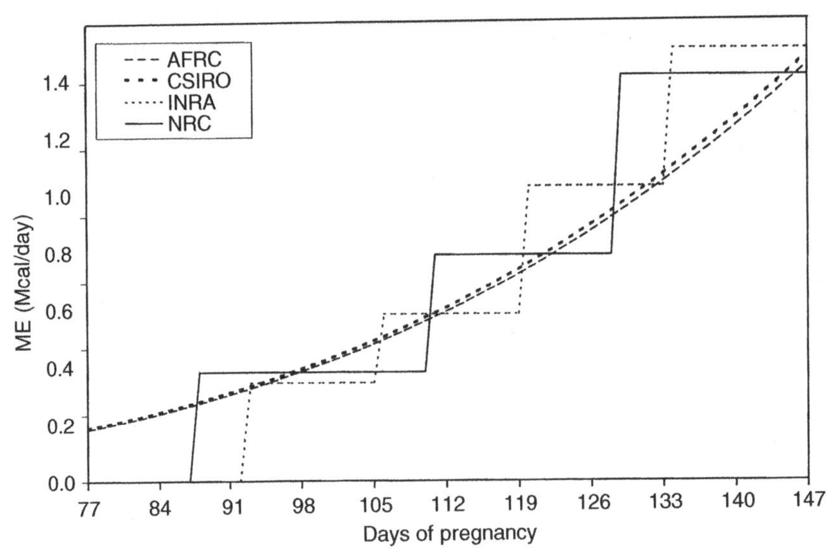

Fig. 3.4. Metabolizable energy requirements for pregnancy, assuming lambs weigh 4 kg at birth.

and the length of the period differed among systems. These results are in contrast to a comparison made with pregnant cows (Cannas, 2000), which showed large differences in energy requirements for pregnancy, especially for NE, among systems both on the last day of pregnancy and over the complete pregnancy. This was mostly an effect of the very high efficiency of conversion of ME to NE adopted by NRC (1988) and INRA (1989) for cattle.

3.3.8 Bodyweight changes

In sheep, after the first lactation, the requirements for growth are very low, and almost all bodyweight change (BWC) is due to variations in either gut contents or body reserves (i.e. fat, protein and water).

The AFRC and INRA systems consider a fixed net energy value for each kg of body tissue mobilized or accumulated by mature animals (Table 3.8). For growing ewes, the CSIRO model devised growth curves that allow the estimation of the energy content and composition of bodyweight gain, depending on the ratio between actual BW and mature size BW and on the rate of gain achieved during growth. However, in mature animals, these factors become marginal and body composition can be considered equal for all sheep having the same degree of fatness. Despite the different approaches considered, AFRC and CSIRO use very similar NE values for BWC (Table

Table 3.8. Energy and protein variations and efficiencies associated with bodyweight changes in mature sheep.

	AFRC	CSIRO	INRA	NRC
Loss				
Energy from body reserves (Mcal NE/kg BW loss)	5.700	5.827	3.800	ng
Efficiency of conversion of NE from BW loss to NEL	0.84	0.84	0.80	ng
NEL for milk production from BW loss (Mcal/kg BW loss)	4.788	4.895	3.040	ng
Milk from BW loss (kg of milk/kg BW loss)[a]	4.50	4.60	2.86	ng
Protein from body reserves (g of NP/kg BW loss)	119	70	ng	ng
Protein from body reserves (g of MP/kg BW loss)	119	100[c]	ng	ng
Body protein used for milk protein production	ng	80% of NP	ng	ng
Gain				
Energy required for body reserves (Mcal NE/kg BW gain)	5.700	5.827	9.200	ng
Energy required for body reserves (Mcal ME/kg BW gain)	9.516	9.712	15.157	ng
Efficiency of conversion of NE to ME for BW gain in lactation[b]	0.599	0.600	0.607	ng
Protein required for body reserves (g of NP/kg BW gain)	83	70	ng	ng
Protein required for body reserves (g of MP/kg BW gain)	140	100	220	ng

ng, not given.
[a] Assuming 1.063 Mcal of NE for 1 kg of milk with 6.5% fat.
[b] 0.95 k_l for AFRC and k_l for INRA; values estimated assuming q_m=0.6 or M/D equal to 11.041 MJ/kg of DM.
[c] This should not be considered a protein supply in non-lactating ewes (CSIRO, 1990).

3.8). In contrast, INRA uses a much lower NE value for BW loss and a much higher NE value for BW gain than the two other systems. When the deuterium oxide method was used to measure body energy content, the fat concentration of gain (90% lipid) was much higher than the fat concentration of loss (50% fat) (Theriez *et al.*, 1987). In its requirement tables, NRC reported BWC but did not give sufficient information to separate the cost of fluctuation in reserves from the other requirements. The NE content of BWC is about 25% higher for either loss or gain for sheep than for cattle, in the AFRC system.

In the CSIRO method, the NE content of BWC for cattle varies depending on the animal's BW and body condition score (BCS), while for sheep a fixed mean value is given. The INRA system for cattle uses a single value for the NE content of BWC, while for sheep much lower values are proposed for live weight losses than for liveweight gains.

The CSIRO and INRA models quantify body reserve changes also in terms of BCS systems, using a 0–5 scale. In the INRA system each change in BCS reflects a 13% change in BW, while in the CSIRO system the standard reference weight changes by 15%. Compared with the approaches used for sheep, the BCS systems adopted for cattle by CSIRO and Cornell Net Carbohydrate Protein System (CNCPS) (Fox *et al.*, 2004) are much more complex, because the energy associated with each BCS varies depending on the BCS considered and on the BW of the animals. The author of the CSIRO publication recognized that their sheep BCS system did not account for differences in composition of body reserves but felt the sheep data were insufficient to make the same adjustments as for cattle. Another aspect to consider relates to breed differences. Indeed, the comparison of Sarda and Lacaune sheep, the two most common dairy breeds, showed that at BCS 2 Sarda sheep had large amounts of visceral fat, while this occurred at BCS 3 in Lacaune sheep (Ronchi *et al.*, 1993). These differences cannot be explained simply in terms of body size, but none of the systems consider the effects of breed on body composition.

3.4 Protein Requirements

The AFRC, CSIRO and INRA systems express protein requirements as metabolizable protein (MP), i.e. the feed or bacterial protein absorbed by the small intestine. The intestinal digestibility of feeds and microbial proteins is considered in terms of *truly* digested protein by AFRC and INRA and in terms of *apparently* digested protein by CSIRO. INRA calls metabolizable protein PDI (protéines digestibles dans l'intestin), CSIRO calls it ADPLS (apparently digestible protein leaving the stomach) and AFRC calls it MP. In contrast, NRC expresses requirements in terms of CP, assuming that net protein (NP) is equal to 0.561 CP. This value was obtained assuming a true protein digestibility equal to 0.85 and a biological value of CP equal to 0.66. This last coefficient can be used to estimate NRC requirements in terms of MP, even though the NRC system does not partition the nitrogen absorbed by the intestine into true protein and non-protein nitrogen. Because of this difference, the estimates obtained in this way are not fully comparable with those of the other systems.

The AFRC, CSIRO and NRC systems based their calculation of total requirements on a factorial approach, in which the different protein losses and the efficiencies of conversion of NP to MP are calculated independently and then added to obtain total requirements. The INRA system did not use this approach, but instead established its requirements on the basis of the results of feeding trials (carried out by INRA or taken from the literature) comparing each important physiological function.

3.4.1 Maintenance requirements

All systems recognize that minimum maintenance NP requirements are represented by the quantity of the endogenous protein lost

Table 3.9. Maintenance and milk production metabolizable protein (MP) requirements.

	AFRC[a]	CSIRO	INRA	NRC
Maintenance				
Wool (g/day)	CW 0.8 /0.26	CW/0.6	0.0492 GF $BW^{0.75}$/0.41	3.81/0.66[b]
UEP (g/day)	2.1875 $BW^{0.75}$/1	(0.147 BW+ 3.375)/0.7		(0.147 BW+ 3.375)/0.66
FEP (g/day)		15.2 DMI/0.7		33.44 DMI/0.66
Total (g/day)			0.828 $BW^{0.75}$/0.41[c]	
Milk production				
MP (g/day)	TP/0.68	TP/0.7	TP/0.59	TP/0.66

UEP, urinary endogenous protein; FEP, faecal endogenous protein; CW, clean wool (g produced per day); TP, milk true protein (g/day); DMI, dry matter intake (kg/day); GF, greasy fleece (kg/year).

[a] AFRC uses a single unit (called total endogenous nitrogen, TEN) that accounts for both UEP and FEP.

[b] Assuming an annual greasy fleece production of 4 kg; a dermal loss equal to 0.1125 $BW^{0.75}$ should be added to wool requirements.

[c] Does not include any requirements to produce wool.

from dermal tissues (scurf and wool), in urine (urinary endogenous protein, UEP) and in faeces (faecal endogenous protein, FEP). However, the approach used to estimate maintenance protein requirements is quite different depending on the system considered (Table 3.9).

The AFRC system adopted the ARC (1984) approach, which partitions maintenance protein requirements into wool losses and total endogenous nitrogen, which, in turn, includes UEP and FEP. These requirements are used for all levels of feeding and only vary depending on wool production and $BW^{0.75}$.

The INRA model adopts the INRA (1978) method, in which protein maintenance requirements include a single component, proportional to $BW^{0.75}$ and thus valid for all feeding levels. These requirements were based on the results of nitrogen balance trials, carried out in animals fed at near-maintenance levels. Requirements for wool growth were added to the maintenance requirement. This relatively simple approach was adopted because the French scientists were unsure of their estimates of FEP and of the conversion of FEP to MP.

The NRC and CSIRO systems use the same equation to estimate UEP losses, which are directly proportional to BW. In the NRC system, FEP losses are assumed to equal 33.4 g/kg of DMI.

This value is the intercept of the linear equation obtained by regressing digestible protein on intake protein. On the basis of this estimate, NRC expresses FEP requirements as a proportion of DMI. In the CSIRO system, based on Nolan's data (cited by CSIRO (1990)), the true endogenous protein losses were equal to 15.2 g/kg DMI, about half of the value used by NRC. The remaining fraction of FEP is considered to be of exogenous origin and is thus not included in the requirements.

The efficiencies used to convert NP to MP vary considerably between systems (Table 3.9), ranging from 0.26 to 1.0 depending on the system and on the physiological function considered. Despite their large effect on MP requirements, the origin of these coefficients is often unclear.

The comparison between systems of the maintenance requirements (Table 3.10) was done by simulating conditions similar to those used for the comparison of the energy requirements. Wool requirements were not included in this comparison. To estimate FEP using the CSIRO system, it was necessary to estimate DMI. This was done using intake prediction equations developed by Pulina *et al.* (1996) for dairy sheep. The MP maintenance requirements were similar for the AFRC, CSIRO and INRA systems (Table 3.10). The NRC values were more than twice those predicted by the other systems. This was the result of the very high FEP

Table 3.10. Metabolizable protein requirements for adult sheep, expressed as g/day and relative to the AFRC requirement for dry sheep, which was set at 100. MP requirements for wool production were not included.

5% true protein milk (kg/day)	50 kg BW						70 kg BW					
	AFRC	CSIRO[a]		INRA	NRC[a]		AFRC	CSIRO[b]		INRA	NRC[b]	
	Total	Total	Maint.	Total	Total	Maint.	Total	Total	Maint.	Total	Total	Maint.
g/day of MP[c]												
0	41	41	41	38	95	95	53	52	52	49	97	97
1	115	126	54	123	191	115	126	137	65	134	203	127
2	188	210	67	207	288	136	200	221	78	218	309	158
3	262	295	80	292	381	154	274	305	91	303	415	188
Indexes												
0	100	100	100	93	232	232	100	98	98	92	183	183
1	280	307	132	300	466	280	238	258	123	253	383	240
2	459	512	163	505	702	332	377	417	147	411	583	298
3	639	720	195	712	929	376	517	575	172	572	783	355

[a] Based on the hypothesis that DMI is equal to 1.2, 1.8, 2.4 and 3.0 kg/day for dry ewes or lactating ewes producing 1, 2 and 3 kg/day of milk, respectively.
[b] Based on the hypothesis that DMI is equal to 1.5, 2.1, 2.7 and 3.3 kg/day for dry ewes or lactating ewes producing 1, 2 and 3 kg/day of milk, respectively.
[c] MP requirements to produce 1 kg of milk with 5% true protein: 74 g for AFRC; 71 g for CSIRO, 85 g for INRA and 76 g for NRC.

level predicted by this system. Although the CSIRO and NRC methods agree on the amount of metabolic faecal nitrogen produced per kg of DMI, there is considerable disagreement on how much of it should be considered truly endogenous.

3.4.2 Milk production requirements

In all systems, NP requirements for milk production are equal to the amount of true protein contained in the milk. However, the efficiency of conversion of MP to NP varies with the system considered (Table 3.9), ranging from 0.59 (INRA) to 0.70 (CSIRO). This variation is responsible for the differences observed in MP requirements for milk production, which ranked: INRA > NRC > AFRC > CSIRO (Note c, Table 3.10).

3.4.3 Maintenance and milk production requirements

Total MP requirements (maintenance + milk production) were by far the highest with the NRC system, lowest with the AFRC system and intermediate with the CSIRO and INRA systems, which, despite the different approaches used, gave similar estimates (Table 3.10).

3.4.4 Pregnancy requirements

As was the case with energy requirements for pregnancy, the pregnancy requirements for protein are calculated using different approaches.

Similar to the energy requirements, AFRC and CSIRO use the ARC (1980) Gompertz model for estimation of the total protein content of the gravid uterus, using lamb birthweight and day of pregnancy as variables. A simpler approach is used by the other systems. The INRA system considers pregnancy requirements to be important in the last 8 weeks of pregnancy. During this period, INRA suggests that NP requirements in pregnancy should be varied every 2 weeks based only on the expected litter weight at birth. NRC considers only two stages of pregnancy: early pregnancy and the last 4 weeks, suggesting that NP supplies vary depending on expected weight at

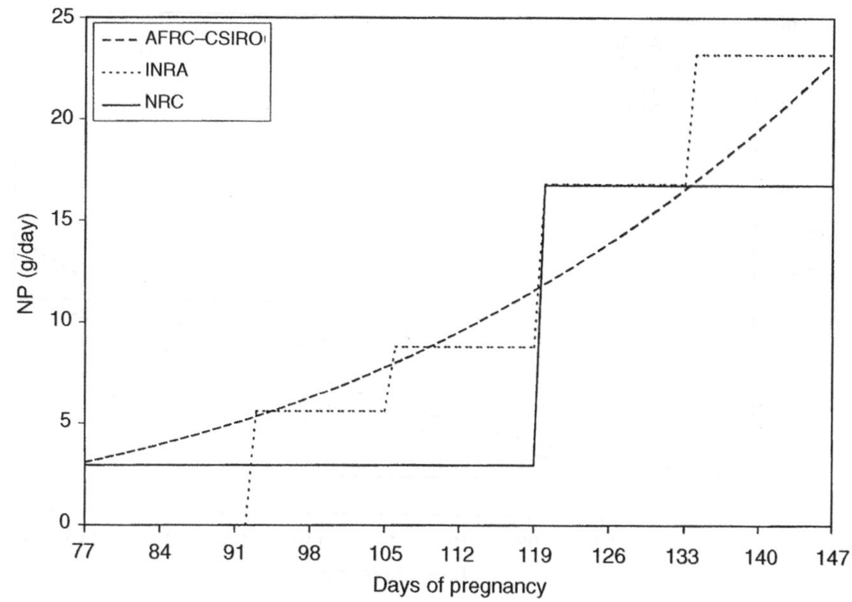

Fig. 3.5. Net protein requirements for pregnancy, assuming lambs weigh 4 kg at birth.

birth. NP requirements are converted to MP requirements with coefficients that range from 0.41 (INRA) to 0.85 (CSIRO).

Pregnancy requirements were compared using a 4 kg lamb at birth. The comparison showed that NP requirements of AFRC and CSIRO systems were identical (Table 3.7 and Fig. 3.5). However, due to the different coefficients used to convert NP to MP, MP requirements differed between AFRC and CSIRO (Table 3.7 and Fig. 3.6). The NP requirements of the INRA system were close to those of AFRC and CSIRO, but much higher than any other system when expressed in terms of MP, because of the low efficiency coefficient adopted (Table 3.7 and Fig. 3.6). Despite the different approaches used, NP and MP pregnancy requirements of NRC were fairly close to the estimates of AFRC and CSIRO.

3.4.5 Bodyweight changes

During lactation, changes in body fat reserves are accompanied by changes in protein reserves (Table 3.8). The AFRC model adopts the ARC (1980) value of 119 g of NP/kg BWC. No information is given on the efficiency of utilization of mobilized protein for milk production or for any other use. In the case of dairy cows, AFRC considered this efficiency equal to that occurring for any other MP source.

The CSIRO system adopts a value of 70 g of NP/kg of BWC for both BW gain and loss. The proposed efficiency of utilization of the protein mobilized for milk protein synthesis is equal to 0.8 NP. The INRA model gives only the MP requirements for BW gain. This requirement is much higher than those proposed by AFRC and CSIRO. The NRC system does not discuss this subject.

3.5 Concluding Remarks on the Systems Compared

Due to the direct connection between requirement estimation and supply of nutrients, the comparison between feeding systems here described does not allow conclusions to be drawn on the accuracy of the requirements predicted. Nevertheless, it highlighted many differences in the

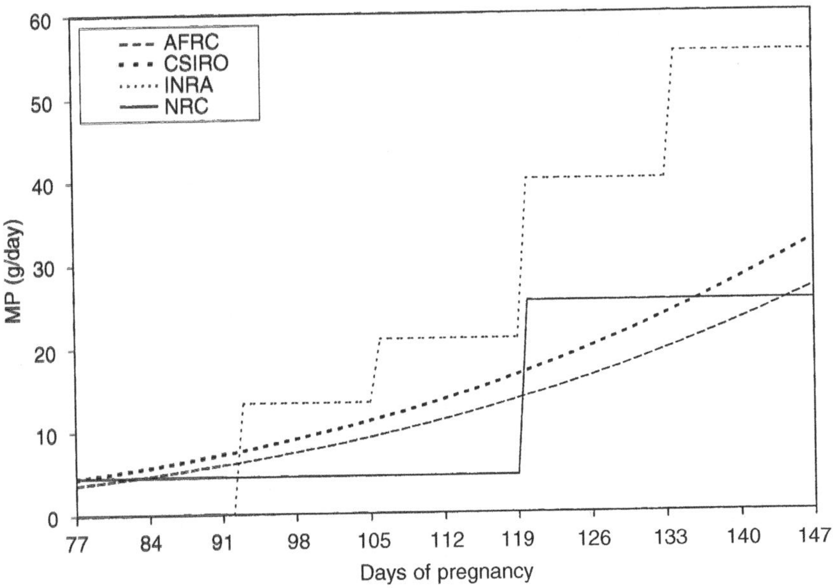

Fig. 3.6. Metabolizable protein requirements for pregnancy, assuming lambs weigh 4 kg at birth.

approaches used and in the estimates given.

The NRC system, probably reflecting the extensive range systems which encompass most sheep production in the USA, accounts for very few variables and does not include some important variables, such as milk production and liveweight changes. Moreover, the information on maintenance energy requirements given in the text contrasts with the allowances reported in the tables. These limitations are somehow counterbalanced by its ease of use. The most striking aspect of the NRC predictions was that the MP required for maintenance was almost twice as high as that of any other system.

The AFRC and CSIRO systems have many similarities, because they based many of their requirements on the ARC (1980) systems. Pregnancy requirements and energy efficiencies, in particular, are identical. The estimates made by AFRC on total requirements for milk production are the lowest of the systems considered, for both energy and protein.

The CSIRO system accounts for many more variables than the AFRC system. The submodels on maintenance energy and protein requirements account for many variables not considered by the other systems. The use of maintenance energy and protein requirements that vary with feeding level, the use of a standard reference weight for estimating body composition, and, above all, the introduction of submodels that account for many variables related to the effects of cold stress are very innovative.

The INRA system is the only one specifically developed for dairy ewes. Compared with CSIRO, it is based on a simpler approach and considers fewer variables in most of the submodels. It does not account for cold stress, which may have dramatic effects on the requirements of sheep. Even though it does not consider the effect of feeding level on maintenance requirements, its estimates of total requirements for milk production are similar to those of CSIRO. This suggests that the underestimation of feed required, attributed to INRA in the introduction, is mostly due to an overestimation of feed values, or that both CSIRO and INRA underestimate energy requirements.

The AFRC and INRA systems seem to have used similar levels of aggregation and flexibility in their cattle and sheep requirement systems, even though most of the equations and values used for sheep are different from those used for cattle. A very important and unique characteristic of the CSIRO system is that most of its equations were developed for use in both cattle and sheep, differing only in the coefficients used or, sometimes, not differing at all. This reconciles the necessity of providing estimates of requirements with biology. With greater application than previously utilized, many of the differences between sheep and cattle can be explained using appropriate scaling factors.

3.6 New Feeding Systems for Sheep

Two new feeding systems for sheep have recently been published (Cannas, 2001; Cannas *et al.*, 2004). These feeding systems basically use the same set of equations for predicting energy and protein requirements but adopt two different approaches to predict dietary nutrient supply. The first system (Cannas, 2001), called ROC 3, predicts feed energy and protein value by using a modified version of the discount system of Van Soest and Fox (1992). The second system (Cannas *et al.*, 2004), called CNCPS Sheep, predicts ruminal function and feed value with a modified version of the Cornell Net Carbohydrate and Protein System (CNCPS) for cattle (Fox *et al.*, 2004).

In both systems, the energy and protein requirement submodels were obtained by integrating equations derived from the ARC (1980), the INRA (1989) and the CSIRO (1990) systems for sheep, and from the CNCPS (Fox *et al.*, 2004) system for cattle. New equations were developed for the prediction of body reserves in adult sheep. Some details of the CNCPS Sheep energy and protein requirements submodel are given below.

ME requirements were developed for sheep primarily from equations of the ARC (1980) and CSIRO (1990) systems. In the CNCPS Sheep, energy requirements for basal metabolism are adjusted for age, gender, physiological state, environmental effects, physical activity, cost of urea excretion, acclimatization and cold stress in order to estimate total ME requirements for maintenance. ME requirements for milk production are estimated from milk NE, based on measured milk yield, fat and true protein (Pulina *et al.*, 1989). NE is converted to ME with an efficiency of 0.644. Pregnancy ME requirements are estimated using ARC (1980) and CSIRO (1990) equations. Maintenance MP requirement is calculated as the sum of wool, urinary and faecal endogenous protein losses, following the CSIRO (1990) model. MP requirement for milk production is predicted from measured milk yield and its true protein content. The coefficient for conversion of MP to NP (0.58) is that suggested for sheep by the INRA system (INRA, 1987). Protein requirements for pregnancy are calculated using the recommendations of ARC (1980) and CSIRO (1990). The energy available for growth (young sheep) or for body reserves changes (mature ewes or rams) depends on the energy balance after maintenance, lactation and pregnancy requirements are satisfied. The same approach is used for MP available for growth or for body reserves changes.

The CSIRO (1990) growth model is used to predict ME and MP requirements for growth of lambs. New equations were developed for estimating the relationships between body condition score (BCS, 0–5 scale), full BW (FBW) and body composition in adult sheep. A model that predicts the relationship between FBW and BCS in various sheep breeds was developed. It allows the prediction of current FBW when BCS and breed mature weight at BCS 2.5 (FBW@BCS 2.5) are known inputs, or the prediction of FBW@BCS 2.5 when current BCS and FBW are known. The relationship between the proportion of fat in the empty body (AF) and BCS was predicted with the equation of Russel *et al.* (1969). The relationship between the proportion of protein in the empty body (AP) and BCS was estimated assuming that the ratio AF:AP for various BCS reported by Fox *et al.*, (2004) for cattle is also valid in sheep. This model was evaluated from published experiments with sheep of diverse body sizes and physiological stages, fed diverse diets at various levels of nutrition (Cannas *et al.*, 2004).

Even though integrated information was derived from different classic feeding systems, the submodels used to predict energy and protein requirements by the CNCPS Sheep (Cannas *et al.*, 2004) and the ROC 3 system (Cannas, 2001) were mainly influenced by the CSIRO (1990) system. Compared with this system, they adopted a more complete and mechanistic body reserve model for adult sheep and are able to account for the effect of mature size on the cost of body reserve variation. In addition, the CNCPS Sheep is able to predict the effect of urea excretion on energy requirements. However, the main differences between these two feeding systems and the previous ones are related to the approach used in predicting nutrient supply. Indeed, the newest feeding systems account for the effect of feeding level on feed digestibility and value. The CNCPS Sheep also includes a mechanistic rumen model.

References

Agricultural and Food Research Council (AFRC) (1990) Technical Committee on response to nutrients. Report no. 5. Nutritive requirements of ruminant animals: energy. *Nutr. Abstr. Rev., Series B*, 60: 729–804.

Agricultural and Food Research Council (AFRC) (1992) Technical Committee on response to nutrients. Report no. 9. Nutritive requirements of ruminant animals: protein. *Nutr. Abstr. Rev., Series B*, 62: 787–835.

Agricultural and Food Research Council (AFRC) (1995) *Energy and protein requirements of ruminants.* CAB International, Wallingford, UK.

Agricultural Research Council (ARC) (1980) *The Nutrient requirements of ruminant livestock.* Tech. Rev. Agric. Res. Council Working Party. Commonwealth Agricultural Bureaux, Farnham Royal, UK.

Agricultural Research Council (ARC) (1984) *The Nutrient requirements of ruminant livestock.* Suppl. No.1. Report of the Protein Group of the ARC Working Party. Commonwealth Agricultural Bureaux, Farnham Royal, UK.

Ball A.J., Thompson J.M., Alston C.L., Blakely A.R., Hinch G.N. (1998) Changes in maintenance energy requirements of mature sheep fed at different levels of feed intake at maintenance, during weight loss and realimentation. *Livest. Prod. Sci.*, 53: 191–204.

Blaxter K.L. (1977) Environmental factors and their influence on the nutrition of farm livestock. In: W. Haresign, H. Swan, D. Lewis (eds) *Nutrition and the climatic environment.* Butterworths, London, pp. 1–16.

Cannas A. (2000) Sheep and cattle nutrient requirement systems, ruminal turnover, and adaptation of the Cornell Net Carbohydrate and Protein System to sheep. PhD. dissertation, Cornell University, Ithaca, New York.

Cannas A. (2001) Le esigenze energetiche e proteiche (Energy and protein requirements). In: G. Pulina (ed.) *L'alimentazione degli ovini da latte,* Avenue media, Bologna, Italy, pp. 33–65.

Cannas A., Fox D.G., Tedeschi L.O., Pell A.N., Van Soest P.J. (2004) A mechanistic model to predict nutrient requirements and feed biological values for sheep in each unique production situation. *J. Anim. Sci.*, 82: 149–169.

Commonwealth Scientific and Industrial Research Organisation (CSIRO) Standing Committee on Agriculture. Ruminants Subcommittee (1990) *Feeding standards for Australian livestock. Ruminants.* CSIRO Publications, East Melbourne, Australia.

Ferrell C.L. (1988) Contribution of visceral organs to animal energy expenditures. *J. Anim. Sci.*, 66 (Suppl. 3): 23–34.

Fox D.G., Tedeschi L.O., Tylutki T.P., Russell J.B., Van Amburgh M.E., Chase L.E., Pell A.N., Overton T.R. (2004) The Cornell Net Carbohydrate and Protein System model for evaluating herd nutrition and nutrient excretion. *Anim. Feed. Sci. Technol.*, 112: 29–78.

Institut National de la Recherche Agronomique (INRA) (1978) *Alimentation des ruminants.* Jarrige, R. (ed.) INRA Publications, Versailles, France.

Institut National de la Recherche Agronomique (INRA) (1987) Alimentation des ruminants: revision du systèmes et des tables de l'INRA. *Bull. Tech. Centre de Recherches Zootechniques et Vétérinaires de Theix, INRA*, 70.

Institut National de la Recherche Agronomique (INRA) (1988) *Alimentation des bovins, ovins et caprins.* Jarrige, R. (ed.) INRA, Paris.

Institut National de la Recherche Agronomique (INRA) (1989) *Ruminant Nutrition. Recommended allowances and feed tables.* Jarrige, R. (ed.) INRA, Paris.

Lanari D., Pinosa M., Tibaldi E. (eds) (1983) *La stima del valore nutritivo degli alimenti: metodi classici e concezioni moderne (Estimation of nutritive value of feeds: classic methods and new approaches).* Associazione Scientifica di Produzione Animale, Rome.

Moe P.W. (1981) Energy metabolism of dairy cattle. *J. Dairy Sci.* 64: 1120.

Moe P.W., Flatt W.P., Tyrrell H.F. (1972) The net energy value of feeds for lactation. *J. Dairy Sci.*, 55: 945–958.

Morand-Fehr P., Sauvant D., Brun-Bellut J. (1987) Recommandations alimentaires pour les caprins. In: Alimentation des ruminants: revision du systèmes et des tables de l'INRA. *Bull. Tech. Centre de Recherches Zootechniques et Vétérinaires de Theix, INRA,* 70: 213–222.

National Research Council (NRC) (1985) *Nutrient requirements of sheep.* National Academy Press, Washington, DC.

Ortigues I., Doreau M. (1995) Responses of the splanchnic tissues of ruminants to changes in intake: absorption of digestion end products, tissue mass, metabolic activity and implications to whole animal energy metabolism. *Ann. Zootech.*, 44: 321–346.

Pilla A.M., Taibi L., Dell'Aquila S. (1993) Consumo di alimenti e composizione della dieta ingerita da pecore da latte alimentate ad libitum per quantità e qualità (Feed intake and composition of free choice diets supplied *ad libitum* to dairy sheep). *Zoot. Nutr. Anim.*, 19: 221–226.

Pulina G., Serra A., Cannas A., Rossi G. (1989) Determinazione e stima del valore energetico di latte di pecore di razza sarda (Measurement and prediction of energetic value of milk of Sarda ewes). *Atti della Società Italiana delle Scienze Veterinarie,* 43: 1867–1870.

Pulina G., Bettati T., Serra F.A., Cannas A. (1996) Razi-O: costruzione e validazione di un software per l'alimentazione degli ovini da latte (Razi-O: construction and validation of a software for dairy sheep feeding). *Proceedings of the XII National Meeting of the Società Italiana Allevamento e Patologia Ovini e Caprini,* Varese, Italy, 22–25 October 1996.

Rattray P.V., Garrett W.N., Hinman N., Garcia I., Castillo J. (1973) A system for expressing the net energy requirements and net energy content of feeds for young sheep. *J. Anim. Sci.* 36: 115.

Robinson J.J. (1987) Energy and protein requirements of the ewe. In: W. Haresign, D.J.A. Cole (eds) *Recent advances in animal nutrition.* Butterworths, London, pp. 187–204.

Ronchi B., Bernabucci U., Bertoni G. (1993) Valutazione comparata del metodo body condition score (BCS) nelle razze ovine Sarda e Lacaune (Comparative evaluation of the body condition score (BCS) method in the Sarda and Lacaune breeds). *Proceedings of the Società Italiana Scienze Veterinarie,* 47: 1985–1989.

Russel A.J.F., Doney J.M., Gunn R.G. (1969) Subjective assessment of body fat in live sheep. *J. Agric. Sci., Camb.* 72: 451–454.

Susmel O., Cuzzit R. (1988) An approach to defining the energy requirements of dairy sheep. In: *Isotope-aided studies on livestock productivity in Mediterranean and North African countries.* International Atomic Energy Agency, Vienna, pp. 301–317.

Theriez M., Bocquier F., Brelurut A. (1987) Recommandations alimentaires pour le brebis à l'entretien et en gestation. In: Alimentation des ruminants: revision du systèmes et des tables de l'INRA. *Bull. Tech. Centre de Recherches Zootechniques et Vétérinaires de Theix, INRA,* 70: 185–197.

Van Soest P.J., Fox D.G. (1992) Discounts for net energy and protein. 5th revision. *Proceed. Cornell Nutrition Conference,* Ithaca, New York, pp. 40–68.

4 Dietary Intake of Vitamins and Minerals, and Water Requirements

Giovanni Annicchiarico and Luigi Taibi
MiPAF, Istituto Sperimentale per la Zootecnica – Sezione di Segezia, Foggia, Italy

4.1 Dietary Intake of Minerals

4.1.1 Classification

The cell may contain traces of all the elements present in its surrounding environment, although only 25 elements of the periodic table are considered to be indispensable for life (Table 4.1). Of these, 11 are known as plastic elements, or macro-constituents, because of the amounts in which they are present. The other 14 elements are also indispensable for life but they are present only in small quantities and are thus called micro-constituents, or trace elements; they have no structural function but are irreplaceable cofactors in biologically active molecules.

In the field of animal husbandry, a further classification may be made on the basis of the amounts required in the rations. Excluding those elements which are part of the organic compounds of cells and are not found in the free elementary state (H,C,O,N), the elements may be classified as macronutrients (> 100 ppm) and micronutrients, or trace minerals (<100 ppm).

Since animals are unable to synthesize mineral elements, their diet has to be supplemented with adequate amounts to meet their nutritional needs. The importance of the minerals is not based on their concentrations within the animal body but stems from the biological function they are involved in. Micronutrients enter biologically active molecules, such as enzymes, enzyme cofactors, vitamins, hormones, etc., which play essential roles in the regulation and control of metabolic pathways and in the synthesis of molecules which may result in disease in the case of micronutrient deficiency. Some major mineral elements, such as Ca and P, have biological as well as structural functions. Ca and P are mostly concentrated in bones and teeth, Ca being a regulatory factor of cell permeability and neuromuscular excitability, while P is involved in energy transport.

4.1.2 Macronutrient content in the main natural feed supplements

In most circumstances, the feeds and forages consumed in feed rations provide all the necessary mineral nutrients to cover nutritional needs. Minerals are introduced in the diets of animals at peak productivity by applying mineral supplementation, or by using integrated commercial feeds specifically formulated for each physiological stage. Often though, particularly in grazing sheep and goats not yet at peak productivity level, a slight correction in the mineral content of their feed, with the addition of salts containing one or two of the main elements (i.e. Ca and P), is sufficient.

There are mineral salts with a high natural content of certain elements, and appropriate dosages of these contribute to a good

Table 4.1. Macro- and micro-constituents of living matter (Bortolami *et al.*, 1985).

Element	%	Symbol	Atomic number	Main recognized functions
Macro-constituents				
Hydrogen	9.80	H	1	Basic building blocks of cells
Carbon	21.15	C	6	Basic building blocks of cells
Nitrogen	3.10	N	7	Basic building blocks of cells
Oxygen	62.43	O	8	Basic building blocks of cells
Sodium	0.08	Na	11	Involved in action potentials and acid–base balance
Magnesium	0.04	Mg	12	Cofactor of numerous enzymes
Phosphorus	0.95	P	15	Energy transport/bone constituent
Sulphur	0.16	S	16	Structure of proteins and sulphur amino acids
Chlorine	0.08	Cl	17	Biological anion; counter-ion for Na and K
Potassium	0.23	K	19	Membrane potential cation
Calcium	1.65	Ca	20	Enzyme cofactor; regulates muscle contraction; membrane constituent; main constituent of bone
Micro-constituents				
Iron	0.005	Fe	26	Cofactor of numerous oxidative enzymes
Zinc	0.0025	Zn	30	Cofactor of numerous enzymes
Copper	0.0004	Cu	29	Cofactor of numerous oxidative enzymes
Manganese	0.00005	Mn	25	Activator of numerous enzymes (arginase, peptidases)
Iodine	0.00005	I	53	Indispensable for T2 and T3 synthesis
Bromide	0.002	Br	35	Action unclear
Fluorine	0.009	F	9	Protects dental enamel
Cobalt	traces	Co	27	Constituent of cyanocobalamin, or vitamin B_{12}
Molybdenum	traces	Mo	42	Cofactor of some enzymes
Selenium	traces	Se	34	Constituent of vitamin E
Nickel	traces	Ni	28	Action unclear
Vanadium	traces	V	23	Action unclear
Tin	traces	Sn	50	Action unclear
Silicon	traces	Si	14	Action unclear

mineral balance in the feed. Tables 4.2, 4.3 and 4.4 show the most commonly utilized salts that supply the main macronutrients.

4.1.3 Dietary needs and factors involved in meeting those needs

Tables 4.5 and 4.6 show the requirements for macro- and micronutrients, expressed as % of dry matter (DM). In practice, it is extremely difficult to balance the real needs of animals by providing such elements. On the one hand, needs vary depending on factors such as the individual animal, the environment and the feed rations (Fig. 4.1); on the other, the exact chemical composition of feeds available to formulate a diet is not always known, therefore the amount of actual nutrients provided is also unclear.

As far as *factors linked to the animals* are concerned, the following points must be taken into account:

- Age – it is well known that young growing animals absorb high quantities of minerals, both for tissue formation and for skeletal development, with the skeleton retaining 99% Ca, 80% P, 65% Mg, 40% Na and 5% K of the body's reserves by

Table 4.2. Salts with high Ca /P content (AAVV, 1985).

Salt	Ca (%)	P (%)	Notes
Calcium carbonate	40		Frequently used in diet
Calcium sulphate	29		Increases urinary excretion of Ca/P
Calcium chloride	36		
Calcium phosphate	15–19	22–24	
Calcium citrate	21		Readily absorbed
Calcium gluconate	9		Readily absorbed
Calcium formate	31		Readily absorbed
Calcium lactate	13		Readily absorbed
Calcium acetate	25		Readily absorbed
Phosphoric acid		31	
Sodium phosphate (mono- or bihydrate)		2–25	Na 16–19%
Potassium phosphate (mono- or bihydrate)		17–22	K 28–44%
Ammonium phosphate (mono- or bihydrate)		23–27	N 12–21%

the time they reach adult life. Mineral absorption rates decrease as animals grow older.

- The animal's physiological status (dry, gestation or lactation) – mineral absorption increases by 20–40% to allow for fetal growth. This also applies to the first months of lactation, reaching very high levels as the animal reaches peak productivity.
- Bodyweight variations.

Factors linked to climate conditions (heat and cold stress) affect the acid–base balance and the amount of minerals absorbed or excreted in urine.

Factors linked to feed rations are directly related to:

- Mineral contents in the soil and plant growing techniques (whether or not fertilizers have been used).
- Digestibility and availability of minerals, which vary depending on the DM content in feeds, the vegetative phase of plants and the season in which they have been harvested (different degrees of lignification and different ratios of minerals).
- Palatability of feeds. Apart from sodium chloride, various mineral supplements are particularly unappealing (although the problem is related more to cattle than to sheep).

4.1.4 Feed absorption through digestion and its influence on the microbial activity of the rumen

In the rumen, macronutrients not only provide nutrition for the microbial flora but they also exert a chemical, physical and regulatory action on ruminal contents (osmotic pressure, pH and dilution levels). Trace ele-

Table 4.3. Salts with high Mg content (AAVV, 1985).

Salt	Mg (%)	Other minerals present (%)	Notes
Magnesium oxide	60		Readily absorbed
Magnesium hydrate	40		Increases urinary Ca/P excretion
Magnesium carbonate	29		
Magnesium chloride	12	Cl (35)	
Magnesium sulphate	20	S (26)	Decreases Ca/P utilization
Magnesium sulphate hydrate	10	S (13)	Decreases Ca/P utilization

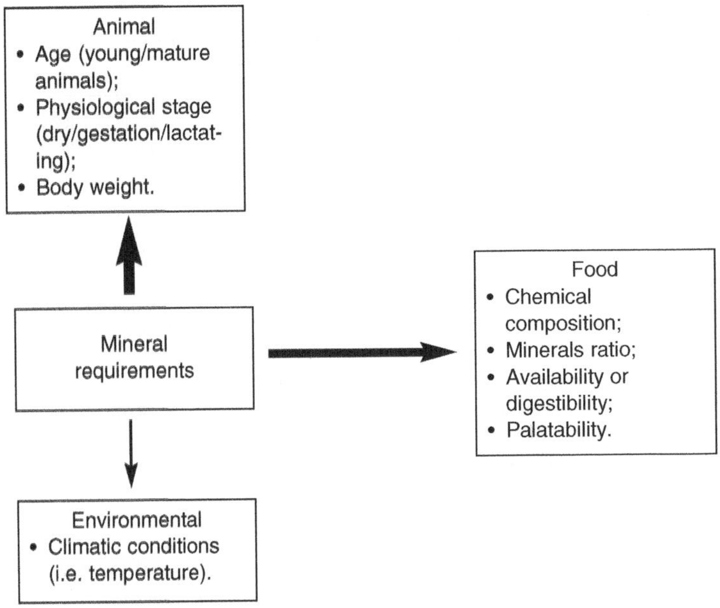

Fig. 4.1. Factors that influence mineral requirements in dairy sheep. Note: the arrows' dimensions describe the weight of the factors.

Table 4.4. Salts with high Na content (AAVV, 1985).

Salt	Na (%)	Others minerals present (%)
Sodium chloride	39	Cl (60)
Sodium bicarbonate	27	—
Sodium sulphate	32	S (23)

ments have a trophic function in microorganisms, as they enter numerous enzymes or vitamin cofactors (e.g. Co in Vitamin B_{12}), which are fundamental for ruminal syntheses and microbial growth.

The ratio of one element to the other (i.e. Ca:P, Ca:Mg, etc.) and the excess and/or deficiency of some minerals may either down- or up-regulate their utilization. For instance, in cases of micronutrient deficiency there is a reduction of ruminal synthesis and consequently of microbial growth. Ruminal pH levels, depending on diet type and dosages of rations, may affect mineral solubility (e.g. Mg becomes less soluble as soon as pH values rise). In turn, minerals may act as buffers (e.g. phosphates and carbonates) in maintaining pH at optimum levels for microbial synthesis.

4.1.5 Macronutrient deficiencies and/or excesses and their effects on nutritional metabolism

An unbalanced intake of the necessary minerals during the productive cycle can cause either primary or secondary deficiencies, or excesses, of one or more minerals. This can immediately affect production potential, later causing symptoms of acute and chronic

Table 4.5. Recommended dietary content of macro-elements (% DM) (AAVV, 1985; NRC, 1985).

Element	Recommended content (%)
Sodium	0.04–0.15
Chlorine	—
Calcium	0.20–0.82
Phosphorus	0.16–0.37
Magnesium	0.12–0.18
Potassium	0.50–0.80
Sulphur	0.14–0.26

Table 4.6. Recommended dietary content of micro-elements (AAVV, 1985, 1992; World Animal Science, 1987).

Element	Recommended content	Toxicity threshold
Iron (mg/kg)	30–50	
Copper (mg/kg)	3–10	15
Manganese (mg/kg)	20–40	1000
Zinc (mg/kg)	30–50	250
Selenium (ppm)	0.1	0.5
Iodine (ppm)	0.5–1.2	8
Cobalt (ppm)	0.1	10
Molybdenum (ppm)	0.5	3
Fluorine	Trace	

diseases, depending on the seriousness and duration of such deficiencies. Table 4.7 shows the main functions of dietary macro- and micronutrients as well as the main symptoms of inadequate or excess intake.

Calcium and magnesium (Ca and Mg) deficiencies

Hypocalcaemia, or Ca deficiency, is a pathological syndrome associated with decreased Ca concentration in plasma. This causes some physiological reactions such as a slowing down of the heartbeat and paralysis of skeletal muscles. Differences have been reported in goats and sheep due to genetic predisposition. This condition usually appears at the beginning of lactation, normally after giving birth to the first litter, and it may occur a few days before or after delivery, when the Ca requirement is at its highest. In sheep, plasma Ca may decrease from a normal concentration of 9–10 mg/100 ml to 2–3 mg/100 ml, while in goats concentrations may decrease to 50% of the normal content.

Hypomagnesaemia, or Mg deficiency, may act as a compounding factor in such instances and may appear in the presence of chelating agents which reduce the physiological activity of Ca and/or circulating Mg. In this regard, an adequate feed ration, which specifically meets the changing needs of animals at peak productivity, is essential. For instance, a concentrated supplement containing such minerals could be given 1 month before delivery.

Hypomagnesaemia, as the name indicates, is associated with a drastic decrease in the plasma or serum levels of Mg, from a normal level of 2.0–2.5 to below 1 mg/100 ml. This condition is also known as lactation tetany, grass tetany, or transient tetany. Neurological signs of Mg deficiency are tetanic seizures and convulsions. Animals get up and lie down frequently and have difficulty standing. In acute forms cardiac arrest is possible. In some cases, hypomagnesaemia is directly related to a Mg-deficient diet, but deficiency may also be induced, for example, by heavy fertilization of pastures with K and N. The presence of chelating agents in feeds may also cause Mg deficiency. In some cases, Ca deficiency may be associated with Mg deficiency, and the clinical signs of Ca deficiency may obscure those of Mg deficiency.

Relationships between calcium, phosphorus and vitamin D

Skeletal pathologies such as rickets, osteoporosis, osteomalacia, or a combination of all these conditions, are frequent in goats and sheep, and usually Ca, P and vitamin D are involved. The symptoms of such diseases cannot be specifically ascribed to Ca deficiency; they can also be caused by a deficiency in P, an abnormal Ca:P ratio or by a deficiency of vitamin D.

Calcium and phosphorus (Ca and P) excess

Urolithiasis is the name given to illnesses

Table 4.7. Biological role and main symptoms of mineral deficiency or excess (ASSONAPA, 1996; Bell, 1997).

Mineral	Main functions	Deficiency symptoms	Excess symptoms
Calcium	Bone; muscle contraction; activates some enzymes	Rickets; osteomalacia; milk fever	Ratio Ca:P 7:1 with phosphorus adequate
Phosphorus	Bone; energy binding; phospholipids and neuronal tissue	Rickets, osteomalacia	Weak bones; urinary calculi
Magnesium	Bone constituent; muscle contraction activator; acid–base equilibrium; protein and lipid metabolism	Loss of appetite; low growth hypersensibility; grass tetany; incoordination; convulsions	Reduced feed intake; diarrhoea
Sodium	Osmotic pressure; acid–base balance; cell permeability; nerve impulse transmission	Depressed appetite; taste deprivation	Diarrhoea; anorexia; salivation; abdominal pain; thirst; muscular spasms
Potassium	Osmotic pressure; acid–base equilibrium; nerve impulse transmission	Rapid decline in food and water intake; loss of strength; pica	Cardiac problems; oedema
Chlorine	Electrolyte balance	Frequently confused with sodium deficiency	
Sulphur	Sulphur amino acids	Low productivity	Diarrhoea; dehydration; acidosis; lung and liver damage
Copper	Many enzymes; haemoglobin; bone and cartilage formation	Hair discoloration; reduced growth; lameness	Anorexia; jaundice; abdominal pain; haemolytic crisis
Zinc	Epidermis; skeletal formation; wound healing	Inter-digital dermatitis; reduced fertility; impaired immune function; slow growth	Rarely anaemia; reduced bone formation; low weight gain
Manganese	Growth, skeleton and reproduction	Reproductive anomalies; skeletal abnormalities; poor growth	Destruction of ruminal flora; reduced appetite; reduced growth; anaemia
Cobalt	Constituent of vitamin B_{12}	Loss of appetite	Low growth; muscular inco-ordination; increased haemoglobin and PCV
Selenium	Antioxidant, enzyme constituent	Reduced fertility; muscular dystrophy; cystic ovaries; reduced immune response	Abortion; hair loss; lameness; loss of appetite; death
Molybdenum	Metabolism of the purine bases	Failure to convert xanthine to urate	Diarrhoea; anorexia; bone malformation; reproductive disorders
Fluorine	Constituent of bones and teeth		Mottling of teeth; lameness; dry skin and hair
Iron	Haemoglobin, many enzymes	Anaemia	Reduced intake and growth
Iodine	Thyroid hormones	Goitre, abortions	Chronic: reduced intake and reduced growth of hair; weepy eyes. Acute: excessive salivation; anorexia; abortion; respiratory problems

involving insoluble deposits such as Ca oxalates, carbonates or phosphates, silicon or magnesium–ammonium–phosphate ($MgNH_4PO_4$) complexes; these are found in the kidneys or bladder and are capable of obstructing the urethra or the ureter, since they can aggregate to form stones, or uroliths. Urolithiasis is seen most often in areas with a dry climate, both cool and warm, where the amount of water being drunk is minimal and/or loss of fluids is high; in these conditions most animals tend to produce more concentrated urine. This condition is also predisposed to by particular dietary situations, such as protein-rich rations with high energy content, or those with a mineral imbalance (high or low Ca:P ratio, high Mg:Ca ratio, high soluble silicon content or low Na and K content). The presence of organic types of kidney stones may be due to the types of grasses eaten during the hot part of the season, or to sheep grazing on land where certain leguminous plants are dominant.

Sodium (Na) excess

The physiological needs for Na in ruminants have been studied in depth and have been precisely formulated for sheep and goats. As a rule, sheep and goats have a very effective Na retention, when needed. The addition of NaCl to feed rations (0.5%) is a common practice and the administration of high quantities of NaCl (more than 2–5% of the diet), when very digestible feed rations are given or there is a risk of urolithiasis, promotes diuresis thus reducing the risk of kidney stone formation.

4.1.6 Micronutrient deficiencies and/or excesses and their effects on nutritional metabolism

Copper (Cu) deficiency

Symptoms: the first sign of Cu deficiency in sheep is lack of pigmentation in black wool. Variable availability of Cu for grazing animals can result in a striped coat with pigmented areas only where there is an adequate Cu supply. Greater deficiency will cause reduced crimp of the coat (in breeds that have curly or wavy wool), and at the same time wool formation will decrease, bodyweight will drop, fetal growth will be stunted (in pregnant ewes) and anaemia will develop. Symptoms in sheep which have a very severe Cu deficiency will include giving birth to fewer lambs, which may present as a disease called enzootic ataxia; the pathology of this disease involves multilocular brain cysts and demyelination of the brain and spinal cord, which is progressive. All of the above symptoms can be corrected by introducing sufficient Cu into the diet, except when symptoms are due to a degenerative process of the central nervous system. In fact, progressive demyelination may be stopped when adequate Cu levels are metabolically available, but it will not regress. High concentrations of molybdenum (20–100 ppm of DM in rations), in the presence of sulphates, interfere heavily with Cu absorption, causing deficiency.

Diagnosis of deficiencies: signs of Cu deficiency are likely to appear when pastures have a content of less than 5 ppm of Cu and 1 ppm or less of molybdenum in the DM. If a substantial quantity of molybdenum is present, Cu deficiency can occur with a concentration of 10 ppm or higher. The livers of sheep showing this deficiency and having given birth to ataxic lambs generally contain less than 10 ppm of Cu, although deficiency symptoms can appear at less than 50 ppm. A normal diet contains a concentration of 100–400 ppm, and blood concentrations of 0.7–1.2 mg/l will protect them from any deficiency.

When pastures have a low Cu content, a distribution of about 7 kg/ha of Cu sulphate raises Cu concentrations to a satisfactory level for a few years, and it also has the positive effect of increasing forage production. Where fertilization is not possible, deficiencies can be treated with Cu sulphate: by oral administration of 10 mg/day for 2 to 3 weeks, or by subcutaneous injections, which, however, may cause damage to local tissues.

Copper (Cu) excess

Unlike cattle, which are relatively tolerant of an excess of Cu, sheep and goats may suffer from poisoning under certain circumstances. Providing sheep with high doses of Cu salts results in closure of the oesophagus and by-passing of the rumen, thus causing a high enough amount of Cu absorption in the intestine to cause the death of the animal in a few hours. In pastures with high concentrations of Cu (>10 ppm of DM), interference with molybdenum absorption may cause the accumulation of high quantities of Cu in the ovine liver (>8000 ppm of the dry diet) and chronic poisoning. Under specific stress conditions, the accumulated Cu may be released, thus increasing its concentration in the peripheral blood. This can cause massive haemolysis followed by blockage of the renal tubules, increased kidney size and in most cases, death. Merino crossbreeds appear to be less sensitive than other sheep breeds. Whenever there is a risk of poisoning, molybdenum supplementation helps in controlling Cu absorption.

Manganese (Mn) deficiency

Mn deficiency in sheep and goats affects both growth and reproduction; symptoms are an overdevelopment of bones followed by hampered movement and lack of balance, as well ataxia in young animals.

Requirements: sheep require at least 32 ppm Mn/kg of their DM, whereas goats need 20 ppm or more. Mn abounds in pastures and forages and may reach concentrations as high as a few hundred ppm; in fact, it rarely decreases to less than 40 ppm, hence grazing animals will rarely lack this element in their diet.

Manganese (Mn) excess

An excessive daily intake may depress appetite and lower growth rates. It has been observed that Mn may antagonize iron absorption by inhibiting haemoglobin formation, for instance in lambs, with an intake of up to 5000 ppm of Mn.

Zinc (Zn) deficiency

Symptoms of Zn deficiency include an arrest in fetal growth after a few weeks, poor or no wool growth with loss of crimp and fibre, abnormal keratin formation and slow wound healing, which makes animals prone to tissue diseases such as foot rot and mastitis. Lambs and kids may show hypogonadism, while in fully developed animals the testicles decrease in size, libido is reduced and spermatogenesis ceases.

Requirements: all alterations due to Zn deficiency are reversible if the animals receive adequate supplements: lambs will grow regularly if fed diets with 17 ppm Zn; testicular growth and spermatogenesis will improve again with 32 ppm Zn. The optimal concentration of Zn in the diet should range from 17 to 33 ppm, but no symptoms of Zn poisoning have appeared in sheep with diets with up to, or in excess of, 500 ppm.

Iron (Fe) deficiency

Fe is abundantly present in vegetable feeds. Grazing animals are rarely deficient in Fe, as leaves, many leguminous plants and seed integuments are good sources of iron, whereas both milk and cereal seeds have low Fe content. Over 90% of iron in the body is incorporated in proteins, the most important of which is haemoglobin, containing 0.34 Fe. Other proteins with Fe content are transferrin, which transports Fe from one part of the body to another, and ferritin, whose function is to create organic Fe reserves. Iron deficiency causes anaemia, which is fairly common in young animals in full growth, due to milk's low Fe content. Anaemia is not common in lambs as milk feeds are often rounded off with a dietary supplement. In adult animals receiving a maintenance diet, high levels of Fe are not required since it is recovered from the destruction and neosynthesis of haemoglobin; only 10% of Fe is lost in this process. An excess of Fe in the diet may cause certain disorders and may diminish phosphorus absorption.

Iodine (I) deficiency

Iodine deficiency leads to the onset of goitre; in sheep (and goats) goitre appears as a large swelling under the lower jaw and around the trachea, which is caused by enlargement of the thyroid gland. Iodine deficiency may be primary or induced, or may be caused by an inherited malfunction of the thyroid gland.

Goitre is very common and easily detected in young animals, whilst in adults an enlarged thyroid gland can be palpated but is not always visible. If I deficiency is severe, goitre is generally accompanied by other symptoms of hypothyroidism such as slow or absent wool and hair growth, reduced milk production and, in lambs, weakness, poor limb coordination, inability to suckle and, in some cases, death.

Affected adult males show loss of libido and low-quality semen, while females show irregular oestrus cycles and low conception rates. Iodine deficiency may be induced by consumption of plants or products containing goitrogens, many of which only become active in the presence of marginal I deficiency. Goitrogens are alkaloids or cyanogenic glycosides which release thiocyanate, isothiocyanate or goitrin in the rumen. These antithyroid alkaloids are contained in *Leucaena* sp., the mustard family in general, and in linseed and soybean-based feeds.

Some common herbage, e.g. perennial ryegrass and white clover, contains goitrogenic monovalent anions (perchlorate, thiocyanate and nitrate) which interfere with I transport in the thyroid. Autosomal recessive congenital goitres have also been observed in some breeds of sheep and goats. Poor thyroid function leads to the formation of thyroglobulin and the subsequent inadequate production of thyroid hormone; this enhances the release of hormones into the gland, giving rise to goitre.

The minimum daily I requirement for sheep is approximately 200 μg. If this requirement is not met, the urinary excretion of I drops to less than 50 μg/day and the milk concentration is less than 8 μg/100 ml. Iodine levels in milk, thyroid hormones and I-binding blood proteins are good markers of nutritional I uptake and thyroid status.

Iodine supplements are recommended for preventing deficiency syndromes, and the recommended dietary allowance is approximately 0.5–1.2 mg/kg of DM. Therapeutic I supplements may be given to counter primary or induced I deficiency: either an oral dose of 360 mg of potassium iodate at the end of the third or fourth month of gestation, or alternatively a single I/M injection 1 or 2 months prior to lambing.

Selenium (Se) deficiency

The biological function of Se, a component of the enzyme glutathione peroxidase, represents the first line of antioxidant defence, for the protection of the lipid component of cell membranes.

Selenium deficiency has been reported in sheep and other grazing animals. It is manifested in lambs by enzootic muscular dystrophy, or white muscle disease, whilst in ewes it reduces fertility and increases neonatal mortality.

Selenium deficiency or Se-responsive diseases can result when the diet contains 0.02 ppm or less of Se. Values of 0.06 ppm are considered sufficient to prevent deficiency disorders, and 0.1 ppm supplies a good safety margin.

In some geographical areas, reproductive disorders have been prevented by combining Se with vitamin E, instead of giving Se alone. Selenium deficiency in animals at risk can be prevented and corrected by scheduling periodical I/M injections or by giving a Se salt orally – generally sodium selenite, or a Se slow-release bullet. The use of Se salt licks is an inadvisable method as it is difficult to regulate Se ingestion.

The addition of salt to improve poor pasture, e.g. 2 kg/ha of Se, may result in suitable Se concentrations for about 3 years. However, this practice may lead to overconsumption of Se due to rapid absorption by plants and superficial contamination of the vegetation during treatment.

Selenium (Se) excess

Excessively high Se concentrations in forage crops – toxic to grazing animals – have been reported in some parts of the USA, Ireland, Australia, Israel and Russia. Some edible plants may accumulate toxic concentrations of Se from Se-rich soils or even from soils with very low Se concentrations. There are soils rich in a particular form of Se that is accumulated only by certain plants (called accumulator plants), which, on dying, release Se into the soil in a form that is accumulated by forage crops. Levels as high as 260 ppm of Se have been reported in pasture, and 6000–15,000 ppm in accumulator plants. Animals that ingest over 5 ppm in their diet show loss of hair coat, emaciation, atrophy of the heart and liver, cirrhosis of the liver and inflammation of the intestine and kidney. Sheep do not appear to be seriously affected by severe Se poisoning, although generally they display poor appetite, weight loss and reproductive disorders.

Several compounds and elements have shown an ability to reduce Se toxicity by reducing absorption, increasing excretion or modifying Se metabolism in the tissues, e.g. bromobenzene, S, As, Ag, Hg, Cu and Cd. Generally speaking, it is advisable to prevent animals from grazing on Se-rich pasture.

Cobalt (Co) deficiency

Cobalt deficiency (pernicious anaemia): following the earliest reports from the UK in the 18th century, many cases of diseases of unknown origin affecting ruminants at pasture were reported throughout the world. In each instance, the symptoms were progressive appetite and weight loss together with severe anaemia, leading to the death of the affected animals unless they were moved to other pastures.

Eventually, a positive response to an oral dose of Co given to a sheep with the above-mentioned symptoms was obtained in 1935 in Australia, and Co deficiency was rapidly confirmed as the principal cause of the disease.

Cobalt deficiency is associated with certain types of soil (sandy limestone, sandy siliceous, soil derived from granite rock, leached pumice stone, shale forests, peat and heavy shale soils) containing less than 5 ppm of Co, or when less than 0.25 ppm of Co is extracted from a 2.5% acetic acid solution.

Functions of Co: cobalt deficiency was studied for at least 15 years before its nutritional function in ruminants was determined. A great deal of attention was focused on the similarities between blood dyscrasia in sheep and pernicious anaemia in humans. In 1948, it was discovered that Co is the constituent of anti-pernicious anaemia factor (vitamin B_{12}), which accounts for the importance of Co in ruminants. It was clarified that the sole function of Co is to guarantee adequate microbial synthesis of vitamin B_{12} in the rumen, and that all the symptoms of Co deficiency syndrome can be explained as a deficiency of vitamin B_{12}.

Parenteral administration of Co is pointless as it does not reach the rumen, whilst vitamin B_{12} administered parenterally, and Co administered orally, clearly demonstrate the biochemical involvement of Co at the level of the rumen.

Diagnosis of Co deficiency: in ruminants, ingested Co is stored primarily in the liver, almost exclusively as a component of vitamin B_{12}. The liver of Co-deficient sheep often contains < 0.1 ppm, commonly between 0.01 and 0.05, while non-deficient animals generally have values between 0.1 and 0.3 ppm.

Cobalt concentrations in blood are in general too low for accurate measurement, but in sheep serum, the levels of vitamin B_{12} provide a good indication of the availability of dietary Co. In severely vitamin B_{12}-deficient sheep, concentrations are consistently < 0.2 ng/ml, while normal levels are between 1 and 3 ng/ml. Livers of vitamin B_{12}-deficient animals generally have Co concentrations < 0.1 mg/g of fresh matter, while in normal subjects the values are around 1 mg/g of fresh matter, or higher.

An adult sheep generally requires a daily allowance of 0.1 mg of Co to optimize productive performance, and a lamb needs even higher amounts. These amounts are generally supplied in a single feed ration, or pasture, containing 0.1 ppm Co of DM ingested.

Molybdenum (Mo) excess

Molybdenum plays a fundamental role as a component of various enzymes, including xanthine oxidase. Molybdenum deficiency has never been reported in any species, but ruminants are highly prone to the effects of Mo poisoning. The effects of Mo on ruminants are largely associated with the effects of Cu storage, Cu metabolism (see the section on Cu) and the availability of sulphate in the diet.

Much has been written on the relationship between Cu, Mo and sulphur (S), which undoubtedly has a great influence on Cu uptake and accumulation in the rumen. It is well known that sulphate is reduced to sulphite by ruminal microorganisms; when it combines with Mo it forms thiomolybdate, and subsequently the insoluble compound Cu thiomolybdate, which is excreted. When dietary Cu is made unavailable by this reaction and body stores become depleted, the animal inevitably suffers from the effects of Cu deficiency.

Fluorine (F) excess

Fluorine does not appear to have a role in ruminant nutrition. Conversely, young sheep exposed to excessive amounts of F display typically stained, irregular and oversized teeth. Adults develop exostoses of the long bones and jaws, whilst sheep which lose their dentition have difficulty grazing and chewing, and consequently develop problems with wool growth and raising lambs.

4.2 Vitamin Nutrition

4.2.1 Classification

Vitamins are organic compounds which animals require in order to live, grow and thrive. In 1912, Funk coined the term 'vitamins' from vital amines, i.e. substances containing nitrogen in amine form. Today it is known that only certain vitamins contain amino nitrogen, but the term has become generally accepted. Since it is impossible to make a rational classification based on their chemistry or biological properties, vitamins are classified as being either fat-soluble or water-soluble (Table 4.8).

Table 4.8. Vitamin classification as a function of solubility.

Fat-soluble	Water-soluble
A,D,E,K and F	B complex (B_1, B_2, B_3, B_5, B_6, B_9, B_{12}) vitamin C and H

Table 4.11 shows the biological function and the main symptoms of vitamin deficiency or excess.

4.2.2 Vitamin requirements and principal dietary sources

Fat-soluble vitamins

VITAMIN A Vitamin A, also called retinol, is a fat-soluble organic alcohol. It generally accumulates in the liver, egg yolk and milk fats. In plant food it exists in the form of carotenoid precursors, which the intestinal epithelium converts into retinol. The greener the plant, the richer it is in vitamin A precursors. Grazing animals, such as sheep, seldom have vitamin A deficiency, due to the presence of its precursors in pasture, and to its accumulation in the liver. A deficiency of vitamin A causes twilight blindness and reproductive disorders. Vitamin A requirements for different ovine categories are shown in Table 4.9.

Table 4.9. Dietary requirements of vitamin A and β-carotene (AAVV, 1985).

Physiological state	Vitamin A (IU/kg BW)	β-carotene (mg/kg BW)
Maintenance and growth	33	60
Gestation	66	120
Lactation	50	90

VITAMIN D Various forms of vitamin D are known, the most important being ergocalciferol (D_2) and cholecalciferol (D_3), whilst their precursors are ergosterol and 7–dehydrosterol. They commence as pro-vitamins and are subcutaneously converted into vitamins by ultraviolet light (290–315 nm). Vitamin D is more resistant to oxidation than vitamin A, and vitamin D_3 is more resistant than D_2.

A deficiency of vitamin D is characterized by the development of rickets and osteomalacia, which are not specific disorders since they may also be caused by deficiency or imbalance of Ca and P.

Excessively high levels of vitamin D may also cause disorders such as Ca deposits in the heart and arteries. Adult animals that graze, or are exposed to normal amounts of sunlight, generally do not suffer from deficiency, while young or pregnant animals should be given supplements when fed hay only, since hay is low in vitamin D. Feeds of animal origin like cod and halibut liver oil and egg yolk are rich in vitamin D. Adult animals require 2000 IU/kg of DM.

VITAMIN E The vitamin E group includes eight forms of active compounds, the most common and most active being α-tocopherol. Vitamin E is abundant in young shoots, especially in the leaves. Vitamin E is easily destroyed by oxidation, so that grass may lose 90% when made into hay, while silage causes somewhat lower losses.

Signs of vitamin E deficiency include disorders of the skeletal muscles: lambs develop unnatural posture and display abnormalities that go under the names of 'white muscle disease' or 'stiff lamb disease' (enzootic muscular dystrophy). Recommended intakes are shown in Table 4.10.

VITAMIN K Discovered in 1935 and known for its role in blood clotting, vitamin K, or antihaemorrhagic vitamin, exists in several forms, the most important being the naturally occurring phylloquinone, which is obtained from plants. Vitamin K is fairly heat-resistant but is easily destroyed by exposure to sunlight. The main sources of dietary vitamin K are plants such as lucerne, cabbage and oilseed rape, as well as (depending on the diet) egg yolk and fish meal.

Table 4.10. Recommended dietary content of vitamin E (AAVV, 1985).

Category	Vitamin E (mg/kg concentrate)
Lamb	50–100
Sheep	25–50

Sheep and goats can also absorb vitamin K synthesized in the large intestine by bacteria (e.g. *Escherichia coli*); deficiencies, therefore, are not reported in these species. Cattle, on the other hand, may be affected by vitamin K deficiency after consuming sweet clover containing dicoumarol, which causes a lowering of thrombin levels in the blood.

Water-soluble vitamins

All the B vitamins and vitamin C are synthesized in the rumen by the microflora; along with the vitamins contained in feeds, they are generally sufficient to cover the nutritional requirements of sheep. The only vitamin deficiency encountered commonly is that of vitamin B_{12} (cyanocobalamin): its microbial synthesis requires an adequate source of dietary cobalt (see paragraph on Co).

4.3 Water Consumption

Animals obtain water from both food and drink. The water balance (Fig. 4.2) is represented by the difference between the amount of water ingested (in food and drink) and that excreted – in any of the following three ways:

- Through the urine and faeces, following the digestive and metabolic utilization of food.
- In milk and other fluid production.
- Through thermoregulation, i.e. in the form of evaporation in expired air, and sweating.

The amount of water required by sheep

Table 4.11. Biological roles of vitamins, and the main symptoms of their deficiency or excess (ASSONAPA, 1996).

Vitamin	Main functions	Deficiency symptoms	Excess symptoms
A	Vision; tissue trophism; glucose synthesis; bone growth	Twilight blindness; hyperkeratosis; skeletal lesions; reproductive anomalies	Hyperkeratosis; skeletal lesions; excess symptoms are similar to deficiency
D	Bone formation; phosphorylation bone growth	Rickets (in young animals); osteomalacia (in mature animals)	Demineralization of bone and mineralization of soft tissue
K	Formation of the thrombin precursor	Spontaneous haemorrhages; delayed blood coagulation	Relatively non-toxic
E	Antioxidant function; protection of vitamin A; affects muscular structure and epithelial tissue of reproductive tract	Muscular dystrophy; degeneration of encephalus; reproductive anomalies	Relatively non-toxic
B_1, or thiamin	Coenzyme of decarboxylase	Polyneuritis; cardiovascular disorders	Relatively non-toxic
B_2, or riboflavin	Coenzyme FMN and FAD; dehydrogenase activity	Dermatitis and loss of hair; enteritis	Non-toxic
B_6, or pyridoxine	Coenzyme PAL	Convulsions; neuritis; anaemia, dermatitis	Convulsions and death
B_3, or PP, Nicotinamide or niacin	Coenzyme NAD and NADP; dehydrogenase activity	Dermatitis; diarrhoea; dementia; hypersensitivity; oral/lingual ulceration	Vasodilation; pellagra; hepatomegaly
B_5, or Panthotenate	Coenzyme A and acyl group carrier	Dermatitis and loss of hair; enteritis; poor growth	Relatively non-toxic
Vit. H, or Biotin	Essential cofactor for carboxylation reactions	Dermatitis and loss of hair	Non-toxic
Inositol	Lipotrophic factor of the liver	Poor growth	Non-toxic
Lipoate	Methyl group donator; lipotrophic factor; and liver protection function	Fatty degeneration of liver	Chronic diarrhoea
B_9, or Folate and pterines	C_1 metabolism; synthesis of purines and pyrimidines	Anaemia; leucopenia	Non-toxic
B_{12} or cobalamin	Isomeric reactions of methyl group	Anaemia; nervous disorders; poor growth	Non-toxic
C	Hydrogen transport; folic acid activation; formation of collagen	Haemorrhages and anaemia	Non-toxic

Table 4.12. Approximate value of water intake by sheep (kg water/kg DM intake) in winter conditions (average ambient temperature 15°C).

Category	Water intake (l/kg DM)[a]
Growth or fattening	2.0
Maintenance or early pregnancy	2.0–2.5
Late pregnancy	
1 fetus	3.0–3.5
2–3 fetuses	3.5–4.5
Lactating sheep	
first month (milk yield 2 l/day)	4.0–4.5
following months	3.0–3.5

[a]The amounts showed respective increases of 30, 50 and 100% for average ambient temperatures of 20, 25 and 30°C.

depends largely on the amount of DM eaten and on the nature and composition of feed (especially the protein and mineral content), as well as on the ambient temperature, liveweight, amount of milk produced and the development of the uterine contents for animals approaching the end of gestation.

Table 4.12 indicates the average water consumption rates of sheep, in litres/kg of DM. It is extremely difficult to estimate the water requirement of grazing animals. Animals that feed on grass generally have a higher intake of water, but other factors include the moisture content of the grass, air temperature and duration of exposure to the

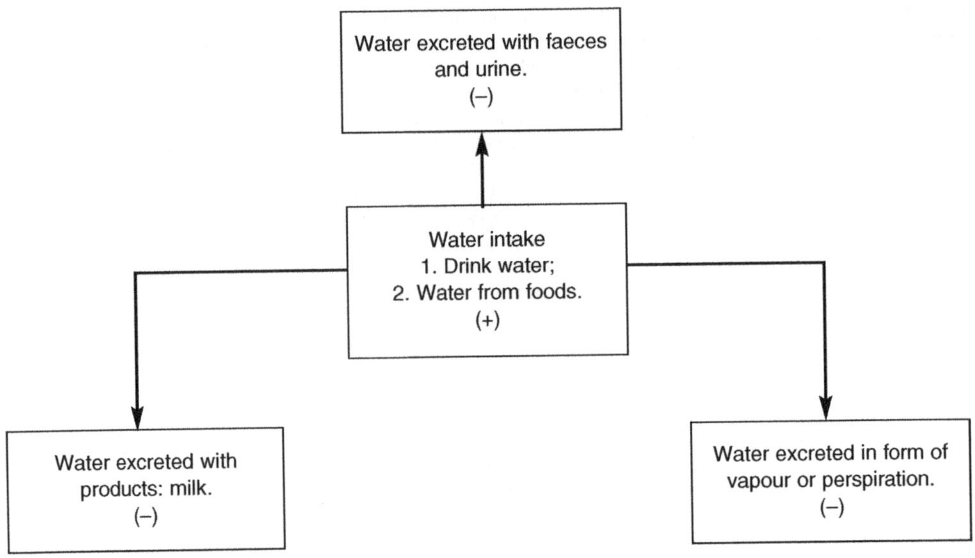

Fig. 4.2. Water balance of the animal.

sun. Furthermore, in semi-arid climates, some breeds of sheep are able to adapt to prevailing conditions, restricting their drinking to twice weekly, and meet their water needs by mobilizing fat reserves (Laticauda or North African breeds).

References

AA.VV. (1985) *Valutazione degli alimenti e dello stato metabolico nutrizionale dei ruminanti.* Associazione Italiana Allevatori, Rome.

AA.VV. (1992) *Ovinicoltura.* UNAPOC.

ASSONAPA (1996) *L'allevamento ovino.* 2nd edition, ASSONAPA, Rome.

Bell B. (1997) *Mineral nutrition in sheep.* OMAFRA.

Bortolami R., Callegari R., Callegari E., Beghelli V. (1985) *La materia vivente, Anatomia e Fisiologia degli Animali Domestici.* Edagricole, Bologna, Italy.

Coop, I.E. (ed.) (1987) Production-system approach C1: sheep and goat production, World Animal Science.

Falaschi A., Verona O. (1977) *Trattato di scienza e tecnica delle produzioni animali,* Vol. III, *Alimentazione degli animali in produzione zootecnica.* UTET, Turin, Italy.

Gueguen L., Lamand M., Meschy F. (1988) *Nutrution minérale – alimentation des bovins, ovins, et caprins.* INRA, Paris.

Jarrige R. (1988) *Ingestion et digestion des aliments – alimentation des bovins, ovins, et caprins.* INRA, Paris.

Masoero P., Giulio L., Ferara B. (1977) *Trattato di scienza e tecnica delle produzioni animali,* Vol. IV, *Fisiologia della nutrizione.* UTET, Turin, Italy.

McDonald P., Edwards R.A., Greenhalgh J.F.D. (1976) *Animal nutrition,* 2nd edition, Longman, Harlow, UK.

National Research Council (NRC) (1980) *Mineral tolerance of domestic animals.* National Academy Press, Washington, DC.

National Research Council (NRC) (1985) *Nutrient requirements of sheep,* 6th Revision. National Academy Press, Washington, DC.

Puls R. (1988) *Mineral levels in animal health.* Sherpa International, Clearbrood, Canada.

Salt Institute (1980) *Salt and trace minerals and Cu toxicity in sheep.* University of Illinois, USA.

Underwood E.J. (1971) *Trace elements in human and animal nutrition,* 3rd edition. Academic Press, New York.

World Animal Science (1987) *Sheep and goat production,* Production System Approach C1. I. E. Coop.

5 Feed Intake

Marcella Avondo[1] and Lucio Lutri[2]
[1]Dipartimento di Scienze Agronomiche, Agrochimiche e delle Produzioni Animali, Università di Catania, Italy; [2]Istituto Sperimentale Zootecnico per la Sicilia, Palermo, Italy

5.1 Introduction

Intake of feed and the consequent absorption processes result in an accumulation of a 'metabolic reserve' that will serve a number of purposes: increase in bodyweight, fetal growth, milk production, etc. Put simply, the greater the requirements of the tissue (that is, the greater the nutritive requirements), the more intensely the metabolites will be removed from the blood (Forbes, 1995). The lower level in the blood allows the animal to continue intake until the accumulation of metabolites is such that the signals from visceral receptors induce the central nervous system to generate a 'discomfort sensation' (Forbes and Provenza, 2000), which induces the animal to stop eating. As the absorption and assimilation processes generally occur long after intake, it has been hypothesized that the animal is capable of intuitively understanding from experience (learning associations) the nutritive capacity of a given food from its organoleptic properties.

Other signals from receptors sensitive to the degree of distension of the ruminal walls have a more immediate effect, so that mature forages, with slow transit times, cause a filling effect that impedes further intake of food. However, it should be noted that the level of blood metabolites, or the filling level of the rumen, that lead to a sense of satiety are often higher than those reached when the animals satisfy their nutritional requirements. This can lead the animal to continue feeding beyond its needs, resulting in metabolic diseases or excessive fattening, as well as the wastage of food with an adverse effect upon production costs. This occurs above all with sheep fed indoors. The latter, more than grazing sheep, often display a greater intake capacity than their true needs, as demonstrated by research conducted on lactating ewes in various environmental conditions (Pauselli *et al.*, 1993; Lanza *et al.*, 1994; Trimarchi *et al.*, 1994). With average-to-low milk production, dry matter intake of over 2 kg can be observed, with a nutritional input obviously exceeding the true needs. This is also confirmed by the high weight increases (in one case of over 10 kg) and by the manifestation, in some cases, of hepatic overload and foot lesions due to metabolic imbalance.

5.2 Factors that Influence Intake

5.2.1 Role of animal factors

It should be remembered that it is very difficult to assess the effects of single variables on intake, as this is the result of the interaction of numerous factors related to the animal, the feed and the environment.

In nature, voluntary intake allows animals to satisfy their nutritive requirements because of an innate physiological tendency to self-regulation. One of the causes of changes in intake is bodyweight, which is correlated, on the one hand, to the basal metabolism that affects nutritive requirements, and on the other to the ruminal volume. An empirical method, though very widely used, for predicting the intake capacity of ruminants is to calculate it as a percentage of bodyweight (for sheep about 4–5.5%). Indeed, the correlation between bodyweight and intake for adult sheep is evident (Table 5.1); however, on analysing the regression for each genetic type, the same tendency is not always observed. Since an increase in weight over the mature bodyweight is an index of fattening, in this case intake tends to diminish. It has been observed that intake in adult sheep increases with weight up to 50 kg, and then decreases (Arnold and Birrell, 1977).

Table 5.1. Dry matter intake and bodyweight in various sheep breeds.

Breed	Bodyweight (kg)	DM intake (kg/head/day)
East Friesian	74.8	2.49
Lacaune	73.2	2.67
Chios	60.0	2.24
Delle Langhe	58.0	1.83
Manchega	57.0	2.24
Massese	52.4	1.98
Comisana	57.4	1.99
Churra	50.0	1.83
Sarda	42.2	1.55

Lactation, which affects nutritive requirements, plays a fundamental role in ovine intake: lactating ewes consume more than dry sheep. However, it has been observed that the correlation between intake and milk production studied by various authors is highly variable (with correlation coefficient values between 0.2 and 0.8 (Serra, 1998)). In fact, the intake capacity after parturition does not always follow milk production increase. As is the case with dairy cows, this causes a phase in which the ewes mobilize fat reserves to satisfy the increased nutritional requirements. Any attempt to estimate the intake capacity of lactating ewes should therefore take into account their physiological tendency to mobilize body reserves in the early months of lactation, and later to restore them. The fall in intake that would therefore be expected in the phase following the peak of lactation is not always so marked. In various experiments (Trimarchi *et al.*, 1981; Pulina *et al.*, 1992; D'Urso *et al.*, 1993; Pauselli *et al.*, 1993) on dairy sheep during lactation, reductions in intake of about 20% have been observed, with corresponding decreases in milk production of approximately 65% (Table 5.2).

For this purpose, an equation has been developed for predicting intake in Italian dairy sheep breeds (Pulina *et al.*, 1996), which takes into account the weight and milk production of the animal, as well as daily weight changes that, in adult animals, can arise due to storage or mobilization of body reserves:

$$I = -0.545 + 0.095\ MW + 0.65\ FPCM + 0.0025\ BWC$$

where: I = DM intake, in kg/head/day; MW = metabolic weight ($BW^{0.75}$), in kg; FPCM = fat (6.5%)- and protein (5.8%)-corrected daily milk production, in kg (Pulina *et al.*, 1989); and BWC = bodyweight changes, in g/day.

Finally, it must be remembered that the state of well-being of the animal affects intake. Any 'negative' stress condition causes intake reduction as one of the first behavioural responses. It has been demonstrated that the following factors can negatively affect the well-being of the animal, reducing the intake capacity:

- Protein/mineral deficiencies or excesses.
- Toxicity.
- Pathological conditions.
- Extreme weather conditions (hot–humid, extreme cold, strong wind).

5.2.2 Role of the chemical, nutritive and physical characteristics of feed

With the exception of cases of availability of feed with an exceptionally high nutri-

Table 5.2. Dry matter intake and milk production.

Lactation week	Breed	Intake (g DM/day)	Milk production (g/day)
3	Comisana	2670	2026
8	(Pauselli *et al.*, 1993)	2634	1025
14		2149	693
7–9	Massese	2200	1068
12–14	(Trimarchi *et al.*, 1981)	2200	588
16–18		2050	346
6	Comisana	1323	1077
10	(D'Urso *et al.*, 1993)	1373	783
14		1820	488
18		1428	424
5th last	Sarda	2539	1095
4th last	(Pulina *et al.*, 1992)	2079	1005
3rd last		2521	976
2nd last		2042	856
Last		2153	721

tional value, very rarely is intake a function only of the animal's requirements. Satiety is linked to the type of feed available: feed with a high energy density results in an arrest in further feed intake before the forestomachs have reached a sufficient level of fill to determine the state of satiety. On the other hand, voluminous forage causes a filling effect that induces the animal to stop eating before the metabolite concentration reaches the threshold level for satiety.

In dairy sheep fed mainly forage, ruminal distension is one of the main causes of intake changes. Most attempts to predict ovine pasture intake based on its chemical–nutritive composition have underlined the important role of cell wall content, well indicated by the presence of neutral detergent fibre (NDF), as this influences ruminal wall distension. Table 5.3 shows some intake prediction equations based on NDF content; correlation is always negative.

Other parameters closely linked to digestibility and speed of ruminal transit, such as lignin or ADF, did not provide similarly satisfactory results. Lignin content of mixed forages can be positively correlated to intake. This is rather anomalous, considering that increased lignin content reflects plant ageing. The cause of this anomaly can be attributed to the presence of legumes, in which high levels of lignin associated with low NDF and high intake levels are typical (Macchioni *et al.*, 1990).

Reduction in the forage particle size, and, as an extreme measure, grinding and pelleting, can significantly increase intake, above all for poor-quality forages. This occurs not only due to the resulting reduction in the filling effect of the forage, but also because its smaller size prevents the selective behaviour that sheep normally exhibit towards the more digestible portions of the plant. This determines a further increase in intake linked to the reduced amount of time dedicated to selection. It also reduces chewing time, and increases feed density and speed of passage through the rumen (Van Soest, 1994).

Silage should be considered separately as its consumption is closely related to its fermentation characteristics. Properly stored silage can result in an intake comparable to that obtained with fresh forage or hay (Bianchi *et al.*, 1990). On the other hand, if silage is highly fermented, the

Table 5.3. Regression equations between dry matter intake (g/kg metabolic weight) and pasture NDF content (Lánari *et al.*, 1993).

Category of forage	Regression equation
All forages (Macchioni *et al.*, 1990)	I = 107.4 – 0.644 NDF
Lucerne hays (Macchioni *et al.*, 1990)	I = 104.6 – 0.488 NDF
Miscellaneous hays (Macchioni *et al.*, 1990)	I = 117.4 – 0.760 NDF
All forages (Reid *et al.*, 1998)	I = 134.5 – 1.10 NDF
Grass forages (Rohweder *et al.*, 1978)	I = 95.3 + 6.70 NDF – 0.0668 NDF^2
Polyphyte hays (Dulphy *et al.*, 1990)	I = 136.5 – 0.12 NDF
Miscellaneous hays (Lanari *et al.*, 1993)	I = 96.5 – 0.38 NDF – 0.0000004 NDF^4

metabolic effect of the high percentage of free fatty acids simulates a state of satiety at low intake levels. Moreover, the modest availability of soluble carbohydrates, in association with high non-protein nitrogen levels, reduces microbial protein synthesis, resulting in reduced ruminal functionality. Very humid silage has been observed to cause lower consumption than the same pre-wilted silage. However, the addition of water to silage does not reduce intake. This demonstrates that humidity has no direct effect on intake, but it causes an indirect effect, setting off anomalous fermentation, with high production of butyric acid, ammonia nitrogen and, in extreme cases, toxic amines. These substances bring about substantial changes to the palatability of silage, reducing its flavour and, as a consequence, voluntary intake.

5.3 Pasture Intake

Prediction of pasture intake for grazing animals is much more complex than for sheep fed indoors. As the animals can freely access the pasture, apart from the animal factors and the qualitative characteristics of the pasture, behavioural factors can also significantly modify intake capacity.

5.3.1 Biomass and structure

Of the pasture characteristics that potentially affect feeding behaviour, herbage availability is one of the most widely studied parameters. In general, as biomass increases, so does intake capacity. However, there are major behavioural factors that render this correlation less evident. Herbage intake at pasture is the result of intake/minute (IR) × grazing time (T); as biomass decreases, IR decreases, as, in these conditions, the animal is only capable of consuming small quantities of herbage with each bite. It has been demonstrated that sheep tend to compensate for reduced herbage availability by prolonging the time dedicated to grazing (Allden and Whittaker, 1970; Theriez, 1983; Brown *et al.*, 1988) and, within certain biomass limits, manage to reach the same intake levels as those reached in good grazing conditions.

Studies on ovine feeding behaviour have highlighted the important role of the structure of the pasture in herbage intake regulation. This represents a combination of various parameters such as plant height, the ratio between leaves and stems, internode length, density in terms of plant number per unit of surface area, and any other parameter defining the spatial distribution of the biomass. In fact, all these parameters have been found to alter the animal bite mode (Hodgson, 1985). The role of biomass and pasture structure will be discussed in more detail in Chapter 11.

5.3.2 Qualitative characteristics of the pasture

The quality of the pasture, expressed in terms of digestibility, filling value, crude protein (CP) or structural carbohydrate content, can affect pasture intake. Very high qualitative levels can enable animals to reach their 'potential' intake, which allows the animal to satisfy its requirements without limitations of a physical nature. Generally, intakes observed in highly digestible pastures are significantly higher than those in more lignified or senescent pastures. It is also true, however, that high digestibility values can lead to animals reaching the state of satiety earlier due to metabolic intake control (Conrad *et al.*, 1964). Our experiments have demonstrated that intake in lactating ewes increases as the herbage level of CP increases and as structural carbohydrates decrease (D'Urso *et al.*, 1993), probably in relation to the greater ease in bite and faster chewing speed linked to these characteristics. We have observed that, of the herbage chemical components, more than any other, the mean level of CP alone is a valid indicator of pasture quality. In particular, we have determined that CP values lower than 16% of DM indicate pasture of poor quality, and this was confirmed by the low levels of intake observed under these conditions (Avondo *et al.*, 2002).

In any case, the correlation between chemical components and intake, considering the diverse pasture typologies, is rather weak. For natural pastures, it is difficult to assess pasture conditions based solely on undifferentiated data on chemical composition. In fact, average chemical composition does not cater for all the qualitative differences between pasture plants and certainly does not consider the spatial distribution of these species, which can be important in affecting the selective behaviour of sheep. For equal chemical–nutritive composition, in

Fig. 5.1. Dense, low and uniform pasture: intake is very high; selective activity is limited.

Fig. 5.2. Highly heterogeneous pasture: intake may be limited due to an increased selective activity.

Fig. 5.3. Pasture with high botanical heterogeneity but structurally compact: the sheep may have difficulty selecting the preferred parts.

a pasture with a heterogeneous structure where selective activity (which subtracts time from intake activity) is very high, intake levels are lower than those observed on a dense and uniformly distributed pasture, which renders selection activity very difficult.

Selection of forage by sheep therefore plays a very important role, not only in diet quality but also in herbage intake, acting as a greater intake-limiting factor than forage availability (Iskander, 1973). A typical example of the limiting role of selective behaviour on intake is illustrated by the case of sheep that have a highly selective behaviour towards forage plants of a high nutritional level or of high palatability. When these plants start to diminish due to selective activity, the animals reduce herbage intake, even if the biomass availability and pasture quality remain high (Ruyle and Dwyer, 1985). As to what induces sheep to select one species or another, it has been hypothesized (Arnold and Dudzinsky, 1978; Provenza, 1995; Forbes and Provenza, 2000) that sheep display a sort of selective 'wisdom' that, while minimizing metabolic discomfort, allows them to eat in relation to their nutritional requirements and to discard toxic plants.

In general, sheep demonstrate preference for the morphological parts of plants and species that are more digestible and richer in CP. As these characteristics are the prerogative of plant shoots and younger plants, selection seems to be determined by the ease of recognition and prehension. Selective tendency towards a protein-rich and digestible diet is therefore more marked the greater the number of botanical species present in the pasture, as long as botanical heterogeneity is associated with differentiation in spatial distribution of the various species (Black and Kenney, 1984). In vetch and barley pasture, which is highly diversified in structure and in growth habit, the advantage of simultaneously having two forage typologies, with chemical–nutritive characteristics that are complementary, is sometimes negated by the highly selective feeding activity of the sheep. In these pasture conditions sheep do not exclusively select one species or another, but they select the two different species at different times. For example, during the earlier part of the day they graze vetch exclusively, with an incredible capacity for discarding any portions of barley they unintentionally bite; later in the day, with the same care, they exclusively select barley. In order to avoid negative consequences from such highly selective activity, it is essential that the pasture be kept low in height (not over 8–10 cm) and at high density, by choosing suitable agronomic practices and adequate stocking rates.

With regard to the choice of morphological parts of the single plant, species with erect growth habit, such as most forage grasses, offer a better chance of selection than creeping species, such as some spontaneous or cultivated legumes (Table 5.4).

Palatal sensitivity to the presence of soluble sugars (e.g. in *Graminaceae*) or to particular substances (e.g. in *Cruciferae* or certain *Compositae*) sometimes leads animals to prefer those with lignified components rather than others with greater nutritional value. An example of this type is illustrated in Table 5.5: in a pasture composed of 35% spontaneous *Graminaceae*, 25% *Vicia* spp., 25% *Trigonella foenum graecum* and 6% *Cruciferae*, we observed that sheep completely rejected *Trigonella* from their diet, well known for its poor flavour, preferring instead lower-protein, less digestible species. The result of this behaviour was a lower protein level in the selected diet with respect to the pasture available, contrary to what normally occurs (D'Urso *et al.*, 1998).

5.3.3 Feed supplementation

The feeding behaviour of grazing sheep can change substantially when a feed supplement is provided. The most marked behavioural response is the change in the pasture intake level. Following administration of a supplement, herbage intake often diminishes noticeably; sometimes it remains unchanged; in extreme cases it can even increase.

We therefore need to define the 'substitution effect' (S), which represents the

Table 5.4. Protein and digestible organic matter content of the whole plant and of selected parts of various pasture species.

Botanical species	Growth habit	Crude protein (% DM)		Digestible organic matter (% DM)	
		Whole plant	Selected part	Whole plant	Selected part
Grasses					
Barley	Erect	18.2	25.4	68.1	72.0
Bromus spp.	Erect	15.8	23.7	69.1	73.7
Mixed pasture	Erect	10.1	16.8	62.6	69.8
Mixed pasture	Erect	8.8	11.3	73.4	76.6
Mixed pasture	Erect	10.8	16.7	–	–
Legumes					
Vetch	Creeping	20.9	22.2	66.4	69.1
Clover	Creeping	21.4	26.5	75.1	75.6
Mixed pasture	Creeping	16.4	18.0	68.4	68.2
Mixed pasture	Creeping	16.8	20.7	–	–

Avondo M., personal data.

change in forage intake per unit of supplement provided, expressed in terms of either dry matter or net energy (Dulphy, 1978). The value of S normally lies between 0 and 1, but, in particular conditions, can even be less than 0 or greater than 1. It is evident that the lower the substitution effect, the higher will be total feed intake.

The substitution effect varies greatly in relation to many variables linked to the characteristics of the pasture and the supplement. The animal response to the supplement can vary in relation to the availability and qualitative level of herbage. When the quantity of herbage is low or quality is poor, the supplement causes an increase in total dry matter intake and improves animal productive performance, as the substitution effect in such conditions is scarce or nil. In contrasting conditions, the feed supplement efficacy is almost nullified due to the high substitution effect (Newton and Young, 1974); in fact, a good pasture allows very high intake levels and, often, sufficient to satisfy ovine requirements. Supplementation (which is normally willingly consumed as it is low in volume and highly palatable to sheep) therefore results in an inevitable reduction in pasture intake.

Changes to the quantity or quality of the supplement administered can cause various responses in relation to pasture conditions as the qualitative–quantitative characteristics of the diet sources interact, affecting ruminal functionality. Generally, as supplementation administration increases, so does the substitution rate (Allden, 1969; Freer *et al.*, 1985). The protein level of the supplement can affect intake (Avondo *et al.*, 1997): our research work led us to hypothesize that sheep are capable of self-regulating intake in relation to their protein requirements. Therefore, when the herbage protein level is high, a supplement with a high protein content results in a marked reduction in forage intake. On the other hand, in a highly lignified pasture, a high-protein supplement (20–22% CP) can even cause increased forage intake (in this case S would be less than 0), due to the improved ruminal function deriving from adequate protein levels.

To conclude, as the pasture is the least expensive portion of the feed, the objective of the farmer, except in exceptional circumstances, should be to maximize herbage intake; if a feed supplement is considered necessary, the substitution effect should be kept to a minimum by adapting the quantity and quality of the supplement to the pasture

Table 5.5. Chemical and botanical composition of a pasture and the relative selected diet by lactating ewes (D'Urso *et al.*, 1998).

Botanical composition (%)	Pasture	Selected diet	Crude protein (% DM)	NDF (% DM)
Vicia spp.	25	23	20.8	33.7
Trigonella foenum graecum	25	0	18.8	27.9
Graminaceae	35	52	8.8	43.9
Cruciferae	6	16	18.3	35.7
Pasture available	–	–	14.4	36.8
Diet selected	–	–	12.8	39.5

characteristics. The concentrate protein level, with the exception of cases in which herbage is at an advanced biological stage, should not exceed 14–15% of dry matter, as it should take into account the tendency of animal selectivity in favour of plants with higher protein levels. Even in grazing conditions less favourable to selectivity (high density, height not over 10–15 cm, structural and/or botanical uniformity), the animal is nearly always capable of selection. This facility enables the grazing animal to feed on a diet characterized by a protein level at least 10–15% higher than the actual protein levels in the available herbage (Avondo *et al.*, 1996).

5.4 Prediction of Pasture Intake in Dairy Sheep in Mediterranean Areas

5.4.1 Dry matter intake

It is now clear that pasture intake is the result of a highly complex behavioural mechanism, affected by the interaction of numerous diverse factors. This makes prediction complicated. During the last 30 years a large number of grazing models have been developed (Elsen *et al.*, 1988). However, the available models are based on genetic types, environmental conditions and, last but not least, production systems very different to those relative to dairy sheep in Mediterranean areas. The last aspect plays a fundamental role in determining intake; most dairy sheep production systems implement short grazing periods that involve taking the ewes to pasture after the morning milking and bringing them back indoors before the afternoon milking (5–8 h grazing per day depending on the season).

Furthermore, as the lactation periods with greater nutritive requirements often coincide with periods of scarce forage availability, the need arises, at least during the initial stage of lactation, for a hay-based or concentrate-based supplement. This underlines the importance of obtaining intake data or utilizing prediction systems deriving from local experimentation.

In an experimental study lasting about 10 years aimed at studying the grazing behaviour of dairy sheep, we have gathered considerable data on individual intake and selective behaviour, which allowed us to propose an intake model that can be applied to Mediterranean systems (Avondo *et al.*, 2002).

In the pasture conditions we analysed, characterized by high botanical and structural heterogeneity, often modest herbage availability and poor chemical–nutritive characteristics, intake is affected by the following parameters:

- Biomass.
- Herbage protein level and dry matter content.
- Pasture height.
- Milk production.
- Feed supplementation.

Data obtained enabled us to identify conditions in which the pasture becomes 'limiting on intake', these are:

Table 5.6. Prediction model for pasture intake (5–6 hours) in dairy sheep (Avondo *et al.*, 2002).

Crude protein ≤ 16% DM
$I = 335.6 + 113.5\ B^{a} + 0.28\ FPCM^{b} - 0.56\ S^{c}$
For crude protein > 16% DM
$y = 997.1 + 73.9\ B^{a} - 27.4\ H^{d} + 20.4\ DM^{e} + 0.16\ FPCM^{b} - 1.24\ S^{c}$

[a] Biomass (t DM/ ha); [b] fat (6.5%)- and protein (5.8%)-corrected milk yield (g/day) (Pulina *et al.*, 1989); [c] supplement (protein, g/day) of CP administered indoors; [d] pasture height (cm); [e] pasture dry matter content (%).

Crude protein > 16% DM and biomass < 1 t/ha, or
Crude protein ≤ 16% DM and biomass < 2.5 t/ha.

Grass height significantly influences intake when the pasture has good qualitative characteristics (CP > 16% DM), whereas it has no effect when the pasture is poor (CP ≤ 16% DM). Indeed, in the latter case intake is inhibited mainly by the unfavourable chemical–nutritive characteristics, rather than by the structure. Processing of data resulted in the formulation of two linear equations for prediction of diversified ingestion for the two qualitative levels of the pasture (CP ≤ 16% or CP > 16% DM) (Table 5.6)

5.4.2 Protein level of the selected diet

In dairy sheep farming, rational feed supplementation should take into account not only herbage intake, but also the qualitative characteristics of the herbage, which, as previously discussed, rarely coincide with those of pasture herbage. Figure 5.4 schematically illustrates the factors that affect selection. To the left, enclosed by the dashed line, are parameters that affect selective grazing: the degree of pasture heterogeneity and its structure. As already seen, sheep, that are faced with a pasture widely differentiated morphologically, have a greater capacity to select more desirable plant parts. Moreover, a low herbage density in association with height over 10 cm makes for easier selection. The right side of the diagram represents variables – circled by the continuous line – that affect selection to a varying extent, in that they induce animal self-regulation in nutrient intake:

- Nutritive requirements, which, if high, lead animals to select less intensively in order to reach higher intake levels.
- Pasture quality, which, if poor in nutritive value, results in more intense search for more digestible or protein-rich plants.
- The CP content provided with the supplement, which, if high, leads animals to select less intensively to limit protein intake.

As sheep demonstrate an intense selective activity above all towards protein-rich plants, we identified a selectivity index for CP defined by the ratio between CP in the selected diet and CP in the available herbage. Based on observations of selective behaviour on about 300 lactating ewes, we developed a linear equation for prediction of the selection index for CP (Table 5.7). The prediction models for dry matter intake at pasture and for CP content in the diet selected are included in the Assis-T rationing programme (see Chapter 12).

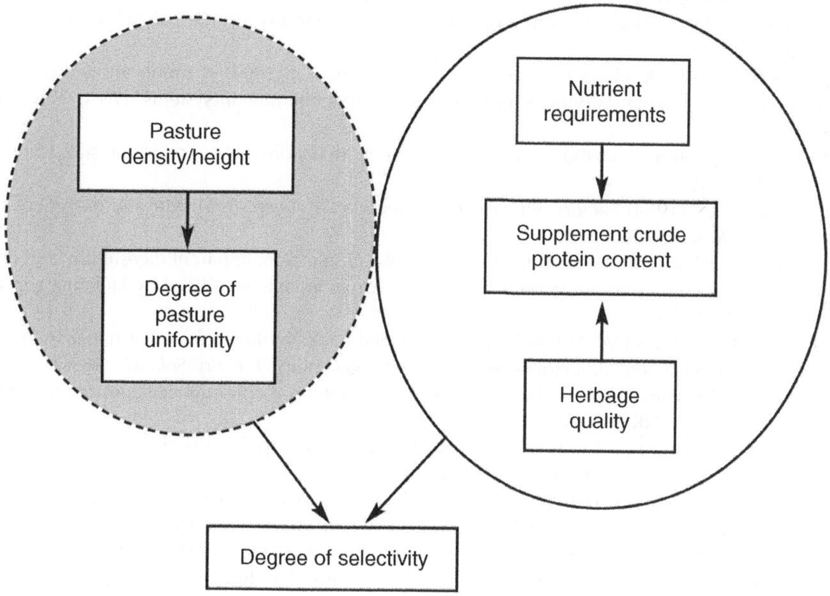

Fig. 5.4. Role of some parameters related to the pasture, management and the animal in sheep grazing selectivity. Parameters in the circle with dashed lines affect selectivity in relation to ease of bite; parameters in the circle with the continuous line affect selectivity in relation to animal requirements (Avondo *et al.*, 2000).

Table 5.7. Protein content prediction of pasture (selected diet).

CP of selected diet = CP × SICP, where:
SICP = 1.65 – 0.00196 S – 0.02089 CP

SICP, selectivity index for CP; S, supplement (g CP/day); CP, CP content of available herbage (% DM).
Avondo M. (unpublished).

References

Allden W.G. (1969) The summer nutrition of weaner sheep: the voluntary feed intake, bodyweight change, and wool production of sheep grazing the mature herbage of sown pasture in relation to the intake of dietary energy under a supplementary feeding regime. *Aust. J. Agric. Res.*, 20: 499–512.

Allden W.G., Whittaker I.A. McD. (1970) The determinants of herbage intake by grazing sheep: the interrelationship of factors influencing herbage intake and availability. *Aust. J. Agr. Res.*, 21: 755–766.

Arnold G.W., Birrell H.A. (1977) Food intake and grazing behaviour of sheep varying in body condition. *Anim. Prod.*, 24: 343–353.

Arnold G.W., Dudzinsky, M.L. (1978) *Ethology of free-ranging domestic animals.* Elsevier, North Holland, New York.

Avondo M., Marletta D., Bordonaro S., Guastella A.M., D'Urso G. (1996) Sheep grazing behaviour in Mediterranean semi-extensive systems. *Proc. pasture ecology and animal intake workshop,* Dublin, pp. 191–196.

Avondo M., Bordonaro S., Marletta D., Guastella A.M., D'Urso G. (1997) Effetti del livello proteico del supplemento sull'ingestione al pascolo di ovini in lattazione. *Proc. XII Congress ASPA,* Pisa, Italy, pp. 251–252.

Avondo M., Bordonaro S., Marletta D., Guastella A.M., Scannavino M., D'Urso G. (2000) Fattori che influenzano l'attività selettiva degli ovini al pascolo. *Proc. XIV Congress SIPAOC*, Vietri sul Mare, Italy, pp. 347–350.

Avondo M., Bordonaro S., Marletta D., Guastella A.M., D'Urso G. (2002) A simple model to predict the herbage intake of grazing dairy ewes in semi-extensive Mediterranean systems. *Livest. Prod. Sci.*, 73: 275–283.

Bianchi M., Errante J., Fortina R. (1990) Prove di razionamento di pecore da latte con fieno silo. *L'informatore Agrario*, 17: 27–30.

Black J.L., Kenney P.A. (1984) Factors affecting diet selection by sheep. II. Height and density of pasture. *Aust. J. Agric. Res.*, 35: 565–578.

Brown D., Salim M., Chavalimu E., Fitzhugh H. (1988) Intake, selection, apparent digestibility and chemical composition of *Pennisetum purpureum* and *Cajanus cajan* foliage as utilized by lactating goats. *Small Rumin. Res.*, 1: 59–65.

Conrad J.R., Pratt A.D., Hibbs J.W. (1964) Regulation of feed intake in dairy cows. 1. Change in importance of physical and physiological factors with increasing digestibility. *J. Dairy Sci.*, 47: 54–62.

Dulphy J.P. (1978) Quantités ingérées et phénomènes de substitution; conséquences pour le rationnement. In: *La vache laitière*, INRA, France, pp. 87–97.

Dulphy J.P., Jailler M., Jamot J., Bousquet H. (1990) Amélioration de prévision de la valeur alimentaire de certains foins au laboratoire. *Fourages*, 121: 65–78.

D'Urso G., Avondo M., Biondi L. (1993) Effects of supplementary feeding on grazing behaviour of Comisana ewes in a Mediterranean semi-extensive production system. *Anim. Feed Sci. Techn.*, 42: 259–272.

D'Urso G., Avondo M., Bordonaro S., Marletta D., Guastella A.M. (1998) Effect of sustained-release somatotropin on performance and grazing behavior of ewes housed at different stocking rates. *J. Dairy Sci.*, 81: 958–965.

Elsen J.M., Wallach D., Charpenteau J.L. (1988) The calculation of herbage intake of grazing sheep: a detailed comparison between models. *Agric. Syst.*, 26: 123–160.

Forbes J.M. (ed.) (1995) *Voluntary food and diet selection in farm animals.* CAB International, Wallingford, UK.

Forbes J.M., Provenza F.D. (2000) Integration of learning and metabolic signals into a theory of dietary choice and food intake. In: P.B. Cronje (ed.) *Ruminant physiology: digestion, metabolism, growth and reproduction.* CABI International, Wallingford, UK.

Freer M., Dove H., Axelsen A., Donnelly J.R., Mc Kinney G.T. (1985) Responses to supplements by weaned lambs grazing mature pasture or eating hay in yards. *Aust. J. Exp. Agric.*, 25: 289–297.

Hodgson J. (1985) The control of herbage intake in grazing ruminants. *Proc. Nutr. Soc.*, 44: 339–346.

Iskander F.D. (1973) Factors affecting feeding habits of sheep grazing foothill ranges of Northern Utah. PhD. dissertation. Utah State Univ., Logan, Utah.

Lanari D., Ribaldi E., D'Agaro E. (1993) Equazioni di stima dell'energia lorda, della digeribilità della sostanza organica e dell'ingestione volontaria di sostanza secca in ovini per alcune categorie di alimenti italiani. *Zoot. Nutr. Anim.*, 19: 57–71.

Lanza M., Biondi L., Pennisi P., Petriglieri R., Keshtkaran A.N. (1994) Sostituzione delle fave con un concentrato appositamente formulato. *Proc. Miglioramento dell'efficienza produttiva degli ovini e dei Caprini*, Bella, Italy, Chapter 3.

Macchioni P., Bosi P., Casini L. (1990) Ricerche sulla previsione dell'ingestione volontaria di foraggio negli ovini. *Zoot. Nutr. Anim.*, 16: 323–330.

Newton J.E., Young N.E. (1974) The performance and intake of weaned lambs grazing S24 perennial ryegrass, with and without supplementation. *Anim. Prod.*, 18: 191–199.

Pauselli M., Morgante M., Duranti E., Casoli C., Ranucci S., Mahrabi H. (1993) Effetto dell'alimentazione sulla produzione e sullo stato metabolico di pecore in lattazione. *Proc. X Congress ASPA*, Bologna, Italy, pp. 325–331.

Provenza F.D. (1995) Post-ingestive feedback as an elementary determinant of food preference and intake in ruminants. *J. Range Manag.*, 48: 2–17.

Pulina G., Serra A., Cannas A., Rossi G. (1989) Determinazione e stima del valore energetico di latte di pecore di razza Sarda. *Proc. XLIII Congr. Soc. Ital. Sci. Vet.*, Pisa, Italy, pp. 1867–1870.

Pulina G., Rossi G., Cannas A., Brandano P., Rassu S.P.G., Serra A. (1992) The use of a pelletted feed as stimulator of chewing activity in dairy sheep. *Proc. of XLIII Annual Meeting of the EAAP*, Madrid.

Pulina G., Bettati T., Serra F.A., Cannas A. (1996) Razi-O: costruzione e validazione di un software per l'alimentazione degli ovini da latte. *Proc. XIII Congresss SIPAOC*, Italy, pp. 11–14.

Reid R.L., Jung G.A., Thayne W.V. (1988) Relationship between nutritive quality and fiber components of cool season and warm season forages: a retrospective study. *J. Anim. Sci.*, 66: 1275–1291.

Rohweder D.A., Barnes R.F., Jorgensen N. (1978) Proposed hay grading standard based on laboratory analysis for evaluating quality. *J. Anim. Sci.*, 47: 747–759.

Ruyle G.B., Dwyer D. (1985) Feeding stations of sheep as an indicator of diminished forage supply. *J. Anim. Sci.*, 61: 349–353.

Serra A. (1998) La valutazione degli alimenti ed il razionamento negli ovini da latte. Dottorato di ricerca thesis, Perugia, Italy.

Theriez M. (1983) Comportement alimentaire et ingestion de l'herbe par les brebis au paturage. *Proc. VIIIième Journées de la Recherche Ovine et Caprine*, Paris, pp. 111–140.

Trimarchi G., Rossi G., Secchiari P., Ferruzzi G. (1981) Influence de la concentration énergétique de la ration sur les performances des brebis laitières. *Proc. International Symposium on Pastoral Sheep Farming Systems in Intensive Economic Environments*, Tel Aviv.

Trimarchi G., Ferruzzi G., Secchiari P., Poli P., Andreotti L., Buonaccorsi A. (1994) Effetti della somministrazione *ad libitum* di fieno e concentrato in pecore da latte. *Proc. Miglioramento dell'Efficienza Produttiva Degli Ovini e dei Caprini*, Bella, Italy, Chapter 6.

Van Soest J.P. (1994) *Nutrition ecology of the ruminant*, 2nd Edition. Cornell University Press, Ithaca, New York.

6 Feeding of Lactating Ewes

Antonello Cannas
Dipartimento di Scienze Zootecniche, Università di Sassari, Italy

6.1 Introduction

Milk production with dairy ewes requires more intensive systems and more nutrients per animal than are usually necessary for meat or wool production systems. During lactation, nutrient requirements may be very high. Inadequate feeding may reduce both the daily milk production and the length of the lactation. Adequate feeding requires proper balancing of rations. This, in turn, requires estimation of the nutrient requirements and feed intake of the animals and of the nutritive value of the feed. Proper feeding strategies for the lactating ewe cannot be based simply on what is known about dairy cows. Even though much of the information available for dairy cattle is valid for dairy sheep, it is necessary to be aware of the differences between the two species to avoid using inappropriate feeding strategies for the lactating ewe.

6.2 Dairy Sheep are Not Just Dairy Cows Ten Times Smaller

Recommendations for feeding dairy sheep are often derived from research on dairy cows, whose nutrition and feeding management have been studied more extensively. Even though both sheep and cattle are ruminants and have many similarities, they tend to have different feeding strategies and are also different in some physiological functions (e.g. wool growth).

Some of the most important differences between the two species are related to their body size. Dairy sheep are, in general, 10–12 times smaller than dairy cows. Many studies have shown that in both species the total volume of the gastrointestinal (GI) tract (as measured by liquid contents) makes up 13–18% of the body volume (Parra, 1978). As adult ruminants increase in size, the wet fermentation contents of the GI tract increase in direct proportion to bodyweight (Demment and Van Soest, 1985). This means that the volume (as defined above) of the GI tract of a 60 kg ewe is, on average, tenfold smaller than that of a 600 kg cow. However, as the bodyweight increases, there is a lower proportional increase in energy requirement for maintenance. Maintenance energy requirements are proportional to the 0.75 power of bodyweight ($BW^{0.75}$, often called metabolic weight, MW). The INRA system (INRA, 1988) reports that maintenance requirements are equal to 56.1 and 70.0 kcal of net energy for lactation (NEL)/kg of MW for sheep and cattle, respectively. This means that the maintenance requirements of a 600 kg (MW = 121.2 kg) cow are only sevenfold higher than those of a 60 kg (MW= 21.6 kg) ewe. An index of the fermentative capacity may be estimated by dividing the liquid contents of the GI tract by the maintenance energy requirements. The resulting fermentative capacity curve (Fig. 6.1) shows that cattle tend to have more GI tract contents per unit of energy required for maintenance than do

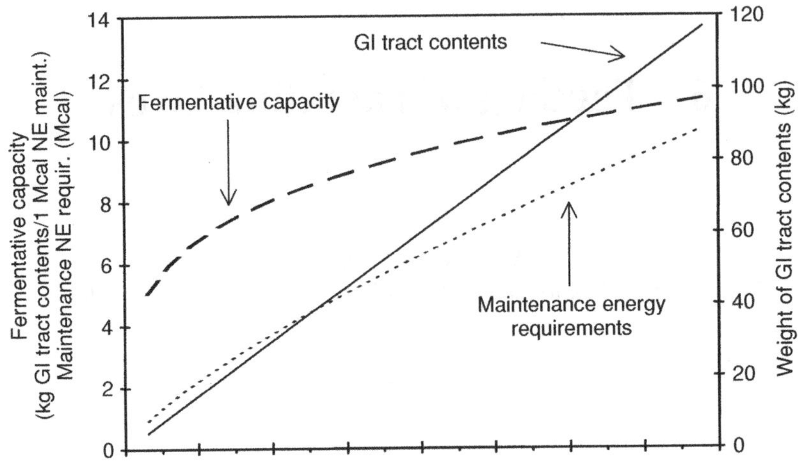

Fig. 6.1. Effect of bodyweight on gastrointestinal tract (GI) contents, maintenance energy requirements (net energy, NE) and fermentative capacity. For greater clarity, energy requirements for maintenance are considered to be equal to 65 kcal/NE/kg of metabolic weight for the whole BW range considered, even though these requirements differ depending on the species (see text).

sheep. This implies that cattle can 'store' more feedstuff in the GI tract for each unit of energy required for maintenance than sheep.

Many feed components (mainly cellulose and hemicellulose, but also insoluble protein and starch) have low rates of degradation. The completeness of fermentation is highly dependent on the time they spend in the rumen (retention time). The longer they are fermented in the rumen, the more they are digested (up to a certain limit). In practice, if sheep and cattle are fed the same fibrous feedstuff, cattle tend to digest it more thoroughly because they have more room, and they can retain the feed in the rumen for a longer period (Table 6.1). This difference in digestibility is maintained even at high feeding levels (Blaxter *et al.*, 1966).

These facts have important practical implications. Sheep have to speed up the passage of feedstuff in the rumen (i.e. decrease the retention time) to compensate for their low fermentative capacity. Thus, sheep need to eat more feed per day (as % of BW) than do cattle to satisfy their requirements, but they digest it less thoroughly. Despite this, the total amount of nutrients digested per day, proportionally, is usually higher due to the higher intake of dry matter. This explains why high-producing dairy sheep may have a level of intake of between 4% and 6% of their bodyweight, while in high-producing cows this figure does not usually exceed 4%. The difference in level of intake explains most of the difference in rumen feed retention time between sheep and cattle (Cannas and Van Soest, 2000).

Another way that sheep deal with this problem is by exercising greater feed selection (Van Soest, 1994). They tend to choose feeds or parts of feeds (young stems, leaves, buds) that are of good quality and whose digestibility is less affected by ruminal retention time. In this, they are helped by their narrow muzzle and by the high mobility of their tongue and lips. However, feed selection cannot be explained in terms only of body size. For example, even though sheep and goats have similar body size, when fed on the same mixed (grassland and shrubs) pasture, they showed quite different behaviour, with goats preferring shrubs, and sheep herbs (Leclerc, 1985).

Sheep and cattle differ also in their chewing activity. Sheep spend between nine

Table 6.1. Apparent digestibility and retention times for ruminants fed the same medium-quality timothy hay *ad libitum* (Uden and Van Soest, 1982; Uden *et al.*, 1982). The intake per kg of BW was higher in goats and sheep than in cattle. As a result, ruminal retention time and dietary total tract digestibility were highest in cattle.

Item	Caprine	Ovine	Bovine
Body weight (kg)	29	30	555
Intake of dry matter			
g/day	700	650	7830
g/kg BW	24.3	21.7	14
g/kg $BW^{0.75}$	56	51	68
Digestibility (%)			
Dry matter	47	47	54
NDF	44	44	52
Retention time of forage particles			
Rumen (h)	28	35	47
Whole GI tract (h)	52	70	79
Ratio: rumen/entire GI tract	54	50	59

BW: body weight; NDF: neutral detergent fibre; GI: gastrointestinal.

and 16 times longer than cows in eating, and ruminate 1 kg of dry matter (De Boever *et al.*, 1990). Sheep have to chew more than cattle because they are smaller animals and their chewing activity is less powerful. Sheep also have to grind the particles more finely than cattle in order that these may pass through the rumen and the other compartments of the foregut (Van Soest, 1994). This behaviour was clearly demonstrated when lactating dairy cows (Holstein) and dairy sheep (Sarda) were fed a pelleted total mixed ration as their only feed (Table 6.2). While sheep needed more than 1 h to ruminate 1 kg of dry matter, cows ruminated very little. Indeed, while sheep were doing well on this diet and were producing a good quantity of milk, the milk yield of the cows dropped, there was less milk fat and clear signs of acidosis occurred.

Since there is a limit to the amount of time a ruminant can spend ruminating (10–11 h/day; Welch, 1982), intake tends to be limited in sheep more than in cattle by the particle size of diets containing long hay. This fact, and the lower fermentative capacity of sheep, explains why grinding often increases the intake of forage and why the response is stronger in sheep than in cattle. A comparison between sheep and cattle fed three different diets: (i) high-quality dehydrated ryegrass (A); (ii) medium-quality dehydrated ryegrass (B); and (iii) a mix of medium-quality ryegrass and barley grains (C) – presented in either long or ground and pelleted form – showed that grinding and pelleting: (i) increased intake more in sheep than in cattle; (ii) increased intake more in young animals than in adult animals; and (iii)

Table 6.2. Intake and chewing activity of cows and sheep fed the same pelleted total mixed ration as sole feed source (Rossi, 1994, cited by Van Soest *et al.*, 1994).

Variable		Dairy cows	Dairy ewes
Intake	(kg of DM/day)	8.4	1.2
Eating time	(min/day)	110.7	56.0
Rumination time	(min/day)	19.4	78.5
Total chewing time	(min/day)	130.1	134.5
Eating efficiency	(min/kg of DM)	13.1	46.3
Rumination efficiency	(min/kg of DM)	2.3	64.9
Total chewing efficiency	(min/kg of DM)	15.4	111.2

Table 6.3. Effects of grinding and pelleting various diets on intake in sheep and cattle (Greenhalgh and Reid, 1973).

			Sheep			Cattle		
		Age (months)	6	18	36	6	18	36
		Bodyweight (kg)	49	72	83	272	464	614
Diet	Form	Intake						
A	Long[a]	(g/kg of BW)	21.9	18.1	23.8	20.5	19.9	15.7
	Ground and pelleted[b]	difference in (%)	+59	+46	+29	+18	−17.1	+5
B	Long[a]	(g/kg of BW)	17.8	15.2	18.0	19.6	15.9	13.7
	Ground and pelleted[b]	difference in (%)	+76	+74	+61	+31	+21	+30
C	Long[a]	(g/kg of BW)	22.0	17.5	24.6	20.5	19.7	17.3
	Ground and pelleted[b]	difference in (%)	+49	+25	+11	+20	0	0

A, perennial ryegrass, 2nd cut, harvested 7 weeks after the 1st cut (NDF 59%, CP 19%, ADL 3.3%); B, perennial ryegrass, 2nd cut, harvested 12 weeks after the 1st cut (NDF 64%, CP 16.6%, ADL 4.1%); C, 60% hay B and 40% milled and pelleted barley.
[a] Long (baled) for cattle, coarsely chopped (5 cm screen) for sheep.
[b] Chopped (1.44 cm screen) and pelleted through a 16 mm die.

increased intake in inverse proportion to dietary quality (B > A > C) (Table 6.3) (Greenhalgh and Reid, 1973). However, even in ground diets the total daily intake of digested dry matter was higher for high-quality than for low-quality diets.

Intense rumination in sheep can also be important when the diet includes grains. Rumination reduces the particle size and increases the ruminal digestibility of grains, and therefore of starch. Sheep tend to chew grains more finely than cattle. This may explain why highly digestible diets (> 66%) tend to be digested better by sheep than by cattle, while with poorly digestible diets cattle are more efficient (Mertens and Ely, 1982).

In conclusion, compared to cows, sheep:

- Have to eat more to satisfy their maintenance requirements. This results in a higher passage rate of feed and lower fibre (forage) digestibility.
- Tend to have a more selective feeding pattern.
- Are more affected in their intake by the particle size and the fibre content of the forage.
- Have to spend more time eating and ruminating each kg of feed.
- Tend to have higher digestibility for grains and high-energy diets.

6.3 Feed Evaluation for Dairy Sheep

Most feeding systems for sheep were developed for meat or wool breeds (NRC, 1985; CSIRO, 1990; AFRC, 1995). Only the French INRA system (INRA, 1987, 1988) specifically considered dairy sheep requirements. This is why it is the most widely used system in the Mediterranean countries, where dairy sheep are particularly common.

The INRA system (INRA, 1988) has been criticized because diets formulated with this system tend to underestimate feed allowances, especially when medium–low-quality forages are used (Dell'Aquila *et al.*, 1978; Pilla *et al.*, 1993, 1994; Bianchi *et al.*, 1994). One of the main problems of the system is that the energy value of feed is estimated assuming that the animals are fed at near-maintenance feeding levels. It is, however, well known that when the level of intake is high, as always occurs in lactating ewes, the rate of passage of feed increases.

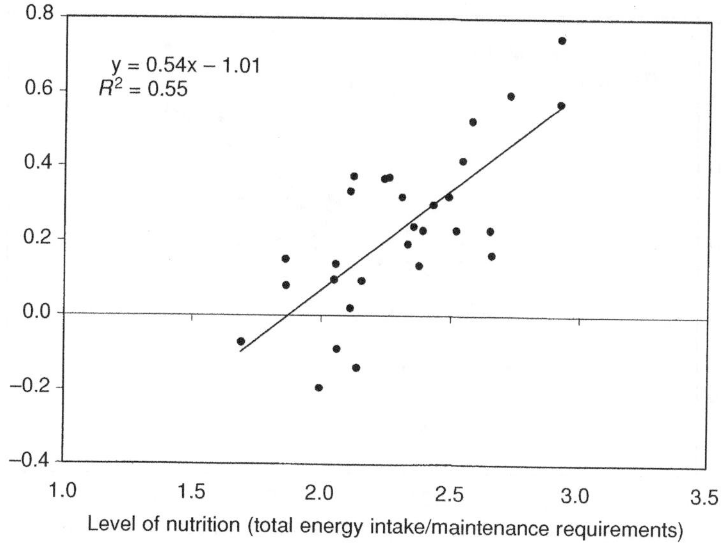

Fig. 6.2. Relationship between level of nutrition and energy balance (daily UFL intake–daily UFL requirements) in ewes fed diets balanced with the INRA (1988) system (Cannas, 2000; Serra *et al.*, 1998). One UFL (Unité Fourragère Lait) = 1.7 Mcal of NEL. Three heavy outliers were discarded.

Thus, the digestibility of slowly fermenting feed fractions (mostly fibre) is less than that for animals fed at near maintenance. In the case of lactating cows the INRA system (INRA, 1988) compensates for this effect by increasing the requirements of the cows in proportion to their feeding levels and their intake of concentrates. No corrections are used for sheep, despite the fact that sheep are often fed at feeding levels similar to or higher than those typical of dairy cows, and that sheep have a faster feed passage rate than cows when fed at the same feeding level (Van Soest, 1994).

The accuracy of the INRA system was tested, as regards calculation of the energy balances, by using published experiments in which 32 different dietary treatments were applied to lactating dairy sheep (Serra *et al.*, 1998; Cannas, 2000). The energy balances were calculated as the difference between net energy intake (UFL/day, as reported in each publication) and net energy requirements of the ewes (UFL/day, estimated on the basis of BW, BW variations, and milk yield and quality reported in the publications). These energy balances were then regressed, after discarding three large outliers, against the feeding level (total energy intake divided by energy requirements for maintenance) observed for each feeding treatment (Fig. 6.2). The results showed that as the nutritional level increased, the energy balance, which in theory should always be equal to zero, increased linearly. This could be the result of either an overestimation of the energy content of feeds or an underestimation of the energy requirements, or both. The energy balances were also expressed as a percentage of each UFL consumed (Fig. 6.3). The results demonstrated that the trend shown in Fig. 6.2 was not due to the accumulation of a fixed error, which increased as energy intake increased, but to an error that increased in proportion to the increase in the feeding level. It seems, then, that the INRA system becomes less accurate as the feeding level increases. Even though (because of the methodology used) it was not possible to be sure whether the error was due to overestimation of the energy value of the feed, or to underestimation of the energy requirements, the first hypothesis seems to be more likely than the second. The effect of feeding level on ovine dietary digestibility is well known (Blaxter *et al.*,

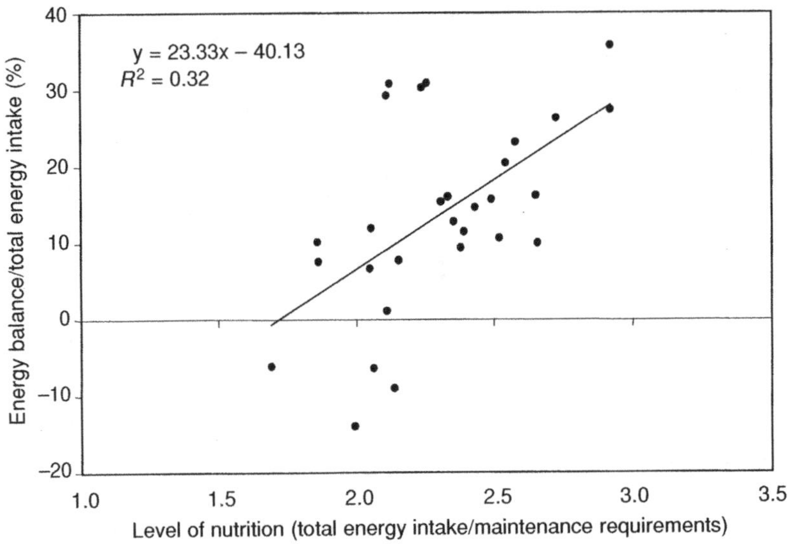

Fig. 6.3. Relationship between energy balance (daily UFL intake–daily UFL requirements), expressed as % of the daily total energy (UFL) intake, and nutrition level in lactating ewes fed diets balanced with the INRA (1988) system (Cannas, 2000; Serra *et al.*, 1998). Three heavy outliers were discarded.

1966; Robertson and Van Soest, 1975). Despite this, data exist for only a few calorimetric and digestibility experiments on lactating ewes fed at high feeding levels.

For this reason, the effect of the feeding level on the nutritive value of the feeds used by sheep may be estimated by following the approach used for cows in some recent feed evaluation systems, such as the Van Soest discount system (Van Soest and Fox, 1992) and the CNCPS (Fox *et al.*, 2004). The CNCPS has recently been adapted to sheep and the results showed that this model quite accurately predicted dietary digestibility at various feeding levels (Cannas *et al.*, 2004).

6.4 Diet Formulation for Dairy Sheep

One of the main objectives of diet formulation is to estimate the optimal dietary concentration of nutrients. This is obtained by dividing nutrient requirements by predicted DM intake. The diet is then balanced by combining available feeds so that the dietary concentration of nutrients matches the concentration of nutrients required by the animal.

In the following section, the existing information on optimal dietary nutrient concentrations for lactating ewes is summarized.

6.4.1 Energy concentration

The optimal energy concentration (UFL or NEL/kg of DM) can be estimated by using energy requirements and feed intake as predicted by any of the published feeding systems. If the French system is used, the previously mentioned overestimation of the energy content of the feeds at a high level of nutrition (Figs 6.2 and 6.3) suggests that feed allowances are often underestimated. For this reason, it has been suggested that the French system could be adapted for lactating ewes by increasing the actual requirements by a factor proportional to the overestimation of feed value observed in Figs 6.2 and 6.3. This correction led to the publication of optimal dietary energy concentrations for lactating ewes with different BW and milk yields (Table 6.4) (Serra *et al.*, 1998). This table was obtained by dividing the requirements, modified as described

Table 6.4. Optimal energy concentration of the diet (UFL/kg of DM) for lactating ewes estimated assuming zero energy balance (Serra *et al.*, 1998). These values were estimated by increasing the energy requirements of the ewes in proportion to their feeding level (see text).

FCM[a]	Bodyweight (kg)								
(kg/day)	30	35	40	45	50	55	60	65	70
0.5	0.65	0.66	0.67	0.67	0.67	0.68	0.68	0.69	0.69
1.0	0.76	0.76	0.76	0.76	0.76	0.76	0.76	0.76	0.76
1.5	0.85	0.84	0.84	0.83	0.83	0.83	0.83	0.82	0.82
2.0	0.92	0.91	0.90	0.90	0.89	0.89	0.88	0.88	0.87
2.5			0.96	0.95	0.94	0.94	0.93	0.93	0.92
3.0					0.99	0.98	0.98	0.97	0.96
3.5							1.01	1.01	1.00
4.0									1.03

[a] 6.5% fat-corrected milk yield (Pulina *et al.*, 1989).

above, by the intake predicted by using an equation developed for dairy ewes (Chapter 5). The values in the table assume that the INRA (1988) method of feed evaluation is used. However, the energy concentrations of Table 6.4 have an anomalous pattern at low milk yields, because the energy concentration increases as bodyweight increases, while the contrary should occur. This may be because only few data were available on ewes with high BW and low milk yields. Despite this limitation, Table 6.4 provides useful reference values for balancing the diets of lactating ewes.

6.4.2 Protein concentration

Optimal dietary protein concentration (g/kg of DM) should be calculated by dividing the required metabolizable protein (PDI or MP depending on the system used; see Chapter 3) by the predicted DM intake. However, when this approach is used the diets balanced for PDI or MP are often quite low in CP concentration, with values ranging between 11% and 15% CP (DM basis). These values are lower than those reported for lactating dairy cows (NRC, 1988).

Moreover, several experiments in which various dietary CP concentrations were tested in lactating ewes found that the highest milk yield was obtained with dietary CP of around 17–18% (DM basis) in both early (Gonzalez *et al.*, 1982, 1984; Robinson, 1987) and mid–late lactation (Pulina *et al.*, 1990; Cannas *et al.*, 1998). In many of these experiments, the diet was rich in escape protein (protein not fermented in the rumen, but digested in the small intestine) of high biological value, such as fish meal. This suggests that in many cases sheep require more protein than predicted. For this reason, optimal dietary CP concentrations for lactating ewes of various milk yields and BW are reported in Table 6.5. The values were based on several published feeding experiments in which different dietary CP concentrations were tested in lactating ewes (Serra *et al.*, 1998). Since lactating ewes usually have higher levels of intake than cows, the concentrations reported in Table 6.5 imply that lactating ewes should have a higher daily intake of CP per kg of BW than lactating cows. The high CP requirement per kg of BW may be explained by the fact that sheep require high quantities of sulphur-containing amino acids to produce wool (Bocquier *et al.*, 1987), suggesting that methionine, the essential amino acid predominately required for wool production, may be a limiting factor. The lack of any specific essential amino acid may result in serious dietary imbalances and waste of protein. Indeed, an experiment carried out on ewes at the beginning of the lactation showed that supplementing suckling ewes with rumen-protected methionine increased

Table 6.5. Dietary CP concentration (% of DM) suggested for different BW and milk yields (Serra *et al.*, 1998).

	Bodyweight (kg)								
5% true protein milk (kg/day)	30	35	40	45	50	55	60	65	70
0.5	16.6	15.8	15.1	14.8	14.5	14.0	13.7	13.3	12.9
1.0	17.7	16.9	16.5	15.9	15.6	15.0	14.5	14.3	13.9
1.5	18.5	17.7	17.4	16.7	16.4	15.9	15.7	15.2	14.8
2.0	19.1	18.7	18.1	17.7	17.2	16.6	16.4	15.9	15.7
2.5			18.9	18.3	17.8	17.5	17.0	16.6	16.4
3.0					18.6	18.0	17.6	17.3	16.9
3.5							18.3	17.8	17.6
4.0									18.0

milk production and the growth rate of their lambs (Lynch *et al.*, 1991). Moreover, sheep tend to have higher rates of passage of feed than cattle (Van Soest, 1994) for similar physiological stages, and consequently greater wastage of feed protein. The fact that feed proteins are usually of lower biological value than bacterial proteins suggests that MP may be used less efficiently in lactating ewes than in lactating cows. However, it is also known that a high rate of passage of feed through the rumen increases microbial yield and efficiency (Robinson *et al.*, 1985; Djouvinov and Todorov, 1994).

Feed protein quality, which is a combination of the composition of essential amino acids and the percentage of escape protein, can markedly affect the milk yield of ewes, as is clearly shown in Fig. 6.4.

6.4.3 Non-structural carbohydrates

Non-structural carbohydrates (NSC) are composed mainly of simple sugars and

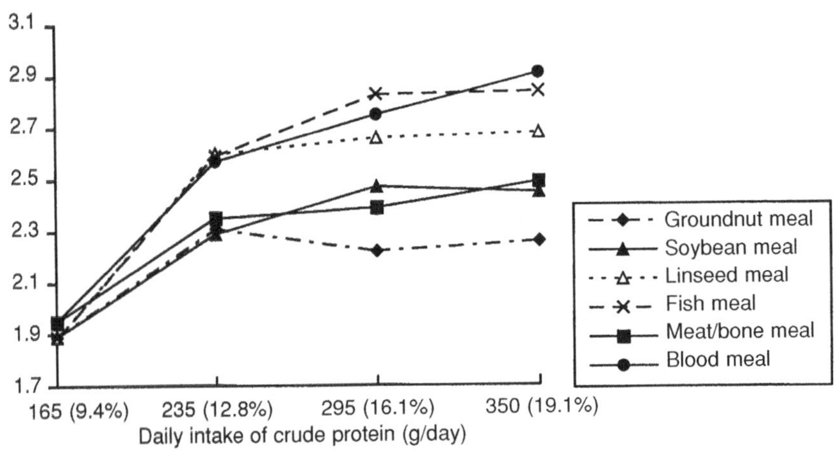

Fig. 6.4. Effect of different amounts and sources of protein in ewes (67 kg BW) in the first month of lactation. The lower protein level corresponds to a basal diet of hay, barley and molasses. Dry matter intake was restricted to 1.8 kg/day for all diets. Numbers in brackets represent CP concentration in the diets (adapted from Gonzalez *et al.*, 1982; Robinson, 1987).

starch. However, if, as is normally done, they are estimated by difference (NSC = 100 – NDF-free–CP – EE – ash), they also include pectins (NDF, neutral detergent fibre; EE, ether extract). In this case it would be more appropriate to call them non-fibre carbohydrates (NFC), to avoid any confusion between chemically measured simple sugars and starch and the estimated sum of simple sugars, starches and pectins (Mertens, 1997). For some types of feed (mostly grasses and cereals) the difference between NSC and NFC is small, but in other cases (legumes and many by-products) the difference can be very large. Dietary NFC and NDF are usually highly and inversely correlated. NFC are a very important energy source for the ewes and their ruminal bacteria. However, an excess of simple sugars and starch may induce acidosis and other digestive and metabolic problems (Ørskov, 1986).

In sheep, the effect of NFC dietary concentration seems to be markedly affected by the stage of lactation. High-roughage diets (60:40 forage to concentrate) resulted in much lower milk yields than low-roughage diets (20:80 forage to concentrate, lower NFC content) in Finn-sheep ewes in the first weeks of lactation (Brown and Hogue, 1985).

In contrast, higher intake and milk yields were found in East Friesian ewes fed diets with 20% NSC from the 5th to the 7th month of lactation, when compared with ewes fed diets with 35% NSC, which in turn had higher positive BW variations (Cavani *et al.*, 1990). In lactating dairy sheep, diets ranging from 14 to 21% CP were compared from the 5th to the 8th month of lactation at two levels of NFC (on average, 29% vs. 40%) (Cannas *et al.*, 1998). The ewes fed the diets with the lower NFC concentration had higher intake (2411 vs. 2195 g/day) and produced more milk (1428 vs. 1252 g/day). This may have been the result of too much starch in the rumen with high NFC diets, which may have caused sub-clinical acidosis. Indeed, with these diets milk fat, milk lactose and milk pH were slightly lower than with the lower NFC concentration diets. It is also possible, however, that with the high NFC diets the energy was used more for body fat deposition than for milk production. This could be due to a high production of propionate, following a mechanism proposed by Ørskov (1986). In a recent experiment, three diets, differing in their forage (chopped dehydrated lucerne) to concentrate (cereal grains and soybean meal) ratio (90:10, 70:30, 50:50) and chemical composition (NFC ranging from 32 to 43% and NDF from 43 to 31%), were fed to dairy sheep in the 8th month of lactation (Cannas *et al.*, 2000). The results showed that as the forage to concentrate ratio decreased, milk production also decreased and bodyweight increased.

The results of these experiments show that NFC has different effects at different stages of the lactation. During early lactation large amounts of grains (NFC up to 35–40%) may help a ewe with a negative energy balance to produce more milk, while at later stages large amounts of grains (and thus of NFC) may be detrimental, because they stimulate fattening but have negative effects on milk synthesis.

However, dairy sheep nutrition in the second stage of lactation has not been investigated as thoroughly as that of the early lactation period. It is clear that dairy sheep breeds have not been subjected to the same intense genetic selection as has occurred in dairy cows. This means that lactation is often not as persistent as in cows. In many dairy sheep breeds the ewes tend to use the nutrients more for body fat deposition than for milk production after the first months of lactation because their natural somatotropin blood concentration decreases quickly. Indeed, in late lactation dairy sheep (Manchega breed) increased their milk yield remarkably (+ 60%) when treated with bovine somatotropin (bST) (Fernandez *et al.*, 1995). Dairy cows treated with bST behave as genetically superior cows do and tend to use the nutrients more for milk production than for body fat deposition (Peel and Bauman, 1987). It is possible that the high positive response of ewes to bST is a sign that there is much scope for improving milk production by genetic methods. Basically, this hormone artificially creates a

Table 6.6. Intake levels and concentrations of NDF in the diet of lactating sheep (BW 51.5 ± 8.3) grazed on pasture (Avondo and Cannas, 2001).

FCM[a] (g/day)	n	DM intake (% of BW)	NDF intake (% of BW)	Dietary NDF (% DM)	Forage NDF (% DM)	Optimal NDF[b] (% DM)
< 500	108	2.78 ± 0.71	1.14 ± 0.29	41.0 ± 3.5	46.3 ± 4.8	44.5
500–799	275	3.29 ± 0.79	1.36 ± 0.36	41.4 ± 3.8	46.9 ± 5.4	45.2
800–1099	156	3.62 ± 1.07	1.43 ± 0.41	39.9 ± 4.6	45.0 ± 5.5	44.5
1100–1399	51	4.05 ± 0.93	1.46 ± 0.31	36.6 ± 4.6	41.5 ± 6.1	41.2
1400–1699	22	3.74 ± 0.89	1.30 ± 0.34	35.0 ± 3.9	38.7 ± 5.0	38.9
1700–2100	9	3.72 ± 0.84	1.20 ± 0.27	32.4 ± 0.8	37.5 ± 2.3	33.2

[a] 6.5% fat-corrected milk yield (Pulina *et al.*, 1989).
[b] NDF concentration in the diet that maintains milk production and maximizes forage intake.

hormonal balance that, in the future, could be achieved by genetic selection.

Regardless of the genetic level of the ewes, it is clear that more research is needed to define optimal levels of NFC during lactation. Proper feeding strategies to account for the effect of NFC on milk yield are suggested at the end of the following section.

6.4.4 Fibre

The quantity of fibre in the diet influences both feed intake and ruminal fermentation. Despite this, there is very little information on the optimal, maximum and minimum quantity of fibre for ovine diets. None of the sheep feeding systems (NRC, 1985; INRA, 1988; CSIRO, 1990; AFRC, 1995) consider this aspect.

In order to study the fibre requirements of sheep, lactating dairy ewes were fed *ad libitum* a pelleted complete diet with fibre concentrations ranging from 32 to 40% of NDF. The ewes with the lowest fibre concentration did not have digestive disorders, despite their high level of intake (4.75% of BW), and had similar milk yield to ewes fed higher dietary fibre concentration (Pulina *et al.*, 1995).

Avondo and Cannas (2001) summarized the results of ten feeding trials conducted in Sicily over a period of 8 years, in which the average concentrations of NDF in the diets chosen by lactating ewes grazed on mixed pastures, with supplements of concentrates and hay, were studied. In total, 621 individual intake measurements on 210 ewes were collected. The animals in the experiments were divided into classes depending on their milk production (Table 6.6 and Fig. 6.5). The results showed that: (i) the average concentration of NDF in the diet was highest (41% of DM) for the two classes of sheep with the lowest production levels, and this gradually diminished to a minimum of 32.4% for the animals which produced more than 1700 g/day of 6.5% fat-corrected milk; (ii) the average concentration of NDF in the forage (grazing + hay) eaten by the ewes was almost constantly 5% higher than dietary NDF concentration in all classes; this suggests that sheep have a remarkable ability to control their fibre intake; and (iii) the DM intake level (as % of BW) was highest (4.05% of BW) with the diets at 37% of NDF. Milk production was highest when the DM intake was 3.7% of BW and the NDF intake was 1.2% of BW. This latter result is very similar to that considered optimal (1.25% of BW) for dairy cows (Mertens, 1983, 1985).

These data can be used to define preliminary optimal dietary NDF concentrations as the sum, for each productive class reported in Table 6.6, of the average dietary NDF concentration plus a standard deviation (last column of Table 6.6). In this way 84% of the measurements of each productive class are included. These values are considered optimal for each milk yield class

Table 6.7. Optimal concentrations of NDF, CP and NFC depending on the productive levels of the sheep. The estimates refer to sheep with BW of 50 kg and assume a total dietary concentration of ash and fat around 12 % of DM.

	Production categories of 6.5% fat-corrected milk yield (g/day)					
	<500	500–799	800–1099	1100–1399	1400–1699	1700–2100
NDF (% DM)	45.0	45.0	44.5	41.2	38.9	33.2
CP (% DM)	14.5	15.0	15.5	16.3	16.7	17.3
NFC (% DM)	28.0	28.0	28.0	31.0	33.0	38.0

because they maximize forage intake and minimize feeding costs. As a result, the optimal NFC levels in the diet (Table 6.7) can be estimated by subtracting the optimal concentrations of NDF (Table 6.6), CP (Table 6.5), ash (8–10%) and lipids (2–3%) from the DM. The maximum NFC and CP and the minimum NDF are probably more likely to occur in the first months of lactation, while lower concentrations of NFC and CP and higher NDF values should be considered during the second half of lactation.

6.4.5 Fibre particle size

As mentioned above, there are substantial differences between cattle and small ruminants with regard to chewing efficiency and the minimum dimensions of the fibre which can stimulate rumination. Thus the general recommendations on the dimensions of the fibre and the optimal quantity of long fibre for dairy cows (Mertens, 1997) cannot simply be applied to dairy sheep.

The minimum feed particle size that stimulates rumination plays a very important role in the ovine diet. Diets that contain too much long fibre limit intake due to their filling effect. In addition, there is a limit to the number of hours per day for which a ruminant can ruminate (10 h/day; Welch, 1982). This means that diets that are too rich in fibre are eaten only in limited quantities because rumination takes too long.

As shown previously, chopping and

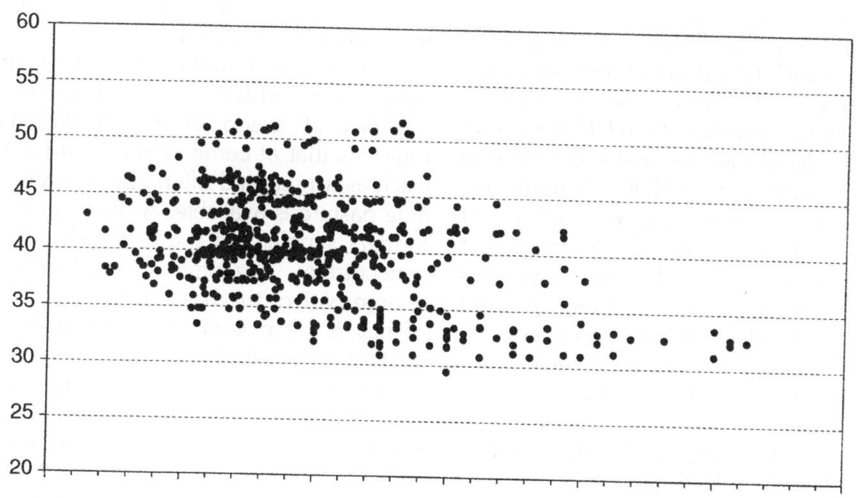

Fig. 6.5. The relationship between milk production and the concentration of NDF in the diet (Avondo and Cannas, 2001).

Table 6.8. The effect of ryegrass silage particle size and timing of cutting on intake and milk production in lactating ewes between the 1st and 4th weeks of lactation (Apolant and Chestnutt, 1985).

Type of silage		Silage					
Length	Timing of cutting[a]	CP (%)	pH	PS[b] (%)	Intake (kg DM/day)	Milk yield (kg/day)	Lamb growth (g/day)
Short	Early	16.6	3.75	74	1.45	2.67	254
	Late	11.8	3.96	82	1.03	2.17	199
Long	Early	16.6	3.75	28	0.93	2.17	200
	Late	11.7	4.22	10	0.76	1.72	156

[a] Early cut: 16.6% CP; late cut: 11.8% CP; the diet was supplemented with 800 g/day of concentrates with 15.5% CP.
[b] PS, particle size (% < 50 mm).

grinding the feed is particularly beneficial for sheep (Table 6.3). Other experimental results showed that fine grinding of the forages can markedly increase intake and milk production in sheep (Apolant and Chestnutt, 1985) (Table 6.8).

In another study, the effect of particle size was studied in Dorset, Finnish and Dorset × Finnish ewes during the first and the second months of lactation (Cannas, 1995). The ewes were fed identical diets with only the length of the cut of the hay being different (ground hay with mesh sizes of 1, 2.4 and 12 mm) (Table 6.9). With the reduction in size of the fibre, both intake and milk production increased, daily production of milk fat did not change and daily production of milk protein increased significantly. The digestibility of the DM in the diet fell only slightly as the particle size was reduced, while that of NDF fell markedly, probably as an effect of an increased rate of passage of feed through the rumen. What was most remarkable in this experiment was that the ewes were able to produce more milk without milk fat depression even with the feed of the smallest particle size.

The reduction of the particle size of the fibre does not always reduce the nutritive value of the diets (Moore, 1964; Paladines *et al.*, 1964). Indeed, it seems that the quantity of heat produced by the animals falls markedly as fibre particle size is reduced. This occurs partly because less physical work is required for eating, ruminating the diet and for the contractility of the digestive system, and also because the ratio of propionate to acetate increases in the rumen as particle size is reduced (Moore, 1964; Paladines *et al.*, 1964).

In summary, grinding the fibrous part of the diet often increases the level of intake, reduces the digestibility of the feed, and increases the efficiency of the transformation of metabolized energy into net energy. In general, this leads to an increase in the amount of energy available for production. In sheep this is particularly important because they are able to use very finely ground diets without the digestive problems usually observed in cattle.

Another important aspect to bear in mind when considering the particle size of the diet of sheep in the Mediterranean region is that at certain periods of the year (such as winter, and often also autumn) very little pasture is available. At these times the farmers supplement the diet with hay and concentrates. The latter often substitute for a large part of the basal diet, as the intake of hay is often very low because its fibre content is usually high. The resulting low forage to concentrate ratio, and the fact that concentrates are usually supplied only twice per day (at milking), often induces subclinical or acute acidosis, with negative effects on milk yield and quality and increases in some pathologies, such as mastitis or enterotoxaemia, which shorten the productive life of the sheep. In an attempt to resolve these

Table 6.9. Effect of hay particle size on the feeding and digestive behaviour and milk production in sheep in the 6th week of lactation (Cannas, 1995).

	Diet[a]			
	Short	Medium	Long	SEM
NDF particle size in diet				
> 1.18 mm (%)	16.0	18.4	54.8	
0.30 mm–1.18 mm (%)	41.0	59.7	34.1	
< 0.30 mm (%)	43.0	21.9	11.1	
Geometrical mean (mm)	0.365	0.535	1.072	
DM intake (g/day)	4005	4132	3767	147
Milk yield (g/day)	2400	2492	1991	192
Fat concentration (%)	7.86[mn]	7.03[m]	9.08[n]	0.56
Fat yield (g/day)	187.7	185.1	178.6	18.9
Protein concentration (%)	4.37	4.13	4.26	0.11
Protein yield (g/day)	105.3[m]	109.7[m]	83.8[n]	8.1
Eating time (min/day)	677[M]	399[N]	419[N]	51
Rumination time (min/day)	45[M]	165[M]	431[N]	38
Apparent DM digestibility (%)	59.8	62.9	63.5	1.6
NDF digestibility (%)	36.5[m]	42.1[mn]	48.61[n]	2.1
Acetate in the rumen (mM/l)	96.6	96.7	87.8	0.08
Rumen propionate + butyrate (mM/l)	92.1[m]	71.4[n]	68.2[n]	0.03
Ruminal pH	6.04[M]	6.05[M]	6.35[N]	0.01

[a] Diet, 54.9% grass hay, 30.1% cracked barley gains, 13.0% soy bean meal, 2% minerals (CP 16.4%, NDF 41.6%, ADL 3.05%). Hay ground with a mesh size of: 1 mm, short diet; 2.4 mm, medium diet; 12 mm, long diet. BW of sheep: 68.1 kg; SEM, standard error of the mean.
[mn] Significant difference ($P < 0.05$).
[MN] Significant difference ($P < 0.01$).

important problems, Rossi *et al.* (1991) have developed a complete pelleted diet (CPD) which can substitute for either the concentrates or the basal diet, and which can replace the whole diet. The basic idea in developing the CPD was to create a pellet which contained enough fibre to stimulate rumination and enough energy to stimulate milk yield. A typical example of the granulometric composition of a CPD is shown in Table 6.10.

In the experiments in which the CPD was used as the only feed, eating time varied between 40 and 110 min/kg of DM, and rumination time between 47 and 73 min/kg of DM, depending on the level of intake, the concentration of NDF and the mean particle size of the pellets (Pulina *et al.*, 1995). When the CPD was used as the only feed, milk yield always increased compared to the pre-experimental period, when the animals were either grazed or fed hay

Table 6.10. Fibre particle size distribution of a complete pelleted diet containing 37.4% NDF (of DM) (Cannas, unpublished data).[a]

Sieve diameter (mm)	NDF retained (%)	NDF not retained (%)
4.74	0.0	100.0
2.36	5.7	94.3
1.18	14.8	79.5
0.60	20.8	58.7
0.30	21.5	37.2
0.15	24.0	13.2
< 0.15	13.2	0.0
Total	100.0	

[a] Pellets treated with 8 M urea before ND treatment to avoid possible starch residues.

Table 6.11. Relationship between dietary crude protein and energy concentration and feed intake, milk production and urea concentrations in the milk of Sarda sheep fed complete pelleted diets (Cannas *et al.*, 1998).

Diet EC	CP (% DM)	Intake DM (g/day)	Intake Energy (Mcal/day)	Milk Yield (g/day)	Milk Fat (%)	Milk True (%) protein	Milk Urea (mg/dl)
HE	21.23	2163	3.57	1329	5.59	5.36^{m}	54.65^{m}
HE	18.84	2331	3.85	1306	5.45	5.35^{m}	47.71^{m}
HE	16.59	2330	3.85	1215	5.68	5.42^{m}	35.85^{n}
HE	14.19	2073	3.42	1165	5.60	5.66^{n}	26.16^{o}
LE	21.09	2516	3.90	1505^{m}	5.85	5.23	56.34^{M}
LE	18.64	2571	3.99	1508^{m}	5.71	5.30	49.64^{N}
LE	16.33	2379	3.69	1425^{mn}	5.72	5.42	37.33^{O}
LE	13.89	2230	3.46	1282^{n}	5.93	5.48	27.80^{P}
HE (mean)	17.71	2204^{p}	3.64	1260^{p}	5.58^{m}	5.45	41.38
LE (mean)	17.49	2409^{q}	3.73	1422^{q}	5.82^{N}	5.36	42.72

EC, Energy concentration; HE, high energy, 1.65 Mcal of ENL/kg of DM; LE, low energy, 1.55 Mcal of ENL/kg of DM.
^{p,q} $P < 0.1$; ^{m,n,o} $P < 0.05$; ^{M,N,O,P} $P < 0.01$.

and concentrates (Rossi *et al.*, 1991; Cannas *et al.*, 1992). Feed intake was always very high, varying between 5 and 6.5% of BW. When the CPD was used in cattle, the result was negative (Table 6.2); the intake of CPD was not limited by the filling effect of the fibre it contained. For example, in the experiment reported in Table 6.11 (Cannas *et al.*, 1998), the intake of DM was higher in the low-energy (48.5% NDF as average) than in the high-energy (38.0% NDF as average) diets and there was no difference in the energy intake between the energy levels. Similar results were obtained in other experiments (Pulina *et al.*, 1995). Using CPD for lactating sheep as the only feed for long periods (4 months) resulted in a slight worsening in the efficiency of the digestive system, but only after more than 2 months of continuous use (Calamari *et al.*, 1991). The liver, however, continued to function normally during the whole experiment. Rapid changes from a diet of CPD alone to grazing with CPD as a supplement did not have any negative effects on milk production (Serra *et al.*, 1990b) or on the health of the animals (Serra *et al.*, 1990a). In addition, the use during the whole lactation of CPD as a supplement to grazing resulted in a marked reduction in the incidences of mastitis and lameness and in the disappearance of ruminal acidosis on the farm where the experiments with CPD took place.

6.5 Diet Balancing for Grazing Dairy Sheep using Nutritional Indicators

There are many practical problems involved in the feeding of grazing sheep. Because it is difficult to estimate the quality or quantity of the grass ingested by the animals it is also difficult to decide what feed supplements should be given to them.

A rational and precise choice of feed supplements should be based on evaluating not only the characteristics of the pasture but also on the nutritional status of the sheep. The easiest nutritional indicators to use are:

- The milk yield and quality.
- The state of health of the animals.
- The concentration of urea in the milk.
- The body condition of the animals.
- The characteristics of the faeces.

Some of these indicators are described and examined below.

6.5.1 The concentration of urea in the milk as a nutritional indicator

In dairy cows, the concentration of urea in the milk (MU) is seen as a good indicator of the metabolism and intake of protein by the animals (Oltner and Wiktorsson, 1983; Bertoni, 1985; Roseler *et al.*, 1993; Bertoni *et al.*, 1999) and is currently used to evaluate and adjust the diet. Blood urea concentration (BU) and MU are closely correlated, even though MU is considered to be more stable (Baker *et al.*, 1992), and is easier to sample. In dairy cows, high concentrations of MU and BU have been associated with reductions in reproductive efficiency (Ferguson and Chalupa, 1989).

MU and BU could also be good indicators of nitrogen intake and used in lactating ewes. However, the available information is mainly on dry sheep, whose dietary requirements, feed intake and rate of passage are much lower than those of lactating ewes. Defining MU and BU reference values for milk sheep would be extremely useful. This is why various experiments on the MU content of milk sheep have been carried out in recent years (Cannas *et al.*, 1997, 1998; Ubertalle *et al.*, 1998).

One of these experiments (Cannas *et al.*, 1998) studied the relationship between dietary crude protein and energy concentration and MU in adult Sarda sheep from the 3rd to the 5th months of lactation (Table 6.11). In this study there was a linear relationship between dietary crude protein concentration and MU at both of the energy levels studied. The difference in MU between the two energy levels was very small and non-significant. The relationship between dietary crude protein concentration or the daily CP intake, on one hand, and MU, on the other, was described by the equations:

$$\text{MU (mg/dl)} = 4.10 \text{ CP concentration (\% of DM)} - 30.30 \ (R^2 = 0.98) \quad [1]$$

$$\text{MU (mg/dl)} = 0.14 \text{ CP intake (g/day)} - 15.23 \ (R^2 = 0.94) \quad [2]$$

The MU values found in this study were compared with the MU or BU reported by various authors in experiments designed to study the effects of diets containing different protein concentrations, or quality, on milk production. These studies investigated 66 different diets and many different sources of protein, both pure and in various mixtures. The regression of the dietary CP concentration on MU or BU suggests that these two variables are closely and linearly correlated (Fig. 6.6), irrespective of the source of the protein or the feed intake. The regression equation (Fig. 6.6) obtained is similar to the above-mentioned equation [1], which was calculated using only data from the experiment just described. The association between MU or BU and daily CP intake was, however, lower (Fig. 6.7). This suggests that the ratios among CP and the other nutrients in the diet are more important in controlling MU or BU than is the total daily protein intake. This is probably because ruminal microbial ammonia utilization is markedly affected by the other nutrients supplied with the diet.

The close relationship observed between MU and dietary CP protein concentration suggests that the different protein sources of the supplements had little influence on MU. The results of the above-mentioned experiments and those of experiments in which maize gluten meal was compared with soybean meal (Cannas *et al.*, 1997), and roasted soybean meal with soybean meal (Ubertalle *et al.*, 1998), seem to confirm that in sheep the degradability of the protein sources has only a limited effect on MU.

6.5.2 Milk sampling for measuring milk urea

The great variability in MU content among animals fed the same diet (Cannas *et al.*, 1998), as has also been seen in cattle, suggests that urea should be measured in pooled samples of milk taken from at least eight to ten sheep at the same stage of lactation. It is also important to consider first-lactation ewes separately from older animals. Indeed, MU was higher by 5–10 mg/dl in first-lactation ewes, compared with ewes in their second and subsequent lactations, in an

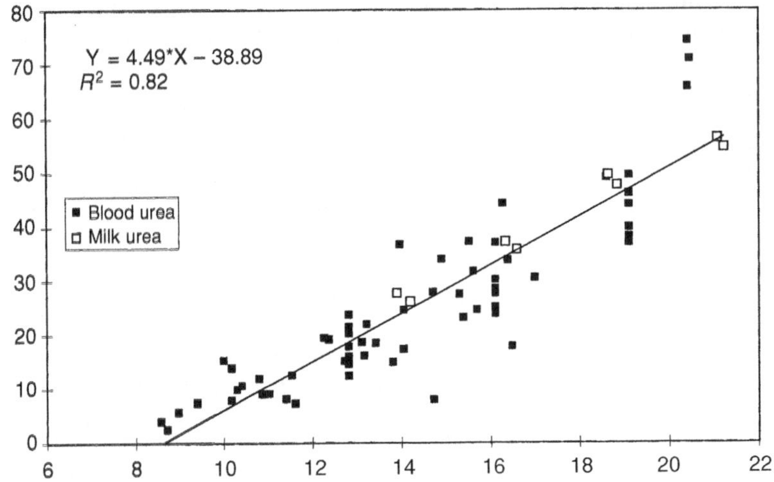

Fig. 6.6. The relationship between protein concentration in the diet and milk urea (MU) or blood urea (BU) in dairy, meat and wool sheep (Cannas *et al.*, 1998). Each point represents the average of an experimental treatment.

experiment conducted on Sarda sheep, despite the fact that the diet supplied was the same (Cannas *et al.*, 1997). In the same experiment, BU values were slightly higher and more variable than MU, as already observed in cattle (Baker *et al.*, 1992).

On many farms mechanical milking has made it difficult to take individual milk samples. Sampling of bulk milk from the refrigeration tank is a possible solution if the flock is made up exclusively of animals at the same stage and order of lactation. Another method is to sample some squirts of milk before the milking clusters are attached to the teats. The differences in MU concentrations obtained at different times of milking are small, and non-significant in practical terms.

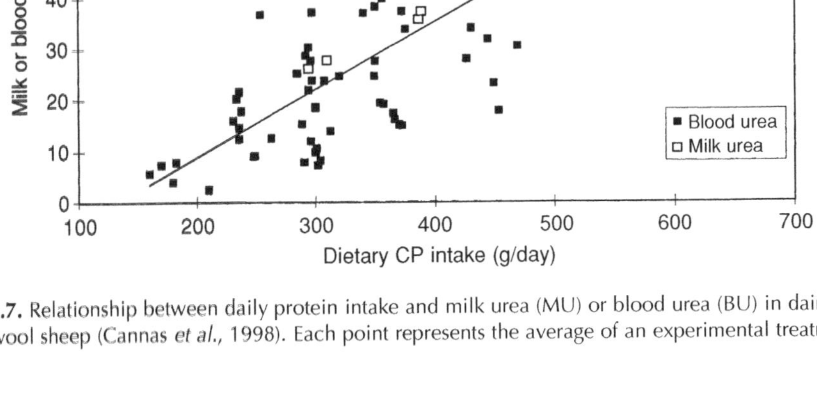

Fig. 6.7. Relationship between daily protein intake and milk urea (MU) or blood urea (BU) in dairy, meat and wool sheep (Cannas *et al.*, 1998). Each point represents the average of an experimental treatment.

Urea can be measured in the laboratory by various analytical methods of very high accuracy. It is quite important not to be confused by the units of measurement. In some countries milk urea is expressed as 'milk urea nitrogen'. Since nitrogen makes up 46.65% of the urea molecule, when analyses are reported as 'milk urea nitrogen' the values are about 50% of those expressed as 'milk urea'; the latter form is used throughout this chapter.

If the diet needs to be changed rapidly it is best to measure the MU directly on the farm. This is why the Azotest® strips are sometimes used on bovine milk. This system is based on the variations in colour that occur in treated plastic strips after they have been dipped in the milk. However, in our experience these strips cannot be used for sheep milk and are also unreliable for bovine milk. Fortunately, portable equipment that measures urea very accurately has recently been introduced to the market.

6.5.3 The relationship between MU, health status and reproductive efficiency in sheep

Another aspect, which makes urea even more important as a nutritional indicator, is that medium to high concentrations of BU (34–50 mg/dl) were associated with negative effects on the initial development and survival rate of sheep embryos cultivated *in vitro* (Bishonga *et al.*, 1994). Concentrations of BU greater than 50–55 mg/dl have been associated with a marked reduction in fertility in artificially inseminated Sarda sheep (Molle *et al.*, 1998) (see also Chapter 7). The information available suggests that the negative relationship between BU and fertility is mainly due to the urea derived from ruminal ammonia, while the effects on fertility are less clear when the urea comes from ammonia produced by the degradation of amino acids in the liver. Other problems associated with excess of feed protein and high milk urea concentrations are:

- Increase in the somatic cell count of the milk.
- Increased incidence of mastitis and mammary oedema.
- Increased incidence of lameness.
- Increased incidence of enteric problems, e.g. diarrhoea.
- Reduction in the efficiency of the liver.
- Reduction in the efficiency of the immune system.

6.5.4 Reference values for milk urea in sheep

Based on what was previously said regarding milk urea and on the optimal dietary protein concentration (Table 6.5), it is likely that MU greater than 40–50 mg/dl is associated with excess dietary protein and lower reproductive efficiency, while values lower than 25–30 mg/dl are associated with insufficient dietary protein and low milk production. These values are higher than those suggested for dairy cattle, but the comparison of dairy ewes with dairy cows showed that MU concentrations were 5–15 mg/dl higher for ewes than for cows when the CP concentration of the diet was the same. The lowest difference between these two species was observed when the CP concentration in the diet was lowest (12–14% of CP on a DM basis), while it increased as dietary CP con-

Table 6.12. Relationship between MU and dietary CP concentrations in sheep (predicted by using the regression equation reported in Fig. 6.6). When dietary CP concentration is unknown, MU may be used for its estimation.

CP (% DM)	12.0	12.5	13.0	13.5	14.0	14.5	15.0	15.5	16.0
Urea (mg/dl)	15.4	17.6	19.8	22.0	24.2	26.4	28.6	30.8	33.0
CP (% DM)	16.5	17.0	17.5	18.0	18.5	19.0	19.5	20.0	20.5
Urea (mg/dl)	35.2	37.4	39.6	41.8	44.0	46.2	48.4	50.6	52.8

centration became higher (> 16% CP) (A. Cannas, unpublished). When the pasture is young and its CP concentration is high, MU concentrations higher than 60 mg/dl are often found in sheep. These levels are certainly associated with excess protein intake, poor health status and reproductive efficiency, and with increases in energy requirements (the energy cost of eliminating 100 g CP in excess of ovine needs is the same as that of producing 200 g of ovine milk).

The relationship found between MU and dietary CP concentration (Fig. 6.6) may be used to estimate the protein concentration in the diet, when it is unknown, by measuring milk urea (Table 6.12). This could be particularly useful for grazing animals, for which some knowledge of dietary CP concentration would greatly help in selecting the type and amount of feed supplements required to complete and balance the diets.

6.5.5 Faeces as a nutritional indicator

Faeces are one of the end-products of the digestive system. Their physical and chemical characteristics depend on the initial characteristics of the diet and on the way in which the digestive system degrades and digests it. Faeces are one of the most representative indicators of the nutritional and health status of the animal. In dairy cows, faeces are used as a nutritional indicator (Bertoni *et al.*, 1999). There has been no such systematic study made for sheep and so the summary of faecal characteristics vs. nutritional level given below should be used with caution.

LIQUID (DIARRHOEAL) OR SEMI-LIQUID

- Dark colour indicates excess of highly fermentable dietary protein (NPN + soluble protein) due to presence of immature grasses or legumes, protein-rich silage or feed with added urea.
- Light colour, strong smell with grain fragments indicates ruminal acidosis due to a diet too rich in NSC and/or low in structured fibre.
- Excess of minerals.
- Dirty feed.

BROWNISH-GREEN IN PASTY BLOCKS

- Well-balanced diet.
- For grazing animals, an indicator of a diet rich in young plants.
- In sheepfolds, an indicator of a ration with a low–medium fibre content.

WELL-FORMED, DARK BROWN PELLETS, LOW–MEDIUM MOISTURE CONTENT (50–60%)

- Well-balanced diet.

WELL-FORMED, COMPACT, VERY DRY PELLETS WITH FIBROUS COMPONENT

- Lack of fermentable protein in the rumen, low ruminal bacterial activity.
- Diet contains too much fibre and/or too lignified (rich in straw, stubble, mature forage).

Whole grains are less frequently found in sheep faeces than in cattle faeces, due to the more intense rumination activity of sheep. This is particularly true for large grains (e.g. maize). Thus, high quantities of whole grains in sheep faeces may indicate ruminal acidosis or sub-acidosis, or at least insufficient rumination. The latter is frequent in older animals with damaged teeth.

6.5.6 Body condition score

Throughout the year sheep usually have large variations of body reserves, with gains during the last part of the lactation and the dry period and losses in the first months of lactation. For this reason, proper dietary balancing requires that the nutritional status of the animals is accurately monitored. Variations in body reserves cannot be estimated adequately by simply measuring BW and its variations (Russel *et al.*, 1969). Indeed, variations in the contents of the digestive and reproductive systems often render BW variations markedly untrustworthy and unpredictable, leading to errors in the evaluation of the energy balance and nutritional status of the animals. Another confounding factor is that in the first 6 weeks of lactation sheep can lose up to 1 kg of protein, which is mostly converted to glucose in the gluconeogenesis pathway, but is also used as a source of amino acids for milk synthesis. With each kg of protein lost, 4 kg of metabolic water is also lost, but when fat instead is mobilized the loss of water is minimal; BW variations measured by weighing the ewes do not discriminate between the mobilization of fat, or of protein and water.

More appropriate techniques for estimating body fat variations are based on the subjective assessment of the fat and muscle thickness on the backbone behind the last rib (body condition score, BCS), following a scale that goes from 0 (very thin, probably sick sheep) to 5 (extremely fat sheep). A more detailed description of the system can be found in the literature (Russel *et al.*, 1969; INRA, 1988; Molle and Sanna, 1992). Subjective estimates by trained personnel can produce sufficiently precise classifications of a large number of sheep in a short time. BCS may be used to estimate the energy costs of fattening or the energy lost when the animal becomes thinner (INRA, 1987) or simply to define the optimal state of the animal's reserves at various stages of the productive cycle (Fig. 6.8). Here, 'optimal' means the specific body condition at which the sheep are most fertile (see Chapter 7) or produce most milk. One practical problem when using BCS is that its relationship with body fat content differs among breeds (Cannas, 2000). As a result, the optimal BCS for each stage of the productive cycle may differ among breeds. For example, the comparison of Sarda sheep with Lacaune sheep showed that Sarda sheep with a BCS of 2.0 had already large visceral fat deposition, while this occurred in Lacaune sheep only when the BCS was around 3.0 (Ronchi *et al.*, 1993). Preliminary estimates have been published on the relationship between BCS and fat reserves in Sarda sheep (Ligios *et al.*, 1995), which differ substantially from those published for other breeds (Teixeira *et al.*, 1989; Zygoyiannis *et al.*, 1997; Cannas, 2000).

6.6 Feeding Sheep in the Various Phases of the Productive Cycle

6.6.1 Last months of pregnancy to the first months of lactation

Milk production in the first 2–3 months of lactation depends on many factors, some of which are linked to the characteristics of the animals themselves (phenotype, body fat reserves, prolificacy) while others are linked to the conditions in which the animals are raised (availability of feed, environmental conditions). As milk production in the first months of lactation profoundly influences the production in later months, this early phase must be treated with particular attention.

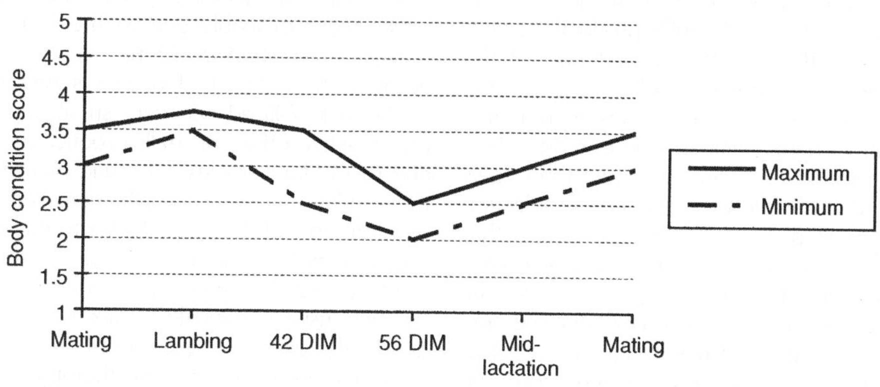

Fig. 6.8. Target body condition score during the production cycle of dairy ewes (based on INRA, 1988). DIM, days in milk.

The body condition of the ewes at lambing is one of the factors that clearly influences their milk production in the first months of lactation, because part of the milk produced at this stage depends on body fat mobilization. Even when sheep are fed high-quality diets, a negative energy balance is none the less inevitable in the first months of lactation. This is because, after lambing, milk production, and thus the energy requirements of the animals, increases more rapidly than the intake of energy from the diet. Furthermore, in Mediterranean regions, the first months of lactation often coincide with winter, when animals require extra energy for thermoregulation, there is little pasture available due to the low temperatures, and grazing is difficult because the weather conditions are often bad and the days are short. Thus, it is inevitable that in the first 2 months sheep produce part of their milk (up to one-third) by mobilizing their body fat and protein reserves. This can be considered normal as long as the loss of BW is not excessive. After this period, the BW of the sheep should begin to increase. In BCS terms, INRA (1988) suggests (Fig. 6.8) an optimal BCS at lambing of 3.25–3.50, which may fall to a minimum of 2.0–2.5 in the 6th to 8th week of lactation. The ewes should not lose more than 1 BCS in 6 weeks. However, these reference values may not be optimal for all breeds. Clearly more research is needed to better define reference values of BCS.

A greater weight loss would certainly cause a reduction in milk production, with negative effects in the second part of the lactation, an increased risk of ketosis and lower fertility. Lactating ewes with highly negative energy balance tend to reduce their milk production more markedly than cows do. Indeed, while in cows genetic selection has resulted in a hormonal status that encourages milk production even with high fat mobilization, in dairy sheep genetic improvement has not been as intensive, and ancestral characteristics designed to protect the life of the animal are often more evident. As a result, when fat mobilization is too intense milk production decreases more readily than in cows.

Thus, it is very important that sheep have sufficient reserves of body fat at the beginning of lactation (Louca *et al.*, 1974; Robinson, 1987a). Robinson (1987a) has clearly shown how body fat reserves and energy intake are of great importance in this period (Fig. 6.9). In his experiments, the milk production of sheep fed a high-energy diet was almost unrelated to their body reserves. However, when the amount of energy in the feed decreased, milk production was greatly influenced by body fat reserves. The thinner the sheep, the less milk was produced. Given that the highest energy intake reported in Fig. 6.9 is hard to achieve in grazing sheep – because concentrates are supplied separately from forages and low forage-to-concentrate ratios are not feasible – it becomes obvious that sufficient fat reserves are essential for high milk production in the first months of lactation. This also explains certain phenomena which, at first sight, appear puzzling. For example, there was a severe drought in Sardinia from January to September 2000. As a result most of the sheep were very thin at lambing time (autumn 2000). However, in the final months of the year (first and second months of lactation) there was a great deal of pasture available, due to the mild temperatures and high rainfall. None the less, milk production was less than in the previous years, even in the first 2 or 3 months of lactation. The average losses on many farms were more than 300–400 g/day per head. These losses were probably due to the fact that normal production could not be obtained because the amount of body reserves that could be mobilized was very small. Part of the negative effect on milk production was due to the large excess of crude protein in the available pasture, which increased the requirements of the animals due to the high cost associated with protein catabolism. It must be said, however, that ewes too fat at lambing usually have low milk production due to the excessive quantity of visceral fat compressing the rumen and thereby reducing feed intake (Stern *et al.*, 1978).

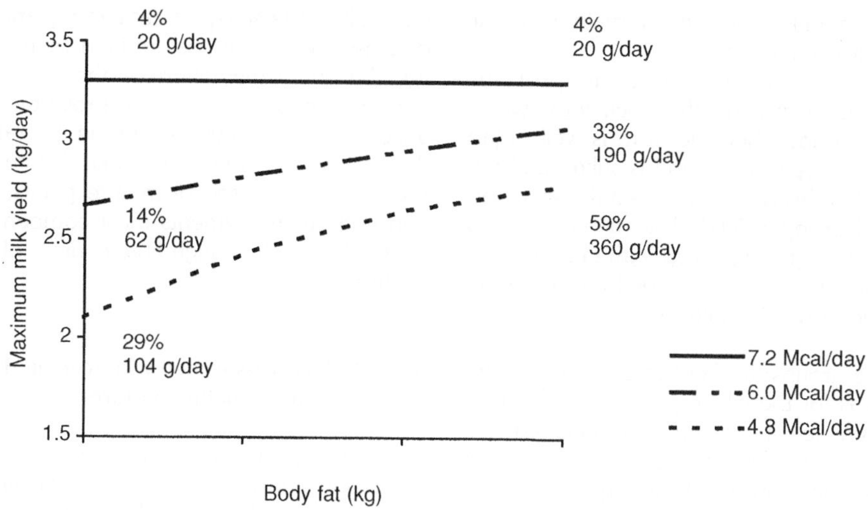

Fig. 6.9. Effect of body fat reserves and daily energy intake on maximum milk production in sheep in the first weeks of lactation (Robinson, 1987a, modified). The numbers in the figure represent the percentage of milk produced by using body fat reserves, and the loss of BW (g/day) in the animals with the lowest and highest fat reserves.

6.6.2 Remainder of lactation

In nutritional terms an exact definition of the moment of transition from initial lactation to full lactation should be based on whether or not the animal has a negative (first phase) or positive (second phase) energy balance, rather than on the month of lactation. In the Mediterranean region this generally coincides with the moment when the growth of pastures resumes after the winter interruption. Thus, for 2–3 months, full lactation coincides with the period of the year when most pasture is available. At these times, the intake of pasture is generally very high and the main technical goals should be to:

- Maximize milk production.
- Allow the body reserves lost in the early phase of lactation to recover.
- Bring the ewes to the body condition that maximizes fertility in the mating season.
- Prevent sheep with low milk production from becoming too fat.

These three objectives are difficult to achieve when sheep with very different productive levels are in the same flock, and thus are fed the same diet. The most frequent result of this is that the most productive animals are underfed, and so become too thin, while the least productive ones are overfed and so become too fat. As the sheep in the flock with above-average production produce most of the milk, underfeeding these sheep has serious economic consequences for the farm. In addition, animals which are too fat or too thin at the moment of mating are less fertile than those at optimum body condition (see Chapter 7). Then, correct feeding strategies require that the flock is divided into at least two groups based on the productive levels of the ewes. These groups should really be separated in the first months of lactation. They should be reviewed periodically and those sheep whose productive level or body condition has changed should be transferred to the appropriate group. Grouping could be decided on the basis of the milk produced or on the body condition score of the animals. In any case, the two criteria are correlated, especially after the first months of lactation. Milk production is probably the easiest to measure and represents the main economic objective of the farm.

There are no particular problems involved in dividing the flock into two or more groups when the farm is well organized and the pasture land is subdivided into plots, but this is not the case if the sheep are grazed on open land. The management of the diet of grazing sheep with different productive levels may be carried out using one of at least two techniques:

- The groups are kept in different grazing areas for the whole period for which different feeding regimes are deemed necessary. The groups are milked separately and the more productive group is given more feed in the parlour. In this way the more productive sheep can also be grazed on better pasture. Dividing the sheep into different groups is already common practice, even though the formation of the groups is often based on their reproductive cycle and is not aimed at improving the diet.
- All the sheep graze together but the different productive groups are identified by marks of different colours on their fleece. At milking time the groups are put into a corral with exits leading to different pens depending on the colours marked on the fleece. The different groups are milked separately and fed the appropriate dietary supplements during milking. After milking the whole flock is taken back to pasture. This technique simplifies the management of grazing but requires drafting facilities and separated yards. There is also the additional labour of separating the flock into different groups before milking.
- A third option could be to have in the milking parlour a feeding dispenser that supplies different amounts of concentrates according to the productive level of the sheep.

Regardless of which technique is used, the aim should be to maximize milk production by utilizing the nutritional principles and dietary requirements mentioned earlier in the chapter. To put it simply, the aim should be to allow the sheep to eat large quantities of grass. When the pasture is of high quality, the supplementation of the diet with concentrates should aim to improve the balance between the nutrients of the diet rather than to increase the total intake. When the pasture is composed of mature plants, the concentrate supplements must compensate for the low protein/high fibre content of the pasture.

6.6.3 Problems associated with feeding sheep on immature pastures

When the pasture is immature, it is also rich in protein and low in fibre. This occurs in both grasses and legumes. In grasses, in particular, the protein concentration is often much higher than is generally believed, often being > 25% of DM. This occurs when the plants are young and are fertilized with nitrogen. Furthermore, a large part of the crude protein is made by non-protein nitrogen or very soluble protein, both of which ferment very quickly in the rumen. In such situations, the intake of protein is often much higher than that required. As already stated, such excess protein is transformed into urea, some of which is later found in the milk. Several years of data of the Regional Association of the Farmers (ARA) of Sardinia have shown that on this Mediterranean island, urea in ovine milk is very high (> 50 mg/dl) during the whole winter and early spring, suggesting that sheep are fed excess protein for at least half of the lactation. It is likely that this occurs in all Mediterranean regions, where most dairy sheep are kept.

This situation is exacerbated when feed supplements rich in protein are used. Since these feeds are expensive their use not only reduces milk production and worsens the health of the animals, but it also increases feeding costs. These high-protein supplements should only be used in late spring, when the pasture is low in protein. Immature pastures are also low in fibre, the ingestion of which does not sufficiently stimulate rumination in sheep. This, in turn, decreases the production of saliva, which contains substances that buffer ruminal pH. In these situations ruminal sub-acidosis often reduces the percentage of fat in the milk; *severe* lack of fibre may also cause *acute* ruminal acidosis.

Some feeding and management techniques can reduce the risks involved in grazing animals on immature pastures:

- At milking, feed supplements should have little protein (no more than 12–13% of DM, and if possible with low ruminal degradability), but fairly high fibre (17–20% of crude fibre, 25–35% of NDF) and starch. A mixture of slow-fermenting (maize or sorghum) and fast-fermenting (barley, oats or wheat) grains should be used, so that there are starches fermenting in the rumen from the beginning of grazing (morning) to the end (late afternoon). This should improve the synchronization between the fermentation of the protein from the pasture and that of the starch in the supplements.
- Nitrogen fertilization should be used carefully and sheep should not graze pastures for several days after fertilization, especially when low temperatures or drought reduce the conversion of fertilizers to plant proteins.
- Grazing time should be reduced and the flock should be brought to pasture in late morning. The animals should not graze on pastures that are wet with dew. If this is inevitable, it is important to avoid sending the sheep to pasture when they are too hungry. Since the animals are not fed for some hours – the period between first milking and the start of grazing – it may be advisable to give them feed rich in energy and digestible fibre, but low in protein, shortly before they go to pasture. For this third meal (assuming that the other two feeds of concentrates occur at the morning and evening milkings) the best supplements are, in order of preference, beet pulp, maize silage and silages made from wheat, barley or oat grains – at milky stage mixed with some grains. Pelleted feeds may also be used as long as they have the characteristics described above.

The third-meal technique has many advantages:

- The ewes are less hungry when they start grazing and therefore they do not eat too much pasture (and proteins) in a short time.
- It supplies the energy that bacteria require for exploiting the nitrogen available in the rumen.
- It allows an increase in the number of energy-rich meals per day, reducing the risk of acidosis.

The adoption of a third meal does not imply that more supplements have to be used; it is simply a matter of subdividing the same amount of supplements into more meals.

- Milk urea should frequently be measured to monitor protein intake.
- The faeces should be routinely observed; if there is diarrhoea or the faeces are very soft, almost liquid and dark, there is probably too much protein in the diet.
- Good-quality hay should be added at night when the ewes are in the barn. However, when the pasture is abundant sheep generally eat little hay, especially when it is too mature or of low quality.

In conclusion, the problems for sheep caused by grazing pasture over-rich in proteins and poor in fibre can be reduced by the use of simple feeding and management techniques. Such techniques are relatively inexpensive and practicable.

6.7 Feeding Housed Sheep

6.7.1 The use of hay and the importance of its quality

At certain times of the year hay may be the only source of fibre in the diet, especially in areas where silage production is not common. Because the fermentative capacity of sheep is not very high, the intake of hay is high only when it is of good quality, i.e. NDF and lignin are low and the hay is prop-

Table 6.13. Practical criteria for evaluating the quality of hay.

Characteristics	Good quality	Poor quality
Colour	Pale green–yellow[a]	Brown[b]
Leafiness	High[c]	Low[c]
Maturity	No ears and/or legumes	Rich in ears and/or legumes
	Elastic culm (little lignification)[d]	Rigid culm (very lignified)[d]
Conservation	No mould	Mouldy
Smell	Pleasant	Sweet or mouldy smell[b]

[a] Indicates that even though the mowed forage has been washed by rain it has been baled at the right humidity; [b] Indicates that the hay was too wet when it was baled and its temperature increased inside the bale (Maillard's reaction), with a consequent reduction in digestibility of sugars and proteins; this problem is often associated with mould development; [c]of particular importance for legumes; [d]for grass hay, if the culms break when twisted round the finger, there is too much lignification, i.e. the hay was made from over-mature plants.

erly harvested (Table 6.13). If the fibre content of the hay is too high, as happens when the cutting is delayed for too long to maximize the quantity of DM produced per hectare, intake is often low, milk production falls and more concentrates are required, with increased risk of feeding disorders.

There are two ways of increasing the intake of mature hay: one is to chop it finely, the other is to allow the animals to choose the better parts of the hay, which means that large amounts of hay will be left as waste in the feeding trough. The quantity of waste increases as quality and digestibility decrease, as has been shown in studies on dairy goats (Van Soest *et al.*, 1994) (Table 6.14). These studies, as well as others on dairy cows (Mertens, 1983, 1985), have shown that it is much more expensive and difficult to feed the animals on poor-quality hay than to use good-quality forage.

6.7.2 Total mixed rations

In recent years, a growing number of dairy sheep farms have started to use total mixed rations (TMR). Increasingly more farmers are considering adopting this system. In most of these cases TMR have been used for only a few years and are generally employed in combination with a few hours of daily grazing.

The two main reasons why shepherds are considering this system are: (i) the stocking rates can be markedly increased by producing haylage, maize silage or by feeding the animals with freshly cut grass; and (ii) the quality of life for the farmers can be

Table 6.14. Estimates of practical refusals for optimal lactational performance in goats (Van Soest *et al.*, 1994).

Forage	Estimated digestibility[a] (%)	Residues in feeding trough (%)	Digestibility of consumed forage (%)	Utilization[b] (%)
Lucerne hay	65	15	69	59
	58	25	66	50
	50	35	60	39
Grass hay	70	20	75	60
	60	35	69	45
	50	50	60	30

[a] Based on the composition of the forage available; [b]intake of digestible substances as % of those available.

improved significantly by reducing the number of working hours dedicated to the animals, as it is no longer necessary to follow the animals to pasture, and to move the animals from one field to the next, or even from one farm to another. It also improves the quality of the work itself as it is more mechanized and is carried out indoors. The latter reason is of particular importance for young farmers, who wish to have the same quality of life as that of farm workers in other sectors of the industry.

However, the adoption of TMR has sometimes resulted in large decreases in milk production. As a result, mixed systems are used, where the animals are fed TMR in the morning, taken to pasture during the middle of the day, and then often, but not always, given second TMR when they return from grazing.

Many of the problems associated with the introduction of TMR are caused by the ration not being prepared in the correct manner, because the approach followed is the same as that used for dairy cows, while some adaptations should be considered in the light of what has been previously said about the differences between sheep and cattle with regard to fermentative capacity, rumination efficiency and size of ruminated particles. Due to these differences, when preparing TMR for sheep it is important to adhere to the following guidelines:

- Grind the forage much more finely than one would for cattle. This increases intake and has positive effects on milk production. While in cattle, particles ground too finely may dramatically reduce rumination and salivation and cause ruminal acidosis, in sheep it is practically impossible, with the equipment currently available to prepare TMR, to reduce the fibrous particles to such dimensions that the diet becomes dangerous. Sheep intensely ruminate feed particles which would be too small to stimulate rumination in cattle. Another advantage of grinding the diet finely is that it reduces the risk that the sheep will select the diet in favour of concentrates. Indeed, when sheep are fed TMR prepared for cattle they often consume the concentrates first – which increases the risk of acidosis – and then the forage.
- Use good forages, of the same quality as those used for lactating dairy cows with high levels of production. As already stated, feed intake is affected more negatively by high fibre content in sheep than in cattle. Compared to traditional feeding techniques, TMR reduces feed selection; it means that if only low-quality forage is used, then a traditional feeding technique, which allows the sheep to discard the worst part of the forage, is better than TMR.
- Favour supplements high in pectins and digestible fibre, such as soy hulls, beet pulps or citrus pulps; this is advocated in order to avoid the use of large amounts of starch-rich feeds.
- Use high-quality silage. Many sheep farmers produce and use silage that is too fibrous and/or badly preserved; this is probably because dairy sheep farmers, at least in the Mediterranean area, still have little experience of producing silage; intake and milk production are reduced both by excessive fibre content and by substances produced by anomalous fermentations. In addition, in poorly preserved silage there are always large numbers of spores of butyric bacteria; when these bacteria are transferred to the milk, they ferment and damage the cheese, especially when the cheese is mature.

In conclusion, if TMR is to be used on dairy sheep farms not only should a careful cost–benefit analysis be made, but the forage must also be of high quality and careful attention must be paid to the way in which the diet is chopped and mixed. These prerequisites are essential for obtaining satisfactory results in practice.

6.7.3 Feeding concentrates

In grazing sheep, concentrates are usually supplied during the two daily milkings. If large quantities of concentrates rich in starch are used, in a short time ruminal pH will be reduced due to the high level of propionate production. This may reduce cellulolytic bacterial activity, fibre digestibility and induce ruminal sub-acidosis or acute acidosis. The latter sometimes occurs even when the average quantity of concentrates in the diet is not high, because the dominant and most aggressive sheep, which are often the most productive, consume more concentrates.

To reduce the occurrence of these problems, the number of meals in which concentrates are supplied should be increased as much as possible. For example, a third meal could be given to the sheep either at night or in the morning shortly before they are taken to pasture. Excessive competition between animals can be avoided by using individual feeding trough systems. Highly fermentable grains such as barley, wheat or oats should not be rolled, flaked or milled because these treatments may further increase their rates of fermentation (Vipond *et al.*, 1985; Ørskov, 1986), even though in some cases flaked barley has been associated with higher milk production (De Vincenzi *et al.*, 1999). Whole grains stimulate rumination and reduce the rate of intake compared to processed grains. It is unlikely that significant quantities of grain will be lost in the faeces as the grains are finely minced during mastication and, in particular, rumination. If many whole grains are found in the faeces the diet should be checked for fibre effectiveness.

Rolling, flaking and milling are advisable only for grains with a low rate of degradation, such as maize or sorghum. This is particularly true for animals with high production levels and a fast passage rate of feeds through the rumen. Whole maize grains should not be used because older sheep, whose teeth are no longer completely efficient, will have difficulty chewing them. However, this problem can be resolved by cracking the grains. Part of the grain component can be substituted by energy-rich feeds which do not stimulate ruminal acidosis, such as those feeds rich in pectins (beet pulp, soybean hulls, citrus pulps) or in cellulose (cotton seeds). This prevents high starch intake. Alternatively, pelleted feeds rich in effective fibre can be used.

The risks associated with high doses of concentrates can be reduced by the use of buffers, either in the concentrate mixtures or in the pellets. For example, mixtures of sodium bicarbonate (64%) and potassium bicarbonate (34%) in amounts equal to 3.5% of DM intake kept the pH at an acceptable level in lambs fed on a diet very rich in barley grains (Mould *et al.*, 1983).

6.8 Conclusions

The feeding techniques for dairy sheep should always consider the differences in the feeding behaviour and fermentative capacity between sheep and cattle. These must be taken into consideration in order to avoid feeding errors. These differences are of particular importance in the case of sheep with high milk production. Balancing the diet of grazing sheep is more complex when TMR are used, because of the difficulties in estimating pasture intake and quality. In this situation, nutritional indicators are essential in deciding which dietary supplements should be given to the animals.

References

Agricultural and Food Research Council (AFRC) (1995) *Energy and protein requirements of ruminants.* CAB International, Wallingford, UK.

Apolant S.M., Chestnutt D.M.B. (1985) The effect of mechanical treatment of silage on intake and production of sheep. *Anim. Prod.*, 40: 287–296.

Avondo M., Cannas A. (2001) NDF intake in lactating ewes fed at pasture. *Proc. XIV Congress ASPA*, Italy, pp. 502–504.

Baker L.D., Ferguson J.D., Ramberg C.F. (1992) Kinetic analysis of urea transport from plasma to milk in dairy cows. *J. Dairy Sci.* 75 (Suppl.1): 181.

Bertoni G. (1985) Controlli ematici di allevamento per saggiare lo stato nutrizionale-metabolico delle bovine da latte. In: *Valutazione degli alimenti e dello stato metabolico nutrizionale dei ruminanti.* Associazione Italiana Allevatori, Rome, Italy, pp. 121–213.

Bertoni G., Calamari L., Trevisi E. (1999) Sistema diagnostico integrato per la valutazione delle lattifiere. *Inf. Agr.*, 55 (35), Suppl.: 5–66.

Bianchi M., Battaglini L.M., Fortina R. (1994) Effetto dell'integrazione con alimenti sotto forma fioccata oppure frantumata nella razione della pecora: indagini comparative sulle caratteristiche quanti-qualitative del latte. *Proc. XI Congress SIPAOC*, Italy, pp. 447–450.

Bishonga C., Robinson J.J., McEvoy T.G., Aitken R.P., Findlay P.A., Robertson I. (1994) The effects of excess rumen degradable protein in ewes on ovulation rate, fertilization and embryo survival *in vivo* and during *in vitro* culture. *Anim. Prod.*, 58: 447.

Blaxter K.L., Wainman F.W., Davidson J.L. (1966) The voluntary intake of food by sheep and cattle in relation to their energy requirements for maintenance. *Anim. Prod.*, 8: 75–83.

Bocquier F., Theriez M., Brelurut A. (1987) Recommandations alimentaires pour le brebis en lactation. In: Alimentation des ruminants: revision du systèmes et des tables de l'INRA. *Bull. Tech. Centre de Recherches Zootechniques et Vétérinaires de Theix, INRA,* Paris 70: 199–211.

Brown D.L., Hogue D.E. (1985) Effects of roughage level and physical form of diet on Finnsheep lactation. *SID Research Digest,* Fall, pp. 11–14.

Calamari L., Rossi G.C., Cannas A., Cappa V. (1991) Influenza della grassatura e del livello proteico di un alimento unico pellettato (Unipellet) sul profilo metabolico di pecore da latte di razza Sarda. *Proc. IX Congress ASPA,* Italy, pp. 335–349.

Cannas A. (1995) Effects of the particle size of the diet on feeding behaviour and milk production in sheep. MSc thesis, Cornell University, Ithaca, New York.

Cannas A. (2000) Sheep and cattle nutrient requirement systems, ruminal turnover, and adaptation of the Cornell Net Carbohydrate and Protein System to sheep. PhD dissertation. Cornell University, Ithaca, New York.

Cannas A., Van Soest P.J. (2000) Allometric models to predict rumen passage rate in domestic ruminants. In: J.P. McNamara, J. France, D.E. Beever (eds) *Modelling nutrient utilisation in farm animals.* CAB International, Wallingford, UK, pp. 49–62.

Cannas A., Rossi G.C., Brandano P., Pulina G. (1992) Effect of the feeding method of a complete pelleted feed (Unipellet) as supplement of grazing in dairy ewes. *Studi Sassaresi, Ann. Fac. Agr. Univ. Sassari* 34: 145–149.

Cannas A., Pes A., Fresi S., Serra F., Pulina G. (1997) Effects of feeding level and type of dietary protein supplement on milk urea content in ewes. *Volume of abstracts XLVIII Annual Meeting of the E.A.A.P.:* p. 304.

Cannas A., Pes A., Mancuso R., Vodret B., Nudda A. (1998) Effect of dietary energy and protein concentration on the concentration of milk urea nitrogen in dairy ewes. *J. Dairy Sci.* 81: 499–508.

Cannas A., Annicchiarico G., Taibi L., Dell'Aquila S. (2000) Effetto del rapporto foraggi:concentrati della razione su produzione di latte e variazioni di peso corporeo in pecore da latte nella fase finale della lattazione. *Proc. XIV Congress SIPAOC,* Italy, pp. 335–338.

Cannas A., Fox D.G., Tedeschi L.O., Pell A.N., Van Soest P.J. (2004) A mechanistic model to predict nutrient requirements and feed biological values for sheep in each unique production setting. *J. Anim. Sci.,* 82: 149–169.

Cavani C., Bianconi L., Mongardi D. (1990) Soybean hulls and cereal distillers in dairy sheep feeding (in Italian). *Proc. IX Congress SIPAOC*, Italy, pp. 6.9–6.10.

Commonwealth Scientific and Industrial Research Organisation (CSIRO) Standing Committee on Agriculture. Ruminants Subcommittee (1990) *Feeding standards for Australian livestock. Ruminants.* CSIRO Publications, East Melbourne, Australia.

De Boever J.L., Andries J.I., De Brabander D.L., Cottyn B.G., Buysse F.X. (1990) Chewing activity of ruminants as a measure of physical structure – a review of factors affecting it. *Anim. Feed Sci. Technol.*, 27: 281–291.

Dell'Aquila S., Malossini F., Settineri D., Carretta A. (1978) Influenza di diversi livelli alimentari sulla produzione di latte delle pecore di razza Comisana. *Ann. Ist. Sper. Zootec. Roma.*, 11: 263–270.

Demment M.W., Van Soest P.J. (1985) A nutritional explanation for body-size patterns of ruminant and non-ruminant herbivores. *Am. Nat.*, 125: 641–672.

De Vincenzi S., Pauselli M., Morgante M., Casoli C., Costantini F., Duranti E., Ranucci S. (1999) Effect of two non-structural carbohydrate sources on the metabolism and productive performance of lactating Comisana ewes. *Zoot. Nutr. Anim.*, 25: 177–186.

Djouvinov D.S., Todorov N.A. (1994) Influence of dry matter intake and passage rate on microbial protein synthesis in the rumen of sheep and its estimation by cannulation and a non-invasive method. *Anim. Feed Sci. Technol.*, 48: 289–304.

Ferguson J.D., Chalupa W. (1989) Impact of protein nutrition on reproduction in dairy cows. *J. Dairy Sci.*, 72: 746–765.

Fernandez N., Rodriguez M., Peris C., Barcelo M., Molina M.P., Torres A., Adriaens F. (1995) Bovine somatotropin dose titration in lactating dairy ewes. 1. Milk yield and composition. *J. Dairy Sci.*, 78: 1073–1082.

Fox D.G., Tedeschi L.O., Tylutki T.P., Russel J.B., Van Amburgh H.E., Chase L.E., Pell A.N., Overton T.R. (2004) The Cornell Net Carbohydrate and Protein System model for evaluating herd nutrition and nutrient excretion. *Anim. Feed Sci. Technol.*, 112: 29–78.

Gonzalez J.S., Robinson J.J., McHattie I., Fraser C. (1982) The effect in ewes of source and level of dietary protein on milk yield, and the relationship between the intestinal supply of non-ammonia nitrogen and the production of milk protein. *Anim. Prod.*, 34: 31–40.

Gonzalez J.S., Robinson J.J., McHattie I. (1984) The effect of level of feeding on the response of lactating ewes to dietary supplements of fish meal. *Anim. Prod.*, 40: 39–45.

Greenhalgh J.F.D., Reid G.W. (1973) The effects of pelleting various diets on intake and digestibility in sheep and cattle. *Anim. Prod.*, 16: 223–233.

Institut National de la Recherche Agronomique (INRA) (1987) Alimentation des ruminants: revision du systèmes et des tables de l'INRA. *Bull. Tech. Centre de Recherches Zootechniques et Vétérinaires de Theix.* INRA, Paris, 70.

Institut National de la Recherche Agronomique (INRA) (1988) *Alimentation des bovins, ovins et caprins.* INRA, Paris.

Leclerc B. (1985) Utilisation du maquis Corse par des caprins et des ovins. II – Comparaison du régime des ovins ed des caprins. *Acta Oecologica Applicata*, 6 (4): 303–314.

Ligios S., Molle G., Casu S., Nuvoli G. (1995) Validation de la méthode de l'eau lourde pour estimer la composition corporelle des brebis au paturage. *Options Mediterranéennes – Series Seminaries*, 75–84.

Louca A., Mavrogenis A., Lawlor M.J. (1974) Effects of plane of nutrition in late pregnancy on lamb birthweight and milk yield in early lactation of Chios and Awassi sheep. *Anim. Prod.*, 17: 341–349.

Lynch G.P., Elsasser T.H., Jackson C.J., Rumsey T.S., Camp M.J. (1991) Nitrogen metabolism of lactating ewes fed rumen-protected methionine and lysine. *J. Dairy Sci.*, 66: 2268–2276.

Mertens D.R. (1983) Using neutral detergent fibre to formulate dairy rations and estimate the net energy content of forages. *Proceed. Cornell Nutrition Conference*, Ithaca, New York, pp. 60–68.

Mertens D.R. (1985) Factors influencing feed intake in lactating dairy cows: from theory to application using neutral detergent fibre. *Proc. 1985 Georgia Nutr. Conf.*, pp. 1–18.

Mertens D.R. (1997) Creating a system for meeting the fibre requirements of dairy cows. *J. Dairy Sci.*, 80: 1463–1481.

Mertens D.R., Ely L.O. (1982) Relationship of rate and extent of digestion to forage utilization: a dynamic model evaluation. *J. Anim. Sci.*, 54: 895–905.

Molle G., Sanna A. (1992) Tecniche di allevamento degli ovini da latte. In: *Ovinicoltura*, Unione Nazionale Associazioni Produttori ovi-caprini, pp. 141–159.

Molle G., Branca A., Casu S., Landau S. (1998) Alimentazione e riproduzione nella pecora: preparazione alla monta e primi tre mesi di gravidanza. *L'allevatore di ovini e caprini*, 15 (12): 6–10.

Moore L.A. (1964) General principles involved with ruminants and effect of feeding pelleted or wafered forage to dairy cattle. *J. Anim. Sci.*, 23: 230–238.

Mould F.L., Ørskov E.R., Gauld S.A. (1983) Associative effects of mixed feeds. II. The effect of dietary addition of bicarbonate salts on the voluntary intake and digestibility of diets containing various proportions of hay and barley. *Anim. Feed Sci. Technol.*, 10: 31–47.

National Research Council (NRC) (1985) *Nutrient requirements of sheep.* National Academy Press, Washington DC.

National Research Council (NRC) (1988) *Nutrient requirements of dairy cattle.* National Academy Press, Washington DC.

Oltner R., Wiktorsson H. (1983) Urea concentrations in milk and blood as influenced by feeding varying amounts of protein and energy to dairy cows. *Livest. Prod. Sci.* 10: 457–467.

Ørskov E.R. (1986) Starch digestion and utilization in ruminants. *J. Anim. Sci.*, 63: 1624–1633.

Paladines O.L., Reid J.T., Van Niekerk B.D.H., Bensadoun A. (1964) Energy utilization by sheep as influenced by the physical form, composition and level of intake of diet. *J. Nutrition*, 83: 49–64.
Parra R. (1978) Comparison of foregut and hindgut fermentation in herbivores. In: *The ecology of arboreal folivores*. Smithsonian Institute Press, Washington, DC.
Peel C.J., Bauman D.E. (1987) Somatotropin and lactation. *J. Dairy Sci.*, 70: 474–486.
Pilla A.M., Taibi L., Dell'Aquila S. (1993) Consumo di alimenti e composizione della dieta ingerita da pecore da latte alimentate ad libitum per quantità e qualità. *Zoot. Nutr. Anim.*, 19: 221–226.
Pilla A.M., Taibi L., Dell'Aquila S. (1994) Alimentazione delle pecore da latte: produzione e composizione del latte, fabbisogni. *Proc. Giornata Scientifica Miglioramento dell'efficienza produttiva degli ovini e dei caprini*, Bella, Italy.
Pulina G., Serra A., Cannas A., Rossi G. (1989) Determinazione e stima del valore energetico di latte di pecore di razza sarda. *Proc. XLIII Congress S.I.S.VET. (Italy)*, pp. 1867–1870.
Pulina G., Rossi G., Cannas A., Papoff G., Campus R. (1990) Effetto del contenuto proteico della razione sulla produzione e sulla qualità del latte di pecore di razza Sarda. *Agricoltura Ricerca*, 105: 65–70.
Pulina G., Cannas A., Rassu S.P.G., Rossi G. (1995) Effect of fibre and protein content of a complete pelleted feed on lactating dairy ewes. *Agr. Med.*, 125: 115–120.
Robertson J.B., Van Soest P.J. (1975) A note on digestibility in sheep as influenced by level of intake. *Anim. Prod.*, 21: 89–92.
Robinson J.J. (1987a) Energy and protein requirements of the ewe. In: W. Haresign and D.J.A. Cole (eds) *Recent advances in animal nutrition*. Butterworths, London, pp. 187–204.
Robinson J.J. (1987b) Nutrition of housed sheep. In: I.F. Marai and J.B. Owen (eds) *New techniques in sheep production*. Butterworths, London, pp.175–188.
Robinson P.H., Sniffen C.J., Van Soest P.J. (1985) Influence of level of feed intake on digestion and bacterial yield in the forestomachs of dairy cattle. *Can. J. Anim. Sci.*, 65: 437–444.
Ronchi B., Bernabucci U., Bertoni G. (1993) Valutazione comparata del metodo body condition score (BCS) nelle razze ovine Sarda e Lacaune. *Proc. XLVII Congress S.I.S.VET. (Italy)*, pp. 1985–1989.
Roseler D.K., Ferguson J.D., Sniffen C.J., Herrema J. (1993) Dietary protein degradability effects on plasma and milk urea nitrogen and milk non-protein nitrogen in Holstein cows. *J. Dairy Sci.*, 76: 525–534.
Rossi G., Serra A., Pulina G., Cannas A., Brandano P. (1991) L'utilizzazione di un alimento unico pellettato (Unipellet) nell'alimentazione delle pecore da latte. I. Influenza della grassatura e del livello proteico sulla produzione quanti-qualitativa di latte in pecore di razza sarda. *Zoot. Nutr. Anim.*, 17: 23–34.
Russel A.J.F., Doney J.M., Gunn R.G. (1969) Subjective assessment of body fat in live sheep. *J. Agric. Sci., Camb.*, 72: 451–454.
Serra A., Calamari L., Cappa V., Cannas A., Rossi G. (1990a) Trial on use of a complete pelleted feed (unipellet) in lactating ewes: metabolic profile results. *Studi Sassaresi, Ann. Fac. Agr. Univ. Sassari*, 34: 13–21.
Serra A., Vallebella R., Cannas A., Rossi G. (1990b) L'impiego di un alimento unico pellettato (Unipellet) nell'alimentazione delle pecore da latte. *Proc. XLIV Congress S.I.S.VET. (Italy)*, pp. 1507–1512.
Serra F., Cannas A., Pulina G. (1998) Alimentazione degli ovini da latte: il razionamento. *L'allevatore di ovini e caprini*, 15(4): 1–5.
Stern D., Adler J.H., Tagari H., Eyal E. (1978) Responses of dairy ewes before and after parturition to different nutritional regimes during pregnancy. II. Energy intake, body-weight changes during lactation and milk production. *Ann. Zootech.*, 27: 335–346.
Teixeira A., Delfa R., Colomer-Rocher F. (1989) Relationships between fat depots and body condition score or tail fatness in the Rasa Aragonesa breed. *Anim. Prod.*, 49: 275–280.
Ubertalle A., Fortina R., Battaglini L.M., Mimosi A., Profiti M. (1998) Effect of protein degradability on urea nitrogen in sheep milk. *Scienza e Tecnica Lattiero-Casearia*, 49(2): 67–81.
Uden P., Van Soest P.J. (1982) Comparative digestion of timothy (*Phleum pratense*) fibre by ruminants, equines and rabbits. *Br. J. Nutr.*, 47: 267.
Uden P., Rounsaville T.R., Wiggans G.R., Van Soest P.J. (1982) The measurement of the liquid and solid digesta retention in ruminants, equines and rabbits given timothy (*Phleum pratense*) hay. *Br. J. Nutr.*, 48: 329–339.
Van Soest P.J. (1994) *Nutritional ecology of the ruminant*. Cornell University Press, Ithaca, New York.
Van Soest P.J., Fox D.G. (1992) Discounts for net energy and protein, 5th revision. *Proc. Cornell Nutrition Conference*. Ithaca, New York, pp. 40–68.
Van Soest P.J., McCammon-Feldman B., Cannas A. (1994) The feeding and nutrition of small ruminants: application of the Cornell discount system to the feeding of dairy goats and sheep. *Proceed. Cornell Nutrition Conference*, Ithaca, New York, pp. 95–104.

Vipond J.E., Hunter E.A., King M.E. (1985) The utilization of whole and rolled cereals by ewes. *Anim. Prod.*, 40: 297–301.

Welch J.G. (1982) Rumination, particle size and passage from the rumen. *J. Anim. Sci.*, 54: 885–894.

Zygoyiannis D., Stamataris C., Friggens N.C., Doney J.M., Emmans G.C. (1997) Estimation of the mature weight of three breeds of Greek sheep using condition scoring corrected for the effect of age. *Anim. Sci.*, 64: 147–153.

7 Nutrition and Reproduction

Salvatore Pier Giacomo Rassu[1], Giuseppe Enne[1], Sebastiano Ligios[2] and Giovanni Molle[2]
[1]*Dipartimento di Scienze Zootecniche, Università di Sassari, Italy;* [2]*Istituto Zootecnico e Caseario per la Sardegna, Località Bonassai, Olmedo, Sassari, Italy*

7.1 The Ewe

Nutrition is one of the main factors influencing the reproductive performance of ewes, from the onset of puberty to the total number of lambs produced during their lifetime.

Interestingly, nutritional requirements are low during breeding and for the first 3 months of pregnancy, contrary to those of late pregnancy and lactation, especially during the first period when animals inevitably lose weight. In particular, extensively bred sheep show cyclic changes in body reserves (accretion and depletion) that affect the achievement of satisfactory productive and reproductive results. Therefore, body reserves evaluation, by body condition score (BCS), is an important tool for correct feeding of dairy sheep during the different physiological stages (Bertoni, 1988; Molle and Sanna, 1992).

7.1.1 Nutrition and puberty

The first main reproductive event of ewes is the onset of puberty (first ovulation), i.e. the attainment of reproductive capacity; the earlier puberty occurs the shorter is the unproductive period, from birth to first lambing. Bodyweight is the main factor influencing age at puberty; although differences exist between breeds, ewe lambs usually reach puberty when 60% of adult bodyweight is reached (Hafez, 1952; Dyrmundsson, 1973; Chappel, 1993).

Nutrition during their growth period affects the timing of puberty in ewe lambs: those that show slow growth during the first 2 months of life are less fertile and prolific than ewe lambs well fed and with satisfactory growth rates; in contrast, during the pre-pubertal period, when mammary tissue formation is greatest, too-rapid growth can cause ovarian disorders which encourage excessive body fat deposition, with negative effects on reproductive activity (Manunta, 1985; Ronchi and Subioli, 1991) and, subsequently, also on milk production (Bocquier *et al.*, 1988).

Feeding of concentrates, added at the rate of 200–500 g/day, to a basal ration of hay, or hay and pasture, from weaning to mating or up to 12 months of age, often improves reproductive efficiency of ewe lambs, via the advancement of puberty, higher fertility and prolificacy (Gunn, 1977; Secchiari *et al.*, 1987; Kassem *et al.*, 1989; Bernabucci *et al.*, 1991). Conversely, delays in the attainment of puberty are associated with deficiency of energy and protein during growth. In fact, the delayed achievement of bodyweight threshold necessary for initiating reproductive activity, causes ewe lambs to reach puberty later than normal (Boulanouar, 1996); fewer LH pulses, without compromising FSH secretion, are evident in underfed sheep (Foster *et al.*, 1985).

Nutrition affects the growth of ewe lambs and the onset of puberty, not only in relation to the amount of food supplied, but also as regards the possible presence of secondary compounds with anti-nutritive action. Some phytoestrogens like genistein and biochanin, contained in subterranean clover, can exert an anabolic effect (Lookhart, 1980; Pace *et al.*, 1994; Pace and Settineri, 1996). These substances seem to stimulate growth hormone secretion, e.g. somatotrophic hormone (STH), growth hormone (GH) and somatomedins or insulin-like growth factors (IGFs), as well as inducing protein cellular synthesis (Meyer *et al.*, 1995).

Higher growth rates were obtained in Sarda ewe lambs grazing subterranean clover compared to those in ewe lambs grazing lucerne or oats. Nevertheless, in ewe lambs 7 months old and weighing about 30 kg, i.e. far above the bodyweight threshold for Sarda ewes at puberty (26 kg) (Cappai, 1977; Martemucci *et al.*, 1980), reproductive activity started only 1 month later (August), with the first lambing occurring at 13 months of age (Pace *et al.*, 2000).

7.1.2 Nutrition and ovulation rate

The ovulation rate, i.e. the number of egg cells (ova) released per ovulating ewe, represents the potential number of lambs. Every egg cell (ovum) does not necessarily result in a lamb; however, higher ovulation rates mean a higher likelihood of having twin lambs, and consequently greater meat and milk production (Carta *et al.*, 1996; Pulina *et al.*, 1996a,b).

From primordial (oogonial) stage to ovulation of an ovarian follicle takes about 180 days. Many factors and complex interactions can affect this process. In fact, follicular growth is characterized by a 'step-by-step' process, during which the follicle acquires certain properties; these represent a sort of 'passport' to the subsequent developmental step until ovulation. Without acquiring these properties, follicles undergo atresia and die.

Follicle-stimulating hormone (FSH) is the main regulating factor of ovulation rate. It acts on follicular growth when follicle diameter is larger than 2 mm. Follicles probably show varying responsiveness to FSH action, which regulates the ovary by paracrine and autocrine control factors (Scaramuzzi and Campbell, 1990; Scaramuzzi *et al.*, 1993; Campbell *et al.*, 1995).

Nutrition acts on the ovulation rate in different ways; liveweight, body condition and the diet of ewes during the last weeks before mating are all factors that can cause evident variations in ovulation rate. These effects can now be considered (Smith and Stewart, 1990; Landau and Molle, 1996):

- 'Static effects', relating to net bodyweight (liveweight minus the digestive contents) and body condition at the time of mating or artificial insemination; these reflect the nutritional status of the animal with reference to energy balance (Figs 7.1 and 7.2).
- 'Dynamic effects', related to bodyweight and bodily condition changes occuring 2–3 weeks before mating or artificial insemination, representing mid-term variations of nutritional status (Fig. 7.2).
- 'Immediate effects' which mirror, within 4–6 days, the 'extra' nutrient supply without bodyweight and condition variations (Fig. 7.2).

Studies on static effects have shown a positive relationship between liveweight and ovulation rate, but only above a weight threshold that varies between breeds (Fletcher, 1971; Lindsay and Martin, 1994). In Merino sheep, Lindsay *et al.* (1975) and Morley *et al.* (1978) have observed that, above this threshold, the ovulation rate increased by 1.2 and 2.0%, respectively, for each additional kg of liveweight.

Dynamic effects, which in practice are the conceptual basis of the flushing technique, are much less repeatable than static effects. Liveweight and the ewe's body condition at mating can influence the number of

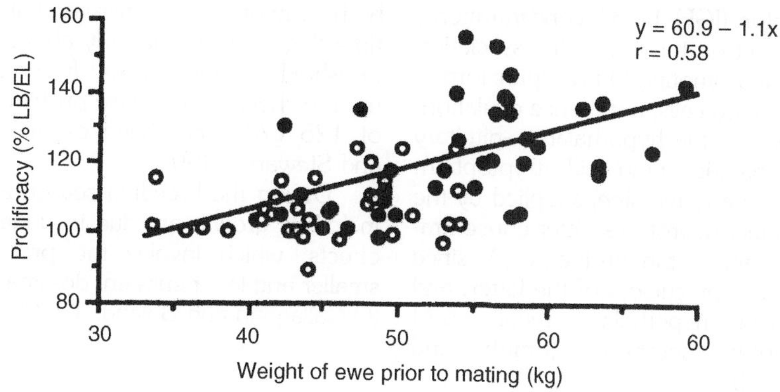

Fig. 7.1. The relationship between weight at mating and the prolificacy of Merino ewes in Western Australia 1971–1987 (Kelly and Croker, cited by Lindsay, 1995).

follicles with a diameter of over 2 mm which are responsive to FSH action, which may allow them to attain maturity; instead, the flushing technique may act on follicles that have already begun this process (Rhind, 1992). Lindsay and Martin (1994) suppose that these effects are regulated by a unique mechanism of controlling ovulation rate, where static and dynamic effects are the same physiological responses, measured at different times with respect to the start of the 'extra' food supply. The nutritional balance – i.e. the sum of endogenous (catabolic processes of the body reserves) and exogenous nutritive sources, minus total requirements – is probably the key factor of this mechanism (Rossi *et al.*, 1996).

Immediate effects (4–6 days), demonstrated by lupin-based supplementation in many studies, can result from direct and indirect effects of this food source, mediated by gonadotrophins.

Increases in energy and protein intake involves an increase in insulin and insulin

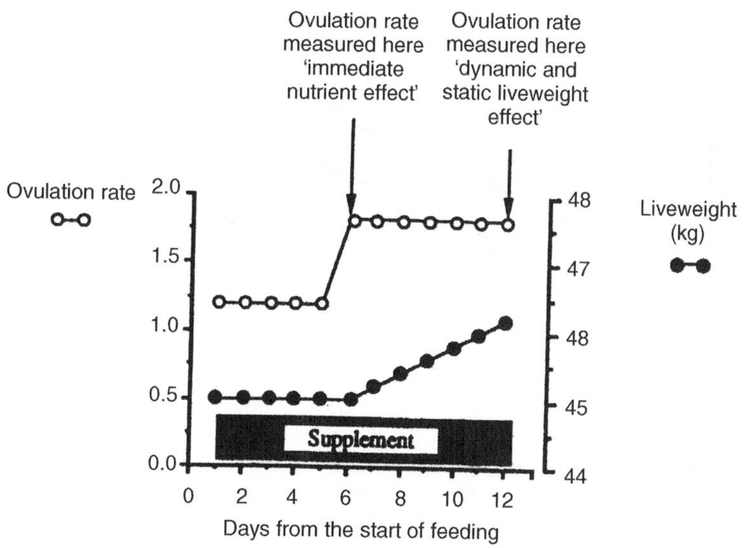

Fig. 7.2. Proposed relationship between the immediate nutrient effect and the dynamic and static liveweight effects (Smith and Stewart, 1990).

growth factor (IGF) blood concentrations, which may act on the follicle in association with FSH by promoting LH-receptor formation in granulosa cells, and hence ovulation. Furthermore, at the hypothalamic–pituitary level, the aromatic amino acids (tryptophan, phenylalanine and tyrosine) supplied by the food can affect neurotransmitter concentrations (dopamine, epinephrine, etc.), since the former are precursors of the latter, and hence influence hypothalamic regulation of gonadotrophin secretion (Smith and Stewart, 1990).

Although there is still a shortage of knowledge about the roles played by both the energy and protein allowances in reproductive efficiency in sheep (Rossi *et al.*, 1996), many researchers have concentrated their attention on proteins. Good results have been obtained with lupins, and recently with soybean meal (Molle *et al.*, 1995, 1997; Branca *et al.*, 2000; D'Allessandro *et al.*, 2000), both characterized by medium to low ruminal degradability. Other studies showed that the ovulation rate is improved by all nutrients that are able to increase energy availability for the ruminant, either as structural carbohydrates, and consequently via acetate, or as protein and non-structural carbohydrates via gluconeogenesis.

Although protein and energy are able to influence the rate of ovulation independently, the best results are obtained when both components are increased at the same time (Fig. 7.3), particularly above a certain threshold protein intake. In non-lactating mutton sheep, a digestible protein threshold of 125 g/day has been estimated (Smith and Stewart, 1990).

During the breeding season alterations in follicle growth are due to undernutrition effects, which involve the production of smaller and less persistent dominant follicles (O'Callaghan and Boland, 1999).

7.1.3 Nutrition and pregnancy

Initial (first month) stage of pregnancy

Pregnancy is the fundamental phase of the reproductive process, due both to its length (5 months) and to the unavoidable risk of fetal losses, which can affect the reproductive efficiency of ewes. Most of these losses occur during the first stage of pregnancy, particularly in the first month, during which embryonic implantation in the uterus takes place (Wilkins and Croker, 1990; Branca *et al.*, 1999).

We can divide this period into two parts: (i) pre-implantation stage (10 days), when egg fertilization is followed by blastocyst development, growth and embryo transfer from the oviduct to the uterus. Total

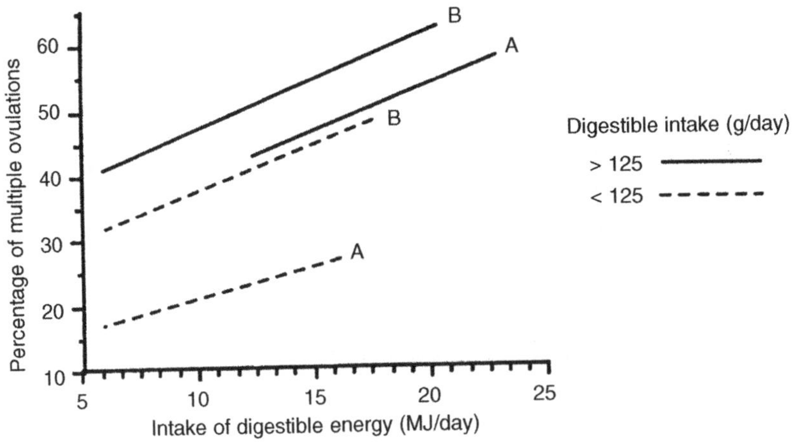

Fig. 7.3. Effect of intake of digestible energy (MJ/day) and digestible protein (g/day) on the percentage of ewes having multiple ovulations in two trials (A and B) (Smith and Stewart, 1990).

blastocystic and embryonic losses in this phase still allow ewes to exhibit signs of heat at the end of the normal oestrus cycle; and (ii) implantation stage (20 days), when the embryo establishes a relationship, by development of the placenta, with the maternal uterus. Embryonic losses in this phase do not allow ewes to exhibit signs of heat at the end of the normal oestrus cycle, but this can occur 19 days after the previous oestrus.

Among the factors that can affect reproductive efficiency, ovulation rate is considered to be a major factor in the variable rate of embryonic survival. There is a negative relationship between the number of corpora lutea (used as an index of potential number of embryos) and the probability of embryonic survival, which is reduced by about 8% for each additional embryo (Table 7.1) (Hanrahan, 1994).

Table 7.1. Probability of embryonic survival as a function of ovulation rate in naturally mated ewes (Hanrahan, 1994).

Corpora lutea (n)	Probability of embryonic survival	Ewes (n)
2	0.82	5069
3	0.74	884
4	0.65	270
5	0.55	91
6	0.45	38

With reference to nutritional effects, studies have concentrated, until a few years ago, on late pregnancy, when the growth of the fetus is more rapid. Based on a summary of published and unpublished information on ovulation rate and the number of fetuses/offspring, most sources only consider pregnancy requirements relative to the final 45–50 days of pregnancy.

However, even if there is much still to be learned, nutritional effects on embryonic survival and placental development at an early stage are very important for obtaining good reproductive results, because most losses occur during the first month of pregnancy. Both energy deficiency (nutritive level < 50% of maintenance) and excess (nutritive level = 200% of maintenance) during the first 3 weeks of pregnancy can induce embryonic mortality (even if < 15%), with higher mortality rates seen in multiple than in single ovulations (Edey, 1976).

In recent years it has been shown that overnutrition more than undernutrition, above all if protracted, has a negative effect on embryonic survival. This is due both to the direct effect on embryonic development caused a by high glucose concentration, as observed during *in vitro* maturation experiments (O'Callaghan and Boland, 1999), and to the indirect effect of the reduction in blood progesterone concentration. The weight and activity of the liver increase due to increases in food intake, causing higher progesterone catabolism. Progesterone concentration reduction is not counterbalanced by an increased secretion of progesterone from the corpora lutea, which would maintain homeostasis and consequently the completion of pregnancy (Parr, 1992; O'Callaghan and Boland, 1999). In sheep the period between 11 and 12 days post-fertilization (implantation phase) is crucial for embryonic survival because, in this period, embryos are very sensitive to reductions in plasma progesterone concentration (Parr, 1992).

Recent studies demonstrated negative effects on embryonic survival resulting from increases in energy and protein supplies, even if they were positive for follicular development, by supporting higher ovulation rates (Fig. 7.4 and Table 7.2).

Regarding the effect of nitrogenous compounds in the diet of ruminants, above all in cattle, great importance has been attached to the negative effects on fertility caused by high blood urea concentrations. High blood and milk urea concentrations, synthesized in the liver from excess ammonia produced in the rumen, usually imply protein excess in the ration or imbalance in the rumen between degradable protein and fermentable energy (Cannas *et al.*, 1998). In reality, most of the negative effects on embryonic survival can be attributed to ammonia itself (as the ammonium ion form) and not to urea (Visek, 1984; Sinclair *et al.*, 2000b). Reproductive efficiency can be

Table 7.2. Effects of low and high crude protein intake on reproductive parameters (Smith *et al.*, 1990; Molle *et al.*, 1997; O'Callaghan and Boland, 1999).

Parameters	Low crude protein intake	High protein intake
Ovulation rate	⇓	⇑
Blood urea	⇓	⇑
Fertility	⇑	⇓
Oocyte/embryo quality	⇑	⇓
Survival of good-quality embryos	⇒	⇒

⇑ increase; ⇓ reduction; ⇒ no variation.

affected by ammonia, which alters the uterine environment, in particular by changing the pH value. Studies carried out on embryo transfer techniques in sheep have demonstrated that ammonia affects follicular development and final development of ova, resulting in very poor-quality embryos (Fahey *et al.*, 1998). This phenomenon results in increases in embryonic mortality, fetal gigantism and higher mortality at birth for both mother and lambs (McEvoy *et al.*, 1997).

In cows in high milk production, where high urea concentration is associated with energy deficiency, negative effects have been observed, particularly during the first 2 months of lactation. In contrast, in milked sheep, a different mechanism underlies this phenomenon: in traditional breeding systems, mating takes place in the final stage of lactation or in the dry period, when nutritional requirements are low and ewes are in positive or nil energy balance.

Recent studies on sheep subjected to artificial insemination have shown a negative relationship between fertility and urea concentration in individual milk samples (> 5.6 mg/ml) (Branca *et al.*, 2000), in ewes grazing grass–legume mixtures under irrigation, or in bulk samples of milk (> 4.5 mg/ml) (Molle *et al.*, 2001) collected from flocks grazing in rainfed and irrigated pastures.

Intermediate (second to third months) stage of pregnancy

The development of the placenta (30th–100th day of pregnancy) is important

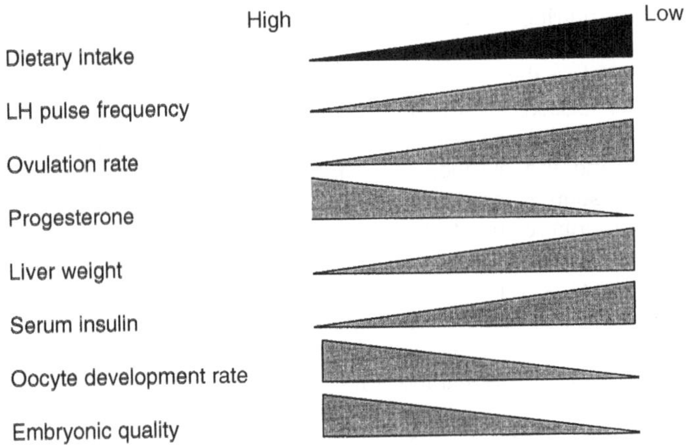

Fig. 7.4. Summary of effects of changing from low to high dietary intake on reproductive parameters (O'Callaghan and Boland, 1999).

because two-thirds of the variation in lamb birthweight are due to variations in placental weight (Robinson *et al.*, 1999). This sensitive period encompasses the second and third months of pregnancy, when the placenta grows rapidly until achieving a weight equal to 10% of the lamb's birthweight. The effect of nutrition on placental development is related to bodyweight, condition score and the age of the ewe. In adult ewes of good bodily condition at mating, slight undernutrition in mid-pregnancy results in greater placental development; conversely, in ewe lambs and in adult ewes of very poor bodily condition at mating, lower placental growth has been observed.

Some studies have shown that an adequate nutritional plan in late pregnancy can effectively compensate for weight losses in mid-pregnancy (up to 20% of liveweight in meat sheep), without negative effects on the birthweight of lambs (Brink, 1990; Kleeman *et al.*, 1993). However, strong correlations have been found between the ewe's liveweight at 100 days of pregnancy and the lamb's birthweight (Fig. 7.5): 48–80% of the variations in lamb birthweight can be explained by variations in the liveweight of the ewe, between the 30th and 100th day of pregnancy (Kelly and Newnham, 1990).

The fact that underfeeding in mid-pregnancy did not cause any birthweight reduction in lambs can be explained by fetal losses (early abortion), allowing improved growth of surviving fetuses (Rattray, 1992), or by the ewe's capacity to mobilize body reserves, provided that they are adequately fed in the last part of pregnancy (Frutos *et al.*, 1998).

Lower fetal survival during the last 2 months of pregnancy has been observed in sheep over-fed in both mid- and late pregnancy; this can cause loss of appetite and result in the 'fatty liver' syndrome and pregnancy toxaemia. Amino acids, rather than energy supply, are particularly important in mid-pregnancy for mitigating the negative effects of undernutrition. In fact, in this period the daily nitrogen requirements of the fetus are 18 mg/kg dry fetal weight, compared with 4–6 mg in the last third of pregnancy (Robinson *et al.*, 1999).

Final (third to fifth months) stage of pregnancy

In the final stage of pregnancy (110–140 days), a positive relationship between nitrogen retention in the pregnant uterus (and in maternal tissues) and daily protein intake (with intakes of crude protein of 80, 140 and 125 g/day) has been demonstrated. Protein depletion in ewes fed at medium- and low-protein levels can partially compen-

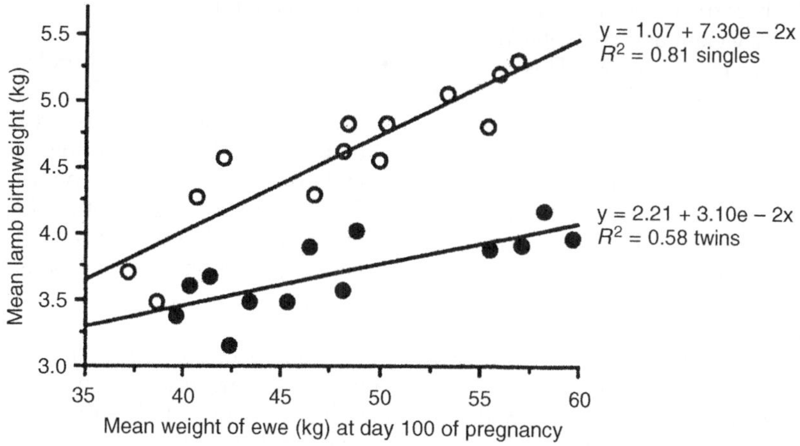

Fig. 7.5. The association between mean liveweight of ewes at day 100 of pregnancy and mean birthweight of their lambs (open symbols, singles; solid symbols, twins) (Kelly and Newnham, 1990).

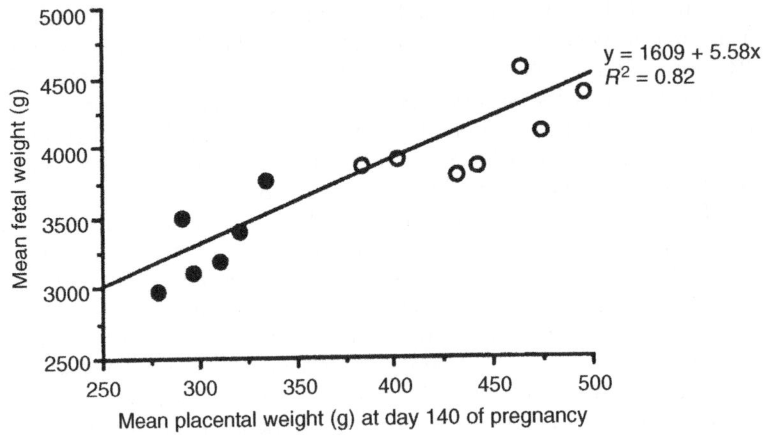

Fig. 7.6. The relationship between mean placental fetal weights at day 140 of pregnancy (open symbols, singles; solid symbols, twins) (Kelly and Newnham, 1990).

sate for nitrogen accumulation in the viscera and udder. Nitrogen is retained in the carcass of ewes fed on 215 g crude protein/day. Skeletal muscles represent the main reserve site of proteins which are used by the ewe to nourish the fetus under deficiency conditions (McNeill *et al.*, 1997).

On this subject, an important role seems to be played by the placental lactogen hormone (produced by the chorion, and having growth hormone-like activity), which, by diverting the flow of maternal glucose, can regulate fetal growth. Its concentration increases with fetal number and placental weight: high correlations have been found between fetal weight and placental lactogen concentration in the fetus. This variable (alone or associated with placental weight) accounts for about 50% and 81% of fetal weight variability, respectively (Schoknecht *et al.*, 1991). Correlations have also been found between placental and fetal weight at 140 days of pregnancy (Fig. 7.6): fetal weight increases by about 560 g for each additional 100 g of placental weight (Kelly and Newnham, 1990).

To adopt adequate feeding plans, it is essential to have a knowledge of the stage of pregnancy (initial, middle or final) and fetal number. This can usually be detected by ultrasound diagnosis at around 30–40 days of pregnancy.

7.1.4 Feeding techniques

For reproduction purposes, feeding techniques will differ between mated and artificially inseminated ewes.

Mated ewes

Body condition score is an important criterion for consideration in animals in order to obtain good reproductive results, particularly in grazing animals. The BCS, easily evaluated on-farm, allows the establishment of some basic rules:

- A reduction in fertility and prolificacy can occur with BCS values < 2.5, as has been demonstrated in the Sarda breed (Molle *et al.*, 1995, 1998), following a period of undernutrition of over 1 month.
- The target level of BCS at mating should be around 3.0–3.5, and should not fall below 2.5 during the first month of pregnancy.

For ewes in poor condition, it is necessary to apply the flushing technique, thereby increasing energy supply (+30–50% of maintenance) for 2–3 weeks prior to mating. This can be achieved by increasing the allowance of green herbage, if available, or by supplying concentrates (or both combined) in relation to their availability on the farm.

For instance, in sheep grazing poor grasses, improvements in ovulation rate and prolificacy can be obtained with protein-based supplementation (with low or medium ruminal degradability). Good results have been obtained with both lupin (400–500 g/day) and soybean meal (200–300 g/day) (Molle *et al.*, 1995), restricted to the synchronization period (short-term flushing) (Molle *et al.*, 1997).

On sheep grazing cereal stubbles during the mating period, a supply of concentrates might be neglected initially because ewes are able to effectively utilize residual grains that have escaped from harvesting. In order to regulate their consumption and to allow the ruminal microorganisms to adapt to the new diet, grazing, initially for a few hours per day, is recommended.

To optimize this feed source, which is widespread in the Mediterranean environment, it is useful to supplement the basal diet with protein and yeasts. If ewes are grazing on green pastures protein supply is usually not a limiting factor, so it should be calculated on the basis of grass quantity and quality as well as on the nutritional requirements of the ewes, which might still be lactating.

Milk urea concentration is a relatively robust index of protein intake and should range between 3.0 and 5.0 mg/ml. In order to avoid asynchronies of ruminal kinetics, which could cause sudden elevations in ammonia concentration without necessarily affecting plasma and milk urea, the food supply should be based on carbohydrates and protein with low to medium degradability (Sinclair *et al.*, 2000a). It follows that, in order to avoid these risks in the mating period, grazing of leguminous plants should represent only a small part of the diet. In any case it is essential to avoid sudden food reductions, often adopted in summer to speed up the drying-off of ewes still producing milk.

Artificially inseminated ewes

The feeding of ewes subjected to artificial insemination is still an overlooked area; there are three main differences between mating and artificial insemination with regard to nutrition:

- The correlations between nutrition and reproductive results are less positive for artificially inseminated ewes, due to the technical (insemination *per se*) and environmental factors, which can markedly affect insemination results (Sanna *et al.*, 1997).
- The ovulation rate (1.8–2.0 in Sarda ewes) and fertilization and embryo losses in the days following insemination are higher than losses observed in natural mating (Branca *et al.*, 1999, 2000).
- The effect of gonadotrophin (PMSG) administration, usually used in conjunction with oestrus synchronization, is additive or synergic (Fig. 7.7) to the nutritive level and body condition (BCS) of the sheep (Scaramuzzi and Murray, 1994).

In fat or overnourished sheep this synergy can cause high ovulation rates, which are not sustainable for some breeds, such as Sarda, and can involve increases in embryonic and fetal losses.

In sheep of BCS $\geq$ 3.0 the flushing technique should be discouraged because it can reduce fertility and result in high frequencies of triple and quadruple lambs, with increased mortality risk for both lambs and ewes. On the other hand, in sheep of BCS $<$ 2.5 (thin ewes) and with high milk production, infertility increases. This negative effect is not attenuated by the enhancement of the nutritional level of the diet in the last 2 weeks before insemination.

The objective of the feeding regimen during the weeks preceding insemination should be to obtain a satisfactory body condition (BCS 2.75–3.0), and the flushing technique should be adopted only in ewes with BCS values of 2.5–2.75. If ewes are too fat (BCS $>$ 3.25) or too thin (BCS $<$ 2.5), adequate feeding must start at least 6 weeks prior to artificial insemination, with two different objectives. In the first case, the aim is to reduce BCS 15 days before insemination, to avoid compromising fertility. In the second case, it is necessary to achieve

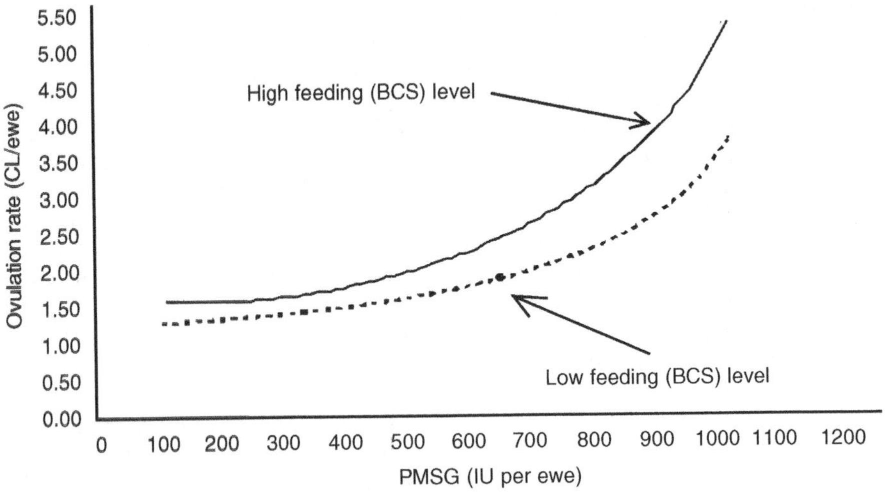

Fig. 7.7. Relationship between PMSG dose and ovulation rate in ewes subjected to different feeding levels or characterized by different body condition scores (BCS) at ovulation; CL: corpora lutea.

BCS values of 2.75–3.00 in order to obtain an ovulation rate not lower than two ova per ewe (Branca *et al.*, 2000). With regard to nitrogen levels, the best criterion is milk urea, which should be in the range 2.5–4.5 mg/ml, in order to avoid problems of low prolificacy due to nitrogen undernutrition, or of embryonic mortality, due to nitrogen excess (Molle *et al.*, 2001). The requirements for energy and protein in sheep subjected to artificial insemination are summarized in Table 7.3.

For ewes subjected to artificial insemination rather than to natural mating, during the first 3 months of pregnancy feeding should be aimed at avoiding variations, either positive or negative (> 0.25 BCS), in bodily condition. Sudden increases in BCS after artificial insemination can lead to high embryonic and fetal losses due to a reduction in blood progesterone concentration associated with energy excess (Parr *et al.*, 1987), irrespective of BCS at insemination (Molle *et al.*, 2001).

7.2 The Ram

The ram is involved in the reproductive process both directly through mating and indirectly by the production of sperm used for artificial insemination. In both cases, knowledge of the quantity and quality of semen and of factors affecting semen pro-

Table 7.3. Guidelines for managing body condition score and milk urea in dairy sheep subjected to artificial insemination.

Body condition score[a]	Milk urea[b] (mg/ml)	Expected reproductive results[c]	
		Fertility	Prolificacy
< 2.50	< 2.50	Low	Low
2.50–3.0	2.50–4.50	Optimum	Optimum
> 3.0	> 4.50	Low	Very high

[a] Individual body condition scores; measurements relate to Sarda dairy ewes. For breeds with higher subcutaneous fat deposition than Sarda (e.g. Lacaune) body score levels should be increased by 0.25 units; [b]bulk milk concentration; [c] refers to the standard performance of each breed at lambing with artificial insemination.

duction is of great importance. Among these factors nutrition plays a primary role, but there is still a shortage of information on the correlation of nutrition and reproduction in the ram, which is an overlooked area of ovine nutrition.

In a flock the ratio of rams to ewes is usually about 1:30–1:40; therefore, bearing in mind the average size of sheep dairy farms, it is very rarely possible to find farms with more than 20 rams. Rams, outside the breeding season, are often kept with barren ewes, replacement lambs etc., so during this period no specific diets are available for males.

Research has often employed wethers for nutritional trials and has neglected studies on rams. The limited information available considers maintenance requirements for rams to be 10–15% more than those for ewes (Bocquier *et al.*, 1988).

The increasing use of artificial insemination in the dairy sheep sector led to the establishment of semen collection centres, where rams are pen-fed. Feeding and management techniques adopted for this class of animal in these centres are based on the empirical experience that, despite the lack of specific experimental results, produces good results in practice. In fact, satisfactory growth in young animals, excellent semen production, increased survival and functional longevity are obtained in these centres, where animals with enhanced breeding value are raised.

7.2.1 Spermatozoa production

Spermatogenesis

Spermatogenesis is the process occurring in the testis, i.e. the formation and liberation of spermatozoa (haploid) from undifferentiated (diploid) germ cells (spermatogonia). Spermatozoal maturation occurs during the passage through the epididymis and deferent ducts; spermatozoa acquire the ability to fertilize (capacitation) only within the female reproductive tract. One of the remarkable features of spermatogenesis is that the duration of the process is constant, and any feeding or hormonal treatment cannot modify it.

Spermatozoa production occurs continuously in the seminiferous tubules, from Sertoli cells, included in the testicular parenchyma. Spermatozoal production begins at puberty, which is reached at 100–150 days of age. Live spermatozoa are present in the ejaculate after this age (Terril, 1975), under the control of: (i) the hypothalamus– hypophysis unit, by GnRH, FSH and LH hormone release; and (ii) the Leydig cells scattered in the interstitial tissue, which release testosterone (Setchell, 1993).

In rams, the time required to produce a single fertile spermatozoon is 49 days, with little variation. The process can be divided into the following stages: (i) 31 days, necessary for development of the spermatozoon from the first division of the germ cells until it reaches the centre of the seminiferous tubule; (ii) 3.5 days, its passage through the tubules, out of the testis and into the epididymis; and (iii) the last 14 days, its journey through the epididymis (Lindsay, 1988).

The first two stages are unaffected by any known treatment, but the third can be reduced by up to 1 or 2 days, increasing the utilization rate of the ram. However, this practice is discouraged because it leads to the ejaculation of immature semen.

A thorough knowledge of the duration of spermatogenesis in rams is very important for practical purposes. In order to improve the quantity and quality of semen it is necessary to intervene at least 7–8 weeks before the expected date of ram usage, whatever treatment is adopted.

Quantity of semen produced

Spermatogenesis is a continuous process characterized by the relatively constant daily semen production per g of testicle (20 million/day), independent of testicular size (Lindsay, 1988; Cameron and Tilbrook, 1990; Martinez *et al.*, 1994). Since semen production is strictly correlated with testicular mass, and in order to evaluate the ram's reproductive capacity, knowledge of this parameter is very important. To overcome the technical difficulty of weighing the testis in the live animal, two indirect systems of

Table 7.4. Equations for estimation of testicular weight (g) in rams (testicular weight = a + bx).

Breed	a	b	x	R^2
Sarda (Rassu *et al.*, 1995)	47.02	0.94	TV (ml)	0.98
	–1331	56.49	SC (cm)	0.90
Varie (Notter *et al.*, 1981)		0.0211	$SC^{2.89}$ (cm)	0.93
		6.12	$TD^{2.32}$ (cm)	0.90
Romney (Knight, 1977)	117 x 0.41	0.41	SV (ml)	0.88
	–373.11	21.66	SC (cm)	0.85
	–218.17	88.64	TD (cm)	0.85
Merino (Lino, 1971)	–323.72	22.592	SC (cm)	0.55

SC, Scrotal circumference measured at the widest point with a tape measure; TV, total testicular volume (devoid of the scrotum) measured by the displacement of water from a plastic cylinder; SV, scrotal volume measured by the displacement of water from a plastic cylinder; TD, testicular diameter measured with a caliper for each testis.

evaluation, both easy and sufficiently accurate, have been adopted:

- Estimation of the volume of testicular tissue using a series of beads of graded sizes and of known volume, and comparing them with the size of each testis.
- Measurement of testicular parameters for volume, diameter and scrotal circumference, which are correlated with testicular weight and, therefore, with semen production (Table 7.4).

Testicular development of rams during their growth phase is very important, in order to evaluate their reproductive potential and to detect the onset of puberty. In fact, puberty appears when testicular growth is at its peak: this occurs at 7–9 months of age in the Awassi breed (Salhab *et al.*, 2001). Since testicular growth parallels body growth, in ram lambs it is possible to estimate the relationship between main testicular parameters and the age and liveweight (Table 7.5).

7.2.2 Nutrition and reproduction in the ram

Nutrition is one of the main factors that can influence the ram's capacity for production of spermatozoa. The testis is very sensitive to nutritional influences, both negative and positive, and as a result its size can alter rapidly. Testicular size, and therefore daily spermatozoa production, changes in keeping with bodyweight variations. An increase of 32% in bodyweight, in rams fed diets of high nutritional value (twice maintenance), produces a 67% increase in testicular volume (Oldham *et al.*, 1978).

Testicular activity in grazing rams, i.e. testicular volume and hormonal secretion, begins in spring–early summer, when food availability and liveweight reach their highest values (Lindsay and Martin, 1994). In contrast, sexual activity and semen production in the ram reach their maxima in autumn (Cappai *et al.*, 1981; Manunta *et al.*, 1981).

Apart from testicular size, feeding can also modify the efficiency of semen production. It has been shown that variations in semen production are greater than variations in testicular mass. For instance, a 25% increase in testicular size corresponds to an 81% increase in semen production (Martin and Walkden-Brown, 1995). Several studies have been carried out to discover which nutrients affect the reproductive capacity of the ram. Food restriction for 3 months resulted in rams with body fat < 12% of bodyweight, as compared to normally fed

Table 7.5. Regression equations for estimation of testicular development by age (months) and bodyweight (kg) in Awassi rams (Salhab *et al.*, 2001).

Parameter	Equation	R^2
Circumference	y = 2.27 + 1.1 age + 0.77 bodyweight	0.90
Length	y = –0.41 + 0.26 age + 0.33 bodyweight	0.87
Width	y = 0.71 + 0.053 age + 0.16 bodyweight	0.45
Volume	y = –167 + 21.8 age + 7.57 bodyweight	0.85

rams with a body fat content of 25–49% of bodyweight. This feed restriction caused:

- A decrease in testicular weight.
- A reduction in the diameter of the seminiferous tubules.
- A reduced number of spermatozoa in the epididymis.
- Lower contractility in ejaculated spermatozoa (Brown, 1994).

Daily semen production is probably affected by digestible energy intake alone or associated with protein intake (Brown, 1994). In particular, the plasma concentration of volatile fatty acids, derived from ruminal degradation, is a key factor in the reproductive capacity of rams (Martin and Walkden-Brown, 1995). These observations have a scientific basis in the relationships (Table 7.6) between some nutritional parameters (energy and protein intake) and changes in liveweight and testicular size (Murray *et al.*, 1990).

As shown in Table 7.6, the parameter of crude protein intake, used as an independent variable, does not improve the accuracy of any of the estimates. Consequently, reproductive parameters of rams (scrotal circumference and testicular volume variations) can easily be evaluated from digestible energy intake only. Most researchers agree on the uselessness of high protein intakes for improving semen production. In fact, the favourable effects from increased lupin ingestion, characterized by high energy (16.4 MJ/kg DM) and protein (CP 337.5 g/kg DM) concentrations, could be due to gluconeogenic amino acids, abundant in lupins (Boukhliq and Martin, 1996).

Energy probably has no direct effects on spermatogenesis or on testosterone production, but it indirectly influences these parameters acting on gonadotrophin secretion (Brown, 1994). The body's response to energy supply occurs at three different levels (Boukhliq and Martin, 1996):

- Increase in LH pulse frequency, evident after 2–3 days from the beginning of

Table 7.6. Correlations between nutrition and reproductive capacity in the ram (Murray *et al.*, 1990).

Parameter	Independent variable	R^2
CBW (g/day)	$-286.7 + 26.7\ \text{DEI} - 0.2\ \text{CPI}$	0.85
CBW (g/day)	$-263.5 + 23.2\ \text{DEI}$	0.85
CSC (mm/day)	$-0.92 + 9.73 \times 10^{-2}\ \text{DEI} - 2.02 \times 10^{-3}\ \text{CPI}$	0.82
CSC (mm/day)	$-0.69 + 6.22 \times 10^{-2}\ \text{DEI}$	0.61
CSC (mm/day)	$4.70 - 4.75 \times e^{-0.0042\text{CBW}}$	0.86
CTV (ml/day)	$-294 + 27.8\ \text{DEI} - 0.26\ \text{CPI}$	0.93
CTV (ml/day)	$-272 + 23.6\ \text{DEI}$	0.93
CTV (ml/day)	$10.15 - 10.26 \times e^{-0.00102\text{CBW}}$	0.52

CBW, Change in bodyweight; CSC, change in scrotal circumference; CTV, change in testicular volume; DEI, digestible energy intake (MJ/day); CPI, crude protein intake (g/day).

Table 7.7. Comparison between data of Istituto Zootecnico e Casario Centre (IZCC) and INRA.

	IZCC Intake (kg/head/day)		INRA Intake	IZCC		INRA		Difference IZCC vs. INRA	
	Hay	Conc.	(kg/head/day)	UFL	g PDI	UFL	g PDI	UFL	g PDI
Group 1	1.31	0.62	1.10	1.29	136	0.90	62	+0.39	+74
Group 2	1.19	0.68	1.20	1.27	134	0.96	66	+0.31	+68
Group 3	1.20	0.48	1.10	1.18	105	0.86	59	+0.24	+46
Group 4	1.23	0.54	1.30	1.26	111	0.96	66	+0.34	+45
Group 5	1.47	0.59	1.47	1.40	174	0.84	58	+0.56	+116
Group 6	1.54	0.68	1.80	1.52	187	0.96	66	+0.56	+121

Bodyweight variation was not observed.

'flushing', and disappearing after 3–4 weeks.

- Increase in FSH concentration, evident after 10 days, but still detectable for several weeks.
- Increase in testicular size, evident after 2 weeks.

Feeding effects on testicular growth depend in part on hypothalamus–hypophysis hormones (GnRH, LH and FSH), and in part on mechanisms as yet unexplained. These probably act in conjunction with other hormones such as insulin, somatotropin, somatomedin and prolactin. These hormones, which have receptors in the testis, could act by promoting cell division and steroid release (Boukhliq and Martin, 1996; Blache *et al.*, 2000).

The limited knowledge available on ram nutrition does not allow careful determination of energy and protein requirements of rams during the reproductive period. Indeed the feeding of the ram is often based on experience rather than on scientifically tested feeding techniques. Available data refer to stall-fed rams of different breeds, which have different maintenance requirements compared with those of grazing rams during the mating period. The different systems of requirements evaluation (AFRC, CSIRO, INRA, NRC) show only data relative to energy and protein requirements for maintenance, e.g. increasing rations for the reproductive activity of rams by 20-40% depending on the reproductive technique (natural mating, controlled mating or penned rams) being followed.

7.2.3 Feeding rams in artificial insemination centres

In this final section, some data derived from centres for semen production for artificial insemination are detailed.

In the 'Le Bourguet' centre in Aveyron, France, the daily diet of Lacaune rams (bodyweight 100–110 kg), at maintenance and out of the breeding season, consists of 1.8 kg of meadow and lucerne hay and 0.5 kg of commercial concentrates. During the period of semen collection the diet is amended to 1.6 kg of lucerne hay and 0.6 kg of commercial concentrate (Brios, personal communication).

Other data are provided by 'Istituto Zootecnico e Caseario per la Sardegna', a centre for ram semen collection for the Sarda breed. At this site, experiments carried out in autumn have shown that rams do not increase in bodyweight while consuming diets with energy and protein levels higher than those recommended by INRA (Table 7.7). This occurs irrespective of feeding regimes (Table 7.8), which differed for included feedstuffs (Table 7.9).

The differences between field experience and INRA recommendations, relative to energy and protein supply, could be due to higher food consumption and to lower alimentary efficiency of rams fed collectively

Table 7.8. Diet for rams housed and fed collectively.

	n	BW	Type of hay	Concentrate (kg/head/day)
Group 1	6	72.4	Italian ryegrass	Mixed feed (0.62)
Group 2	5	78.8	Italian ryegrass	Mixed feed (0.68)
Group 3	6	67.5	Italian ryegrass	Lupin (0.48)
Group 4	5	78.5	Italian ryegrass	Lupin (0.54)
Group 5	5	65.8	Lucerne	Mixed feed (0.59)
Group 6	6	78.5	Lucerne	Mixed feed (0.68)

BW, bodyweight; hay, *ad libitum*; TMR, twice daily.

compared to rams fed individually. The latter are often the subject of studies from which evaluations of nutritive requirements originate (D'Hour *et al.*, 1991; Ingvartsen and Andersen, 1993). This phenomenon has been observed both in beef cattle, which when kept collectively consumed about 4.7% more food than animals kept individually (Broadbent *et al.*, 1970), and in dairy cattle, which consumed 7% more food when kept in an open stall system, without showing weight and milk production differences (Coppock *et al.*, 1972).

Since rams used for artificial insemination are almost always housed indoors, in order to maximize photoperiods in an artificial environment, the use of total mixed ration (TMR) can represent a valuable feeding system. For instance, a typical TMR is composed of 3.0 kg maize silage, 0.15 kg lucerne hay, 0.25 kg ryegrass hay, 0.4 kg commercial concentrate and 0.2 kg peas. The residual represents 5–10% of the offered food, even if a large part consists of maize cobs and lignified stems.

The TMR can also be used for ram lambs, whose diet must have energy and protein concentrations of 0.85 UFL/kg DM and 100g PDI/kg DM, respectively (Table 7.10). Because they are growing animals it is necessary to integrate their diet with minerals and vitamins (20 g/day) and to progressively increase the quantity of food offered relative to the increase in intake capacity of the animals, but without exceeding 750–800 g/day of concentrate.

The future sires, usually born in November from ewes belonging to the genetic improvement plan and which were artificially inseminated in May–June, are transferred to the centre when they are 2.5 months of age and have reached 20 kg bodyweight. Before being subjected to semen collection ram lambs must reach 42–43 kg bodyweight, which is achieved after approximately 120 days at the centre, with 170 g/day of daily growth.

With regard to reproductive features, evident differences in semen quantity and quality have been observed between young and adult rams. In young rams the incidence of discarded ejaculates, poor spermatic concentration and contractility after 6–8 h from semen collection is high. These parameters are often below the sufficiency threshold and accordingly prevent the use of semen for artificial insemination. Therefore, semen collection starts at 6–8 months of age with

Table 7.9. Chemical composition and energy values of feeds.

	DM(%)	OM(%)	CP(%)	UFL	PDIN (g)	PDIE (g)	UEM
Ryegrass hay	85.0	89.9	9.7	0.67	60	74	1.60
Lucerne	85.0	90.1	18.9	0.71	121	98	1.05
Mixed feed	87.5	90.0	16.0	1.00	97	99	–
Lupin	91.9	89.2	32.7	1.12	188	67	–

Table 7.10. Composition of TMR fed to ram lambs.

Food	As fed (g)	Dry matter (g)	UFL	PDIN (g)	PDIE (g)
Lucerne hay	110	95	0.06	11	9
Ryegrass hay	110	95	0.06	9	8
Maize silage	750	230	0.20	16	17
Soybean meal	60	50	0.05	18	12
Mixed feed	600	530	0.48	51	52
Total	1870	1000	0.85	105	99

a frequency of service equal to three times per week. Since semen vitality decreases rapidly from collection to the time of insemination (Bogliolo *et al.*, 2000), semen quality at collection should be high. Therefore, utilization of adult rams, as mentioned previously, is limited to two consecutive services on alternate days. Table 7.11 shows the values of some of the parameters used for the evaluation of semen quantity and quality.

Table 7.11. Semen characteristics of different types of Sarda ram (Sanna *et al.*, 1995).

	< 18 months	18–20 months	> 54 months
Volume/collection (ml)	0.83[a]	1.63[b]	1.84[b]
Motility (scale 0–5)	3.27	3.21	3.21
No. of spermatozoa	3.77	3.92	3.78
Paillettes/collection (n.)	6.6	14.7	16.1
Paillettes/breeding season (n.)	30.0	207.0	300.0

[a] One ejaculate; [b] two consecutive ejaculates.

References

Bernabucci U., Ronchi B., Bertoni G. (1991) Il livello nutritivo dallo svezzamento al primo parto in agnelle di razza Sarda.1. Effetti sull'accrescimento e sull'attività riproduttiva. *Proc. 9th Congress ASPA (Italy)*, pp. 380–381.

Bertoni G. (1988) Come va alimentata la pecora da latte. *L'allevatore di ovini e caprini*, 5(9): 1–6.

Blache D., Chagas L.M., Blackberry M.A., Vercoe P.E., Martin G.B. (2000) Review: metabolic factors affecting the reproductive axis in male sheep. *J. Rep. Fert.*, 120(1): 1–11.

Bocquier F., Theriez M., Prache S., Brelurut A. (1988) Alimentation des ovins. In: *Alimentation des bovins, ovins et caprins*, INRA, Paris.

Bogliolo S., Branca A., Stocchino M.C., Sanna S.R., Del Francia F. (2000) Effetto del trattamento omeopatico costituzionale sulla produzione di materiale seminale di arieti di razza Sarda. *Proc.: Allevamento ovino biologico e veterinaria omeopatica*, Asciano, Siena, Italy.

Boukhliq R., Martin G.B. (1996) Nutrition and reproduction in the ram in a Mediterranean environment. *Proc. Sem. FAO-CIHEAM, Recent Advances in Small Ruminant Nutrition*, Rabat, Morocco, pp. 227–232.

Boulanouar B. (1996) Dietary protein or energy restriction both delay age at puberty in ewe lambs. *Proc. Sem FAO-CIHEAM, Recent Advances in Small Ruminant Nutrition*, Rabat, Morocco, pp. 217–222.

Branca A., Gallus M., Pares R., Molle G., Landau S. (1999) Estimation de la perte embryonnaire precoce chez la brebis Sarde après insemination. *Proc. annual meeting Rencontres Recherche Ruminant*, p. 222.

Branca A., Molle G., Sitzia M., Decandia M., Landau S. (2000) Short-term dietary effects on reproductive wastage after induced ovulation and artificial insemination in primiparous lactating Sarda ewes. *Anim. Reprod. Sci.*, 58: 59–71.

Brink D.R. (1990) Effects of bodyweight gain in early pregnancy on feed intake, gain, body condition in late pregnancy and lamb weights. *Small Rum. Res.*, 3: 421–424.

Broadbent P.J., McIntosh J.A.R., Spencer A. (1970) The evaluation of a device for feeding group-housed animals individually. *Anim. Prod.*, 12: 245–252.

Brown B.W. (1994) A review of nutritional influences on reproduction in boars, bulls and rams. *Reprod. Nutr. Dev.*, 34: 89–114.

Cameron A.W.N., Tilbrook A.J. (1990) The rate of production of spermatozoa by rams and its consequences for flock fertility. In: Oldham, Martin and Purvis (eds) *Reproductive physiology of Merino sheep*. School of Agriculture (Animal Science), University of Western Australia, pp. 131–141.

Campbell B.K., Scaramuzzi R.J., Webb R. (1995) Control of antral follicle development and selection in sheep and cattle. *J. Reprod. Fert. Suppl.*, 49: 335–350.

Cannas A., Pes A., Mancuso R., Vodret B., Nudda A. (1998) Effect of dietary energy and protein concentration of milk urea nitrogen in dairy ewes. *J. Dairy Sci.*, 81: 499–508.

Cappai P. (1977) Sincronizzazione dei calori negli ovini di raza Sarda. *Proc. Round Table La Pastorizia Oggi, Verona (Italy)*. Associazione Nazionale della Pastorizia, Rome.

Cappai P., Manunta G., Branca A. (1981) Variazioni stagionali delle caratteristiche dell'eiaculato in arieti di razza Sarda, Romanof e meticci Frisona X Sarda. *Proc. 4th Congress SIPAOC (Italy)*, 2: pp.115–119.

Carta A., Sanna S.R., Casu S. (1996) Relazione tra numero di agnelli allattati e produzione di latte alla mungitura in pecore Sarde. *Proc. 12th Congress SIPAOC*, Varese, Italy, pp. 211–214.

Chappel G.L.M. (1993) Nutritional management of replacement sheep utilizing southern forages: a review. *J. Anim. Sci.*, 71: 3151–3154.

Coppock C.E., Noller C.H., Crowl B.W., McLellon C.D., Rhykerd C.L. (1972) Effect of group *versus* individual feeding of complete rations on feed intake of lactating cows. *J. Dairy Sci.*, 55: 325–327.

D'Alessandro A.G., Neri M.G., Nardi A., Corino C., Martemucci G. (2000) Impiego del flushing alimentare nel controllo ormonale degli estri con trattamento di corta durata (5 giorni) in pecore da latte. *Proc. 14th Congress SIPAOC*, Vietri, Italy, pp. 389–392.

D'Hour P., Petit M., Garel J.P. (1991) Components of grazing behaviour of 3 breeds of heifers. *Ann. Zootech.*, 44 (Suppl. 1): 270.

Dyrmundsson O.R. (1973) Puberty and early reproductive performance in sheep. *Anim. Breed. Abstr.*, 41: 273–289.

Edey T.N. (1976) Nutrition and embryo survival in the ewe. *Proc. N.Z. Soc. Anim. Prod.*, 36, pp. 231–239.

Fahey J., Boland M.P., O'Callaghan D. (1998) Effect of dietary urea and embryo development in superovulated donor ewes and on embryo survival following transfer in recipient ewes. *Proc. BSAS (Abs)*, 182.

Fletcher I.C. (1971) Effects of nutrition, liveweight and season on the incidence of twin ovulation in South Australian strong-wool Merino ewes. *Austr. J. Agric. Res.*, 22: 321–330.

Foster D.L., Yellon S.M., Olster D.H. (1985) Internal and external determinants of the timing of puberty in the female. *J. Reprod. Fert.*, 75: 327–344.

Frutos P., Buratovich O., Giraldez F.J., Mantecon A.R., Wright I.A. (1998) Effects on maternal and fetal traits of feeding supplement to grazing pregnant ewes. *Anim. Sci.*, 66: 667–673.

Gunn R.G. (1977) The effects of two nutritional environments from 6 weeks' pre-partum to 12 months of age on lifetime performance and reproductive potential of Scottish Blackface ewes in two adult environments. *Anim. Prod.*, 25: 155–164.

Hafez S.E. (1952) Studies on the breeding season and reproduction of ewes. *J. Agric. Sci. Cam.*, 42: 255–265.

Hanrahan J.P. (1994) Embryo survival in small ruminants: incidence and sources of variation. *Proc. 45th Annual Meeting EAAP*, Edinburgh, UK, p. 255.

Ingvartsen K.L., Andersen H.R. (1993) Space allowance and type of housing for growing cattle. A review of performance and possible relation to neuroendocrine function. *Acta Agric. Scand. Sect. A, Sci.*, 43: 65–80.

Kassem R., Owen J.B., Fadel I. (1989) The effect of pre-mating nutrition and exposure to the presence of rams on onset of puberty in Awassi ewe lambs under semi-arid conditions. *Anim. Prod.*, 48: 321–335.

Kelly R.W., Newnham J.P. (1990) Nutrition of the pregnant ewe. In: Oldham, Martin and Purvis (eds) *Reproductive physiology of Merino sheep*. School of Agriculture (Animal Science), University of Western Australia, pp. 161–168.

Kleeman D.O., Walker S.K., Waleley J.R.W., Ponzoni R.W., Smith D.H., Grimson R.J., Seamark R.F. (1993) Effect of nutrition during pregnancy on fetal growth and survival in FEC Booroola x South Australian Merino ewes. *Theriogenology*, 39: 623–630.

Knight T.W. (1977) Methods for indirect estimation of testes weight and sperm numbers in Merino and Romney rams. *New Zealand J. Agric. Res.*, 20: 291–296.

Landau S., Molle G. (1996) Nutrition effects on fertility in small ruminants with an emphasis on Mediterranean sheep breeding systems. In: Lindberg, Gonda and Ledin (eds) *Proc. Sem FAO-CIHEAM, Recent Advances in Small Ruminant Nutrition.* Rabat, Morocco, pp. 203–216.

Lindsay D. (1988) Breeding the flock. In: Inkata (ed.) *Modern research and reproduction in sheep.*

Lindsay D.R. (1995) The role of management in the control of the estrous cycle. *Proc. 30th Symp. SIPZOO*, Milan, Italy, pp. 251–263.

Lindsay D.R., Martin G.B. (1994) Feeding and reproduction in sheep. *Proc. 11th Congress SIPAOC*, Perugia, Italy, pp. 469–478.

Lindsay D.R., Knight T.W., Smith J.F., Oldham C.M. (1975) Studies in ovine fertility in agricultural regions in Western Australia: ovulation rate, fertiliy and lambing performance. *Austr. J. Agric. Res.*, 189–198.

Lino B.F. (1971) The output of spermatozoa in rams. II Relationship to scrotal circumference, testis weight and the number of spermatozoa in different parts of the urogenital tract. *Aust. J. Biol. Sci.*, 25: 359–366.

Lookhart G. (1980) Analysis of cumestrol, a plant estrogen, in animal feed by high-performance liquid chromatography. *J. Agr. Food Chem.*, 28: 666–677.

Manunta G. (1985) L'attività ovarica nella pecora. *Proc. 39th Congress S.I.S.Vet.*, pp. 27–47.

Manunta G., Floris B., Cappai P. (1981) Attività circannuale delle gonadi dell'ariete di razza Sarda. *Arch. Vet. Ital.*, 32: 103.

Martemucci G., Bellitti E., Toteda F., Manchisi A. (1980) Studio delle prime manifestazioni estrali estrali delle agnelle in relazione al tipo genetico. *Ann. Fac. Agr. Bari*, 31: 185–200.

Martin G.B., Walkden-Brown S.W. (1995) Nutritional influences on reproduction in mature male sheep and goats. *J. Reprod. and Fert. Suppl.*, 49: 437–449.

Martinez J., Limas T., Peron N. (1994) Daily production and testicular and epididymal sperm reserves of Pelibey rams. *Theriogenology*, 41: 1595–1599.

McEvoy T.G., Robinson J.J., Aitken R.P., Findlay P.A., Robertson I.S. (1997) Dietary excess of urea influences the viability and metabolism of preimplantation sheep embryos and may affect fetal growth among survivors. *Reprod. Sci.*, 47: 71–90.

McNeill D.M., Slepetis R., Ehrhardt R.A., Smith D.M., Bell A.W. (1997) Protein requirements of sheep in late pregnancy: partitioning of nitrogen between gravid uterus and maternal tissues. *J. Anim. Sci.*, 75: 809–816.

Meyer H.H.D., Stoffel B., Hagen-Mann K. (1995) b-agonist, anabolic steroids and their receptor: new aspects in growth regulation. Ruminant physiology: digestion, metabolism, growth and reproduction. In: Engelhardt, Leonard-Marek, Brevs. and Giesecke (eds) *Proc. 8th Int. Symp. Rum. Phys.*, Ferdinand Verlag, Stuttgart, Germany.

Molle G., Sanna A. (1992) Tecniche di allevamento degli ovini da latte. In: *Ovinicoltura*, UNAPOC, Rome, pp. 141–159.

Molle G., Branca A., Ligios S., Sitzia M., Casu S., Landau S., Zoref Z. (1995) The effect of grazing background and flushing supplementation on reproductive performance in Sarda ewes. *Small Rum. Res.*, 17: 245–254.

Molle G., Landau S., Branca A., Sitzia M., Fois N., Ligios S., Casu S. (1997) Flushing with soybean meal can improve reproductive performance in lactating Sarda ewes on mature pasture. *Small Rum. Res.*, 24: 157–165.

Molle G., Branca A., Casu S., Landau S. (1998) Alimentazione e riproduzione nella pecora: preparazione alla monta e primi tre mesi di gravidanza. *L'allevatore di ovini e caprini*, 12: 6–10.

Molle G., Sanna S.R., Ligios S., Branca A., Oppia P., Caria A., Corda A.R., Demuru G., Fressura G., Ruiu G. (2001) Influenza dell'alimentazione sui risultati riproduttivi della pecora Sarda. *L'Informatore Agrario*, 7: 75–81.

Morley F.W.H., White D.H., Kennedy P.A., Davis I.F. (1978) Prediction of ovulation rate from liveweight in ewes. *Agric. Syst.* 3: 27–45.

Murray P.J., Rowe J.B., Pethick D.W., Adams N.R. (1990) The effect of nutrition on testicular growth in the Merino ram. *Austr. J. Agr. Res.*, 41: 185–195.

Notter D.R., Lucas J.R., McClaugherty F.S. (1981) Accuracy of estimation of testis weight from *in situ* testis measurements in ram lambs. *Theriogenology*, 15: 227–231.

O'Callaghan D., Boland M.P. (1999) Nutritional effects on ovulation, embryo development and the establishment of pregnancy in ruminants. *Anim. Sci.*, 68: 299–314.

Oldham C.M., Adams N.R., Gherardi P.B., Lindsay D.R., Mackintosh J.B. (1978) The influence of level of feed intake on sperm producing capacity of testicular tissue in the ram. *Austr. J. Agr. Res.*, 29: 173–179.

Pace V., Settineri D. (1996) Subterranean clover extract and phytoestrogens as anabolyzing agents for growing mice. *Proc. 4th Int. Feed Prod. Conference,* Piacenza, Italy, pp. 451–452.

Pace V., Settineri D., Marzoli C. (1994) Effects of natural and synthetic plant oestrogens on rumen fluid degradation of some feedstuffs. *Annales de Zootecnie,* 43: 31S.

Pace V., Settineri D., Rassu S.P.G. (2000) Trifoglio sotterraneo sulla crescita e la riproduzione di agnelle di razza Sarda. *Nota I. 35th Proc. Int. Symp. SIPZOO: Produzioni Animali di qualità ed impatto ambientale nel sistema mediterraneo,* Ragusa, Italy, pp. 257–265.

Parr R.A. (1992) Nutrition–progesterone interaction during early pregnancy in sheep. *Rep. Fert. Dev.*, 4: 297–300.

Parr R.A., Davis I.F., Fairclough R.J., Miles M.A. (1987) Overfeeding during early pregnancy reduces peripheral progesterone concentration and pregnancy rate in sheep. *J. Repr. Fert.*, 80: 317–320.

Pulina G., Bencini R., Rassu S.P.G. (1996a) Relation between birthweight of lambs and milk production in ewes. *Agr. Med.*, 126: 316–319.

Pulina G., Cappio-Borlino A., Rassu S.P.G., Salis A.G., Brandano P. (1996b) Effetto della gemellarità e dell'ambiente di allevamento sulla produzione di latte in pecore di razza Sarda. *Proc. 12th Congress SIPAOC (Italy)* pp. 215–217.

Rassu S.P.G., Vallebella R., Doro P., Pulina G. (1995) Relationships between bodyweight and reproduction in Sarda breed sheep. In: G. Enne, G.F. Greppi, A. Lauria (eds) *Reproduction and Animal Breeding Advances and Strategy. Proc. 30th Int. Symp. SIPZOO,* pp. 437–438.

Rattray P.V. (1992) Nutrition of the ewe during gestation and lactation. In: A.W. Speedy (ed.) *Progress in sheep and goat research.* CAB International, Wallingford, UK, pp. 85–106.

Rhind S.M. (1992) Nutrition: its effects on reproductive performance and its hormonal control in female sheep and goats. In: A.W. Speedy (ed.) *Progress in sheep and goat research.* CAB International, Wallingford, UK, pp. 25–51.

Robinson J.J., Sinclair K.D., McEvoy T.G. (1999) Nutritional effects on fetal growth. *Anim. Sci.*, 68: 315–331.

Ronchi B., Subioli G. (1991) Alimentazione corretta, riproduzione efficiente. *Informatore Zootecnico,* 17: 61–65.

Rossi G., Cannas A., Macciotta N.P.P., Pulina G., Rassu S.P.G. (1996) Base genetica della prolificità e della stagionalità dei cicli riproduttivi nella specie ovina. In: G. Enne and A. Lauria (eds) *La riproduzione in zootecnia, RAIZ (Italy),* pp. 101–127.

Salhab S.A., Zarkawi M., Wardeh M.F., Al-Masri M.R., Kassem R. (2001) Development of testicular dimensions and size, and their relationship to age, bodyweight and parental size in growing Awassi ram lambs. *Small Rum. Res.*, 40: 187–191.

Sanna S.R., Carta A., Casu S. (1995) L'utilizzazione dell'inseminazione strumentale nella razza ovina Sarda. *Proc. Seminar Miglioramento genetico degli Ovini e dei Caprini: aspetti scientifici e problemi applicativi,* SIPAOC, Perugia, Italy.

Sanna S.R., Carta A., Cappai P., Branca A., Festante G., Bitti P.M. (1997) Environmental effects on fertility of artificial inseminated Sarda dairy ewes. *Proc. 48th Meeting of EAAP,* Vienna, Abstr: S5.5.

Scaramuzzi R.J., Campbell B.K. (1990) Physiological regulation of ovulation rate in the ewe: a new look at an old problem. In: Oldham, Martin and Purvis (eds) *Reproductive physiology of Merino sheep.* School of Agriculture (Animal Science), University of Western Australia, pp. 71–84.

Scaramuzzi R.J., Murray J.F. (1994) *The nutrient requirements for the optimum production of gametes in assisted reproduction in ruminant animals.* Reunion AETE, Lyon, France, pp. 55–103.

Scaramuzzi R.J., Adams N.R., Baird D.T., Campbell B.K., Dowing J.A., Findlay J.K., Henderson K.M., Martin G.B., McNatty K.P., McNeilly A.S., Tsonis C.G. (1993) A model for follicle selection and the determination of ovulation rate in the ewe. *Reprod. Fert. Dev.*, 5: 459–478

Schoknecht P.A., Nobrega S.N., Petterson J.A., Ehrhardt R.A., Slepetis R., Bell A.W. (1991) Relation between maternal and fetal plasma concentration of placental lactogen and placental and fetal weight in well-fed ewes. *J. Anim. Sci.*, 69: 1059–1063.

Secchiari P., Trimarchi G., Ferruzzi G., Martini A., Pistoia A., Berni P., Luisi M. (1987) Effetti dell'apporto nutritivo della razione sull'inizio della carriera riproduttiva di agnelle Massesi. *Zoot. Nutr. Anim.*, 13: 223.

Setchell B.P. (1993) Male reproduction. In: G.J. King (ed.) *Reproduction in domesticated animals.* Elsevier Science.

Sinclair K.D., Sinclair L.A., Robinson J.J. (2000a) Nitrogen metabolism and fertility in cattle: I. Adaptative changes in intake and metabolism to diets differing in their rate of energy and nitrogen release in the rumen. *J. Anim. Sci.*, 78: 2659–2669

Sinclair K.D., Kuran M., Gebbie F., Webb R., McEvoy T.G. (2000b) Nitrogen metabolism and fertlity in cattle: II. Development of oocytes recovered from heifers offered diets differing in their rate of nitrogen release in the rumen. *J. Anim. Sci.*, 78: 2670–2680.

Smith J.F., Stewart R.D. (1990) Effects of nutrition on the ovulation rate of ewes. In: Oldham, Martin and Purvis (eds) *Reproductive physiology of Merino sheep.* School of Agriculture (Animal Science), University of Western Australia, pp. 85–101.

Terril C.E. (1975) Sheep. In: Hafez (ed.) *Reproduction in farm animals.* Lea and Febiger, Philadelphia, Pennsylvania.

Visek W.J. (1984) Ammonia: its effects on biological systems, metabolic hormones and reproduction. *J. Dairy Sci.*, 67: 481–498.

Wilkins J.F., Croker K.P. (1990) Embryonic wastage in ewes. In: Oldham, Martin and Purvis (eds) *Reproductive physiology of Merino sheep.* School of Agriculture (Animal Science), University of Western Australia, pp. 169–177.

8 Nutrition and Milk Quality

Anna Nudda[1], Gianni Battacone[1], Roberta Bencini[2] and Giuseppe Pulina[1]

[1]*Dipartimento di Scienze Zootecniche, Università di Sassari, Italy;* [2]*School of Animal Biology, Faculty of Natural and Agricultural Sciences, The University of Western Australia, Australia*

In dairy sheep, as in other ruminants, feeding is a major factor affecting the quality of milk. Nutrition directly influences the synthesis and rates of secretion of milk fat and total protein ($N \times 6.38$) and it may also affect the concentrations of minerals and vitamins. Aromatic compounds or toxic and poisonous substances fed to sheep may be transferred to the milk. Nutrition may influence the somatic cell count (SCC) and microbial concentration of milk. Therefore, nutrition influences all aspects of the processing performances of the milk including milk clotting properties, cheese yield, ripening time, preservability and flavour of cheese.

8.1 Fat, Total Protein and Total Utilizable Substances (TUS)

As most ovine milk is used to make cheese, the yield of cheese from each litre of milk depends mainly on its fat and protein concentrations. The 'scotta', which is the whey obtained after ricotta cheese is made by heating the cheese whey, contains non-fermented lactose in concentrations greater than those found in milk (5.0–5.2%). Because of the evaporation that occurs during the processing of cheese, it also contains roughly 1% of total protein (50% protein, 50% non-protein nitrogen) and traces of fat (Table 8.1).

It is known that protein contributes more than fat to the cheese yield (Casu and Marcialis, 1966; Pirisi *et al.*, 1994; Barillet *et al.*, 1998). Because the 'scotta' contains roughly 16–18% of the original milk protein, we propose to consider the two components as having the same weighting and to adopt a unique parameter for evaluation of the quality of milk in terms of cheese yield. This parameter, called total utilizable

Table 8.1. Average composition (%) of whey and 'scotta' from ovine milk (Comendador *et al.*, 1996).

	Water	Fat	Protein ($N \times 6.38$)	Lactose	Ash
Cheese whey[a]	91.77	1.25	1.63	4.75	0.62
Scotta[b]	93.14	0.19	1.10	5.14	0.48

[a] Cheese whey is a by-product of cheese-making obtained after the milk is clotted and the curd is separated from the whey. The protein it contains is mainly represented by whey proteins. These can be further separated by precipitating them at high temperature to produce ricotta; [b]the whey left after the ricotta is made is called scotta.

substances (TUS) of milk, is the sum of the fat and protein content of milk, expressed in g/l. For example, 1 litre of milk, containing 6.5% of fat and 5.8% of protein, corresponds to 123 g TUS.

Combining fat and protein levels to evaluate the productive potential of ovine milk is not a new idea. Casu (1989) and Caroli *et al.* (1993) referred to an analogous parameter used in the breeding scheme of Lacaune sheep, but in their case the fat and protein were weighted differently. However, since cheese yield is highly variable depending on the kind of cheese, the manufacturing process and the possible utilization of cheese whey for manufacturing ricotta, etc., using the same weighting for fat and protein is not too distant from the actual estimate of milk quality for cheese-making, and it has the advantage of being very simple to calculate. The TUS parameter has been used by Barillet *et al.* (1998), who considered the 'Taux de Matière Seche Utile (TMSU)' of milk a useful and accurate predictor of Roquefort cheese yield from the milk of Lacaune sheep.

The effects of nutrition on the concentrations of fat and protein fall under a current paradigm which is based mainly on research conducted on dairy cows. This paradigm, frequently transposed and applied to dairy sheep (Polidori *et al.*, 1991; Bencini and Pulina, 1997) for explaining the relationship between feeding and yield of TSU, is based on four assumptions:

- Milk yield and its fat and protein concentrations are inversely correlated.
- Milk fat is positively correlated with the concentration of neutral detergent fibre (NDF) in the ration.
- Non-fibre carbohydrates (NFC) are positively correlated with protein concentration and negatively correlated with fat concentration in the milk.
- Feeding fat increases fat and depresses protein concentrations in milk.

We shall now examine how these assumptions apply in the case of the dairy sheep, and in Section 8.1.2 we will see that not all of these assumptions hold true.

8.1.1 Fat and protein yield and quality

Milk fat content

Nutrition affects the concentration of fat in the milk and its fatty acids content. Fat concentration (%) is normally correlated positively with the concentration of NDF in the diet. Pulina and Rassu (1991), using data from experiments based on traditional feeding strategies (hay + supplement of concentrate to cover nutritional requirements) in dairy sheep, calculated the relationship between milk fat (%) and NDF (% on DM basis) in the diet:

$$\text{fat} = 4.59 + 0.05\ \text{NDF}\ (R = 0.48)$$

However, a higher percentage of NDF in the ration will cause a decrease in DM intake and digestibility, with a drop in milk production and consequent increase in milk fat concentration. Intake of a low-NDF ration will cause inadequate ruminal fermentation, resulting in a drop in milk yield and fat concentration (Poulton and Ashton, 1972; Oddy, 1978; Chiofalo *et al.*, 1993), especially if it induces ruminal acidosis (Rossi *et al.*, 1988). To avoid this problem, recommended levels of NDF should not fall below 30–32% of the DM in the ration: above this level our experimental data did not show a further increase in milk fat concentration, even if an increase in chewing activity time was observed (Pulina and Rassu, 1991).

Important aspects of dietary fibre are its quality, composition of structural carbohydrates, density and size of fibrous particles. Regarding its quality, leguminous fibre, such as NDF derived from *Medicago* spp. (lucerne), is more digestible and results in greater voluntary feed intake (Van Soest, 1994). The size of the fibrous particles influences chewing time and ruminal retention time. In Dorset and Finn ewes fed a diet with a forage to concentrate ratio of 56:44, an average concentration of NDF of 41.6% and a reduction in particle size of hay, through a grinder with mesh sizes from 12.7 to 1 mm, resulted in a reduction in ruminal activity, an increase in feed intake and an increase in the ruminal concentration of propionate. As a consequence, there was an increase in milk production and protein yield without a depression in milk fat concentration (Cannas, 1995). This suggests that sheep are less sensitive than cows to reductions in the length of fibre particles, even though the forage intake index (FII = average particle size × NDF), previously proposed for cows by Woodford *et al.* (1986) for adjusting forage intake according to particle length, showed a positive correlation with the concentration of milk fat in Sarda dairy ewes (Pulina and Rassu, 1991).

Milk fat may be influenced by other factors such as the lipid content in the diet, the level of fat supplementation, the quality of fat ingested, the use of buffers and the feeding frequency. Generally, feeding sheep diets with added fat results in increases in milk fat of up to 4–5% of fat on DM. Higher concentrations of fat in the diet lead to a depression in milk fat through a fall in microbial activity in the rumen, a reduction of short-chain fatty acid synthesis in the udder and a decrease in the rate of metabolites (mainly fatty acids from lipoproteins) that the mammary gland takes up from the blood.

Rumen-protected fats are naturally present in some feedstuffs (soybean, cotton, sunflower) or can be obtained artificially. The most frequently used are the calcium soaps of fatty acids (CSFA). Their use in dairy sheep increases the fat concentration of the milk (Fig. 8.1) and changes the milk fatty acid profile, which reflects the composition of the CSFA added to the diet (Sklan, 1992), confirming previous findings in dairy cows. The use of CSFA is normally accompanied by a reduction in milk protein concentration (Sklan, 1992; Bencini and Pulina, 1997), partly due to 'dilution effects', since milk yield tends to increase, partly due to the reduced capacity of the mammary gland to utilize amino acids. However, protected fat added to the diet has no effect on the relative proportion of nitrogen compounds and on the clotting properties of the milk. The effects of CSFA on milk composition depend on the dose of calcium soaps and on the stage of lactation. Bocquier and Caja (2001) reported that the optimum amount of CSFA for maximization of milk fat yield is 70 g/day.

Casals *et al.* (1999) observed that greater fat concentrations in milk can be obtained using CSFA at the beginning of lactation because of the lower activity of

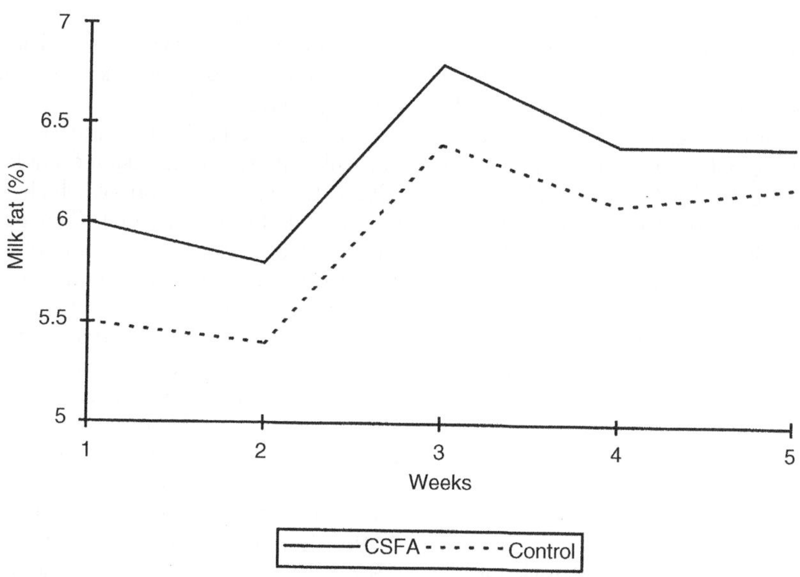

Fig. 8.1. Effect of calcium soaps of fatty acids (CSFA, 4.5% of DM) on the concentration of fat in the milk of Sarda ewes (Rossi *et al.*, 1991).

lipogenic enzymes, which favours the uptake of dietary fatty acids by the mammary gland, whereas a depression in the concentration of protein in milk occurs, particularly in the final stage of lactation.

When sheep are fed rations containing large amounts of concentrates or readily fermentable carbohydrates (e.g. soluble starch or sugars), the addition of buffers to the diet, e.g. sodium bicarbonate, calcium carbonate or magnesium oxide, may prevent a decrease in milk fat concentration because they: (i) maintain ruminal pH within the optimum range for cellulose fermentation; and (ii) increase the efficiency of lipid transit from the gastrointestinal tract to the mammary gland. Another theory regarding the action of ruminal buffer salts has been proposed by Russel and Chow (1993), who reported that the action of bicarbonate could be explained by increased water intake and increased ruminal fluid dilution. This would, in turn, result in a greater outflow of undegraded starch from the rumen with a consequent decrease in the production of ruminal propionate.

There are contradictory reports regarding the effect of protein concentration in the diet on milk fat concentration. Some authors (Cowan *et al.*, 1979; Rossi *et al.*, 1991) reported an increase in milk fat when protein concentration was increased, but others (Cowan *et al.*, 1979; Pulina *et al.*, 1990a) did not find any effect. The use of urea in the diet caused a depression in milk fat, which resulted in a greater concentration of unsaturated fatty acids (Farid *et al.*, 1979).

The diet of high-producing ewes often contains high levels of concentrates. In Sarda dairy ewes we observed that animals that were offered the supplement three times during the day, rather than only once a day, produced milk with a greater fat content as a consequence of reduced fluctuations in ruminal pH.

Milk fatty acid composition

The fatty acid composition of milk fat can also be modified by feeding strategies that influence the pattern of the precursors that the mammary gland removes from the blood for fat synthesis. The main factors that affect fatty acid composition are the energy concentration of the diet, the (acetate+butyrate):propionate ratio in the rumen and the level and quality of fat in the ration.

If sheep are underfed they mobilize body fat, then subsequently long-chain fatty acids (LCFA) – particularly stearic, oleic and linoleic acids – increase in the milk fat (Payne and Rattray, 1980; Storry, 1988; Piva and Fusconi, 1989), while short- and medium-chain fatty acids (S/MCFA) decrease (Rossi and Pulina, 1991). Adding supplementary rumen-protected amino acids – methionine and lysine – to the diet increased the concentration of milk fat (Perna *et al.*, 1994), the levels of C16:0–C18:3 by 5% and the unsaturated:saturated fatty acid ratio by 4% in the milk of Comisana ewes (Sevi *et al.*, 1998). By contrast, the synthesis of S/MCFA is favoured by a high (acetate+butyrate):propionate ratio, which increases as the fibre concentration in the diet increases, decreases as fibrous particle size decreases and depends also on the composition of the cell walls of the plants ingested by the sheep. Adding supplementary rumen-protected fat to the diet can increase the proportion of unsaturated fatty acids in ovine milk (Table 8.2) and desirable fatty acids, such as omega-3, can be included in the milk.

Among the lactic unsaturated LCFA, of particular interest is conjugated linoleic acid (CLA), the concentration of which can be markedly influenced by diet. The acronym CLA is used to describe a mixture of isomers of octadecaienoic acid with conjugated double bonds. The major isomer of CLA in milk is cis-9, trans-11 and it represents about 80% of the total CLA (Parodi, 1977). A range of beneficial effects on health (anticarcinogenic, growth promotion, anti-arteriosclerotic) has recently been reported for this compound in research with animal models (Ha *et al.*, 1989; Kelly and Bauman, 1996; Banni and Martin, 1998; Parodi, 1999). CLA in the milk of ruminants derives from the biohydrogenation of unsaturated fatty acids in the rumen (Parodi, 1977).

Table 8.2. Fatty acid composition of milk fat in dairy ewes fed calcium soaps of fatty acids (CSFA).

	Pulina *et al.* (1990b)		Perez Alba *et al.* (1999)	
	Control	CSFA	Control	CSOFA
Milk yield (kg/day)	1.09	1.21	1.40	1.56
Fat yield (g/day)	84	101	181	198
Fatty acids (% of FA)				
C8:0	1.7	1.1[b]	2.7	1.4[b]
C10:0	6.0	3.6[b]	9.6	4.2[b]
C12:0	3.8	2.3[b]	5.7	2.7[b]
C14:0	11.2	8.4[b]	12.1	8.3[b]
C16:0	29.8	34.2[b]	24.1	21.6[a]
C16:1	1.3	1.3	1.1	1.0
C18:0	10.1	9.0[a]	7.2	9.2[a]
C18:1	20.1	25.7[b]	15.9	31.0[b]
C18:2	4.9	5.4	1.8	1.3[b]
C18:3	5.0	4.0[b]	–	–

CSFA, calcium soaps of fatty acids; CSOFA, calcium soaps of olive fatty acids.
Significant differences: [a] $P < 0.05$, [b] $P < 0.01$.

However, Griinari *et al.* (2000) hypothesized that cis-9, trans-11 CLA would originate in the mammary gland from desaturation of trans-11 C18:1. Many studies have been published on CLA in bovine milk (Jiang *et al.*, 1996; Dhiman *et al.*, 1999; Parodi, 1999; Griinari *et al.*, 2000), but there are very few on ovine milk, despite the fact that the content of CLA found in ovine milk, cheese and especially in the ricotta was greater than in bovine dairy products (Table 8.3) (Banni and Martin, 1998; Prandini *et al.*, 2001).

In any case, in ovine (Banni and Martin, 1998), bovine (McGuire *et al.*, 1997; Banni and Martin, 1998; Griinari, 1998; Dhiman *et al.*, 1999) and caprine (our unpublished results) milk, CLA concentrations are greater when the animals have access to grazing pasture. This is due to the higher ingestion of polyunsaturated fatty acids (PUFA) in pasture compared to that with dry forages; forage maturity is also an important factor affecting CLA concentration in milk. In cows the intake of diets containing forage at its early growth stage increased milk CLA content compared to diets that included late-growth or second-cut forage (Chouinard *et al.*, 1998).

Furthermore, rations containing different fat concentrations and varying relative amounts of roughage can also induce variations in the concentration of CLA (Jiang *et al.*, 1996; Kelly and Bauman, 1996; McGuire *et al.*, 1997). In dairy sheep, if the fibre:carbohydrate ratio is reduced, biohydrogenation is decreased and there is accumulation of the precursors of the cis-9, trans-11 C18:2 in the rumen (Gerson *et al.*, 1985). CLA content also depends on the fat content of the diet: McGuire *et al.* (1997) found that in bovine milk CLA concentration increased from 2.3 to 2.9 mg/g of fat if the concentration of fat in DM was increased from 3.0 to 7.2% using maize oil. This result suggests that protected fat with a particular fatty acid composition can be fed

Table 8.3. C18 fatty acid content (mg/g of fat) of bovine and ovine milk and in some Italian dairy products (Banni and Martin, 1998).

Product	18:1t	18:2	CLA
Bovine milk	19.22	15.79	7.10
Yogurt	43.37	25.12	7.98
Parmesan	32.21	17.67	8.65
Ovine milk (green forage)	85.74	9.78	29.78
Ovine milk (dry forage)	13.14	24.40	11.72
Pecorino cheese	49.92	17.93	13.03
Ricotta	91.04	17.50	24.19

to dairy animals for increasing the target acid fraction. Almost all ovine milk is transformed into cheese, so it is important that transformation and conservation of dairy products do not affect CLA content (Lin *et al.*, 1995). This stability of CLA may be due to its association with the whey proteins, particularly β-lactoglobulin, which may prevent isomerization and oxidation (Banni and Martin, 1998). Since ovine milk has a greater concentration of β-lactoglobulin than bovine or caprine milk, this would explain the greater concentration in CLA found in dairy products derived from ovine milk (Casper *et al.*, 1998; Pena *et al.*, 1998).

Interference by ruminal CLA in fat synthesis in the mammary gland is one of the theories adopted to explain the low fat syndrome in cows (and probably also in dairy sheep) following the ingestion of high doses of concentrates (Bauman and Griinari, 2000). However, the depression of milk fat seems to be caused by trans-10, cis-12 linoleic acid (Baumgard *et al.*, 2000), rather than by the cis-9, trans-11 isomer.

Milk protein content

The protein content of milk is influenced by the concentrations of energy, protein and lipid in the ration and also by the quality of the protein and the lipid. Protein in milk is much harder to manipulate nutritionally than fat. Milk protein concentration is positively correlated with the energy content of the ration, particularly when the additional energy comes from soluble carbohydrates (Gonzalez *et al.*, 1984). This can be explained because: (i) ruminal ammonia decreases and bacterial protein increases, due to improved conditions for the development of ruminal microbes, resulting in improved utilization of nitrogen in the rumen; and (ii) ruminal production of propionate – which increases the availability of non-essential amino acids that are no longer required for gluconeogenesis – can therefore be taken up by the mammary gland and transformed into milk protein.

Generally, dietary crude protein (CP) concentration affects milk yield, but not milk protein percentage, and so protein yield increases. However, when the diet is deficient in CP, which often occurs when sheep are fed poor-quality forage, the protein content in milk decreases (Calderon-Cortes *et al.*, 1977). A negative relationship between dietary CP and milk protein percentage can also be an indicator of excessive rumen-undegradable protein in the diet; this has been associated with a decrease in microbial protein synthesis and amino acid flow to the small intestine. On the other hand, feeding excessive degradable CP, such as urea, can reduce milk protein content.

Supplemental dietary fat and CSFA can decrease protein content in milk (Fig. 8.2). The underlying mechanism is not fully understood, but it could be due to a lower availability of fermentable energy in the rumen for the synthesis of bacterial protein, or to the lower response by the mammary gland to insulin, resulting in the reduction of amino acid uptake, or to a dilution effect resulting from increased milk yield. In cows, increasing the total amount of dietary amino acids reaching the small intestine by using rumen-undegradable protein (RUP), or rumen-protected amino acids, may partly alleviate the milk protein depression associated with supplemental dietary fat. In sheep, supplementary RUP demonstrated effects on milk protein content only when the CP of pasture was low (Casals *et al.*, 1999).

Milk protein fraction

The composition of total protein (caseins, whey proteins, NPN) in the milk depends on the energy:protein ratio in the diet, on its CP concentration and on the use of rumen-protected protein.

To optimize the dietary utilization of nitrogen, the diet should be balanced for energy and protein for both quantity and ruminal degradation kinetics. This results in maximum microbial growth per unit of DM fermented, in lower urinary losses of energy and nitrogen and in the optimum utilization of dietary nitrogen for casein synthesis.

Urea is the nitrogen compound that is

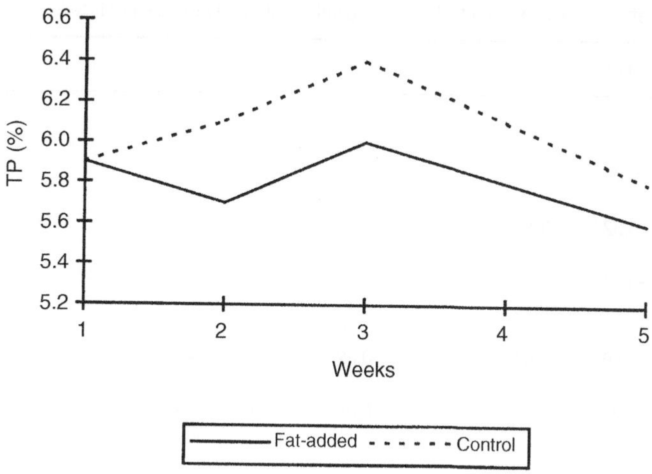

Fig. 8.2. Effect of use of rumen-protected fat (4.5% of DM) on protein content of ovine milk (Rossi *et al.*, 1991).

most sensitive to the concentration of protein in the diet. Cannas *et al.* (1998) showed that there is a close correlation between milk urea nitrogen (MUN) and CP concentration in the diet (see Chapter 6). Effectively, MUN can be used as an indirect indicator of protein in the diet, especially for animals on pasture, for which evaluating protein in the diet is particularly difficult.

8.1.2 Dietary NDF, crude protein and milk TUS yield

So far, we have dealt with the current interpretation of the relationship between the components of the diet and the quality of ovine milk in the light of the 'dairy cow paradigm' briefly summarized in Section 8.1.1. However, we have recently partially refuted this paradigm as we are accumulating experimental data showing that an increase in dietary fibre does not always result in an increase in milk fat concentration (Pulina *et al.*, 2000). We have analysed a series of data derived from over 120 scientific publications on dairy sheep (each datum is the average result from an experiment).

Amongst the correlations that link nutrition and milk composition (Table 8.4), two are particularly important:

- The so-called 'dilution effect', which causes a reduction in milk fat when milk production increases.
- Increase in milk fat concentration with the increased content of NDF in the diet.

It is evident that the first phenomenon confounds the interpretation of the second. With higher fibre content in the diet, less energy is ingested (due to the lower digestibility of the diet and lower feed intake) and therefore there is a reduction in milk production and fat concentration increases. However, with a reduction in milk production, we observed a reduction in the yield of fat and protein (expressed in g/day). Therefore, the positive effect of NDF on fat percentage is due more to the reduced production of milk and consequent increase in fat concentration than to the direct effect of fibre on this component.

The dendrogram reported in Fig. 8.3, derived from cluster analyses of the variables in Table 8.4, highlights the importance of daily energy intake in determining milk production and its quality. However, energy intake is affected positively by the CP content of the diet, and negatively by the content of NDF.

The hypothesis that best explains these results is this: the increase in energy available to the animals following the ingestion

Table 8.4. Correlation matrix between feeding factors and quantity and quality of milk produced.

Variables	DMI	EI	CP	NDF	DBW	MY	Fat	Pr	FY
Dry matter intake (DMI) (kg/day)	–								
Energy intake (EI) (UFL/day)	0.94	–							
Dietary crude protein (CP) (%)	0.32	0.49	–						
NDF (%)	–0.37	*	*	–					
Bodyweight change (DBW) (g/day)	*	0.41	*	*	–				
Milk yield (MY) kg/day	0.59	0.61	*	–0.62	–0.49	–			
Milk fat (Fat) (%)	–0.26	*	*	0.38	0.41	0.64	–		
Milk protein (Pr) (%)	*	*	*	*	*	*	0.48	–	
Fat yield (FY)	0.64	0.69	0.32	–0.57	–0.42	0.93	–0.38	*	–
Protein yield	0.56	0.57	*	–0.65	–0.58	0.99	–0.63	*	0.92

UFL, milk forage unit, energy unit of the INRA system corresponding to 1.70 Mcal of ENL.
Coefficients differ from zero at $P < 0.05$.

of diets with low NDF is primarily utilized in producing lactose as the increase in metabolizable energy coincides with an increase in the blood glucose available for lactose synthesis (Freetly and Ferrell, 1999).

At the same time, blood flow to the mammary gland increases due to local factors controlled by the metabolic hormones (insulin, IGF and neuro-hormones), the mammary gland uptake of milk precursors increases, lactose production increases, protein production increases as some amino acids are released from gluconeogenesis (hepatic or mammary) and their uptake becomes more efficient. Fat production increases because: (i) there is a greater inclusion of LCFA in milk; and (ii) a greater quantity of excess glucose after lactose synthesis can be utilized directly for fat synthesis.

In practice, the correlation between milk fat and NDF depends on the energy status of the animals. Bocquier and Caja

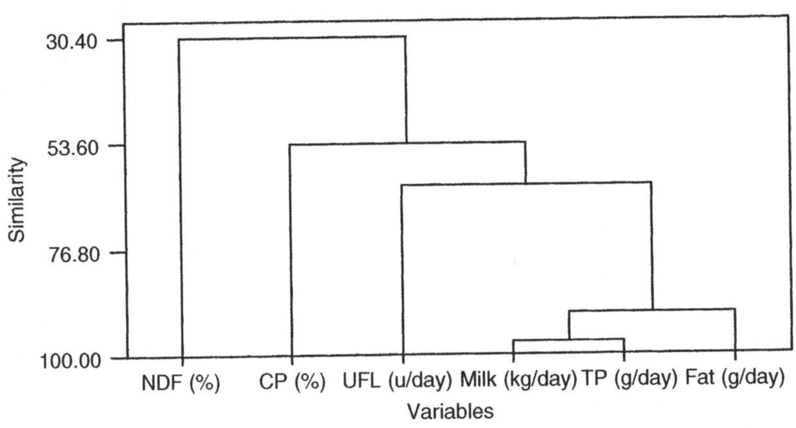

Fig. 8.3. Dendrogram of correlation between the main feeding and production variables reported in Table 8.4.

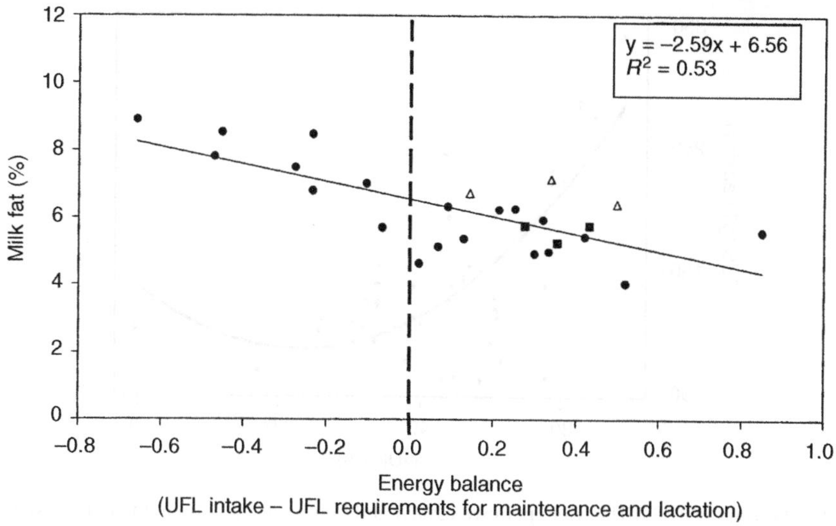

Fig. 8.4. Relationship between energy balance (UFL/head/day) and fat concentration in milk (Cannas *et al.*, 2001).

(2001) and Cannas *et al.* (2001) observed a decrease in milk fat concentration as the energy balance of dairy sheep became positive, independent of milk yield (Fig. 8.4).

When the energy balance is positive, NDF concentration in the ration and milk fat concentration are negatively correlated, but if the energy balance is negative, the two parameters are positively correlated. Cannas (personal communication) suggested that if energy balance is negative, body fat mobilization leads to an increase in LCFA in blood plasma. Because the ovine mammary gland is very efficient at extracting circulating LCFA, milk fat concentration increases, and the proportion of LCFA to C16 in the milk fat increases too. This phenomenon is clearly shown in Table 8.5: as

Table 8.5. Fatty acid composition of ovine milk fat at different energy balances (Rossi and Pulina, 1991).

	Bodyweight change (kg/week)		
Fatty acids (%)	+1.5	−1.1	−3.8
C4:0	3.31	2.49	2.21
C6:0	2.81[a]	1.29[b]	0.84[b]
C8:0	2.87[a]	1.09[b]	0.65[b]
C10:0	5.62[a]	2.70[b]	1.52[b]
C12:0	4.07[a]	1.88[b]	1.10[b]
C14:0	9.84[a]	6.96[a]	3.43[b]
C16:0	22.86	24.67	24.15
C16:1	1.50	1.56	1.57
C18:0	7.14[a]	10.93[a]	13.58[b]
C18:1	16.91[a]	21.52[a]	28.47[b]
C18:2	5.42	5.86	6.47
C18:3	0.31[a]	0.27[a]	0.65[b]

[a,b] Means with different superscript differ $P < 0.05$.

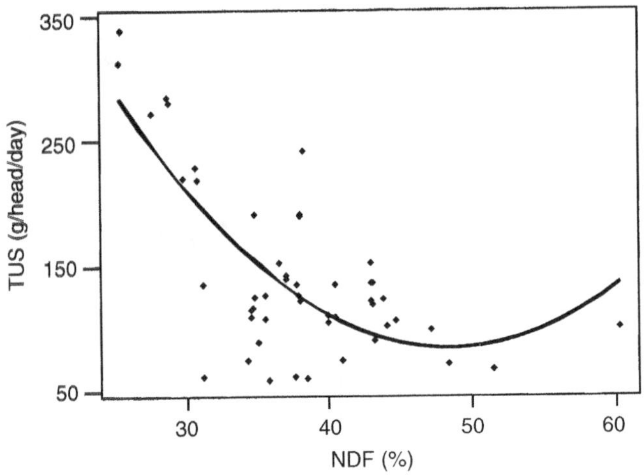

Fig. 8.5. Relationship between TUS yield per ewe (g/day) and fibre concentration in the ration (NDF as % of DM).

sheep move from negative to positive body-weight change, the fraction of LCFA increases dramatically. If energy balance is positive, LCFA concentration in the blood decreases, the proportion of S/MCFA in the blood plasma increases and milk fat concentration decreases because those fatty acids are synthesized directly by the mammary gland.

It would be expected that the production of TUS in milk would be determined by the same characteristics of the diet that separately control the production of fat and total protein. Using the same series of data, we calculated the second-order relationships that link the production of TUS to the concentration of fibre in the ration:

$$TUS = 957.7 - 36.0\ NDF + 0.372\ NDF^2 \quad R^2 = 0.54,$$

where NDF is expressed as % of DM and TUS is in g/day. In any case, data are grouped along the descending part of a parabolic curve (Fig. 8.5).

Nevertheless, the most informative function is the relationship between TUS and the CP:NDF ratio of the diet. The most appropriate equation describing this phenomenon is of second order too, but data are mainly grouped along the ascending part of the parabolic curve (Fig. 8.6):

$$TUS = 96.4 - 252.7\ (CP{:}NDF) + 824.0\ (CP{:}NDF)^2 \quad R^2 = 0.544.$$

The above relationships confirm that NDF is the best index of the transformation of nutrients into milk compounds that are useful for cheese-making. In particular, the first equation shows that diets containing less fibre are more efficient because of the effects of higher energy concentration and higher intake. The second equation, moreover, confirms the positive effect of CP on TUS caused by the increase of feed intake and NDF digestibility, and not by the increase in amino acid availability for protein synthesis at mammary level.

8.2 Minerals and Aromatic Compounds

8.2.1 Minerals

Normally, mineral macroelements (Ca, P, Mg, K, Na, Cl and S) in milk are not modified by feeding. However, feeding concentrate supplements with low-quality grass may increase the concentrations of Ca and P in milk (Ashton *et al.*, 1964). Ca, Na and Cl may be reduced in the milk of sheep fed rations lacking in those elements (Naitana *et al.*, 1992), but an excess in the diet does not

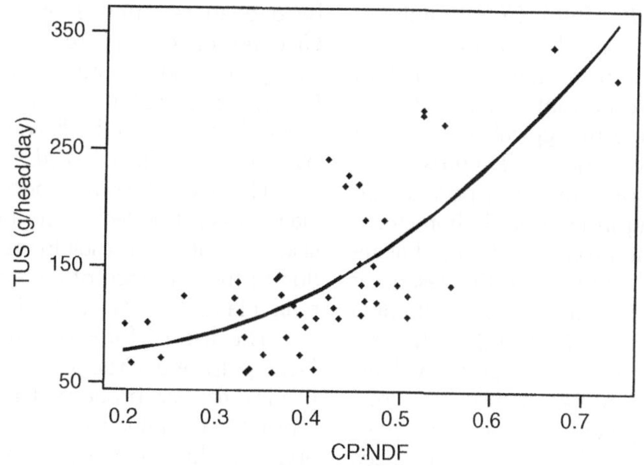

Fig. 8.6. TUS yield per ewe (g/day) as function of dietary CP:NDF ratio.

alter their concentration in milk (Luquet, 1985). Legumes, for example, are usually higher in Ca content than grass forages, which, in turn, are generally higher in Ca content than cereals. Mature forages, and crop residues (e.g. maize stubble) generally contain low levels of P, while cereal grains and oilseed meals are moderate to high in P. K content is lower in cereals than in forage.

On the other hand, there are strong relationships between mineral microelements (Mn, Co, Al, F, I, Mo and Se) in the diet and their concentrations in milk. In particular, feeding sheep with Se supplement during the pre-partum period results in its increase in colostrum and in milk (Paulson *et al.*, 1968; Cuesta *et al.*, 1995). The use of Zn supplements in the same period does not increase its concentration in milk, but results in a greater immunological response in the newborn, probably because of the function of Zn on the interaction of T-helper cells and B lymphocytes (Prasad and Kundu, 1995). Changes in the equilibrium of minerals can be linked, in addition to their concentration in the food, to their modification and absorption in the gut as a result of dysmetabolic syndromes, or of competition among ions.

The mineral content in milk is particularly important for its influence (together with lactose, citrate and urea) on the freezing point of milk. This is used as an indicator of milk adulteration – usually due to the addition of water. In animals with acute mastitis in later lactation, a slight depression of freezing point is normal. Sometimes an alteration in freezing point has been observed in the milk of animals fed with different diets, probably related to the mineral content of the feed. For example, milk from dairy cattle fed with soybean meal showed a lower mean freezing point than milk from those fed rapeseed.

A reduction in the concentration of minerals in milk can be a negative influence on the nutritional value of milk and its clotting properties: for example, low levels of Se and/or vitamin E in milk can cause nutritional muscular dystrophy in suckling lambs; low levels of Ca in milk can compromise clotting properties.

8.2.2 Aromatic compounds

Aromatic compounds have to be given special attention because of their influence (sometimes considerable) on cheese flavour (Ha and Lindsay, 1991; Rubino and Claps, 1999). These substances are generally volatile compounds, e.g. polyphenols or branched fatty acids. In ovine milk there are more than 100, although only 16 have major effects on cheese bouquet (Moio *et al.*, 1993a,b).

The concentration and the type of volatile compounds in milk are closely linked to diet (Moio *et al.*, 1996). Grazing animals on pasture characterized by a specific flora can result in the introduction to the milk of the compounds responsible for the characteristic bouquet of some typical cheeses. For example, the polyphenols of *Thymus herba barona* (Loisel grass) are transported into the milk, resulting in its typical characteristics and pleasant aroma (Lai and Falchi-Delitala, 1982; Lai *et al.*, 1983). Désage *et al.* (1996) fed ewes with a ration containing cumin (*Cuminum cuminum* L.) as the aromatic source, and observed that the volatile compounds most representative of cumin seed aroma (β-pinene, p-cymene, γ-terpinene and cuminaldehyde) were introduced to the milk. Some dicotyledons, such as *Asperula odoratum* and *Geranium molle*, markedly influence the aroma of dairy products because there is a transfer of terpenic compounds into the milk (Rubino and Claps, 1999).

Sesquiterpenes have been found in milk produced in mountain areas (Dumont and Adda, 1978) and in the milk of grazing sheep (Moio *et al.*, 1996); these compounds, therefore, may be useful in distinguishing between cheeses obtained from pasture systems and those derived from the milk of sheep fed on conserved feeds. Table 8.6 shows the influence on ovine cheese of some feeds, frequently utilized in dairy sheep, which may result in off-flavours (Urbach, 1990). Off-flavours in the cheese have been reported after sheep were fed *Cruciferae*, certain cereals and legumes, citrus pulp and poorly conserved silage. Conversely, well-preserved, good-quality silage did not affect the flavour and aroma of ovine cheese (Cavani *et al.*, 1991).

The use of some concentrates (especially oats), if added to the ration in excessive amounts, can result in unpleasant smells due to the presence of certain SCFA in the milk (Ha and Lindsay, 1991).

The use of forages containing betaine (beet pulp and *Cruciferae*) can result in a fish-like flavour, probably due to oxidation of choline – contained in lecithine – to glycol and trimethylamine (Alais, 1988).

The introduction of undesirable substances in the milk depends on the quality and quantity of feedstuff ingested. To avoid these problems it is important to use feedstuffs of good quality, and if feedstuffs responsible for off-flavours have to be fed, these should be limited in their quantity. For example, when animals consume feeds such as lucerne, the dimethyl sulphide content in milk increases. In low concentrations, dimethyl sulphide contributes to the characteristic flavour of milk, but in high concentrations it causes a malty defect (Urbach, 1990).

The characteristic aroma of ovine cheese is mainly due to its high SCFA content, (Palo, 1983) and, in particular, to caprilic (C8:0) and caprinic (C10:0) acids.

Table 8.6. Some feeds responsible for anomalous flavours in cheese (Urbach, 1990).

Feed	Compound	Off-flavour
Beet by-products	Trimethylamine, formed from betaine in the digestive tract	Fish flavour
Rye and wheat	Trimethylamine	Fish flavour
Legumes (e.g. *Medicago*, *Vicia*)	Not identified	Bitter flavour
Cruciferae (*Brassica* spp.)	Mustard oil released during ruminal fermentation	Pungent smell, due to presence of S in flowers
Sunflowers	Not identified	Oxidized flavour
Poor-quality silage	Not identified	Taste of silage
Fruit and residues of its processing	Not identified	Off-flavour
Cress (*Lepidum sativum*)	Products derived from benzyl-glucosinolate	Burnt flavour and pungent smell
Excessive amounts of lucerne	Dimethyl-sulphide	Malt flavour

These fatty acids may affect cheese aroma directly or indirectly by acting as precursors of other aromatic compounds (Luquet, 1985), such as 4-methyloctanoic and 4-ethyloctanoic acids, which are responsible for the 'ram' and 'goat' smells that cannot be found in bovine cheese (Ha and Lindsay, 1991). Again, nutrition, by affecting the relative quantities of fatty acids in the milk, affects the aroma of its derived dairy products (see Section 8.1).

8.3 Somatic Cell Count and Microorganisms

The somatic cell count (SCC) in milk is affected by nutrition if errors occur in the formulation of feeding regimes that may predispose the mammary gland to inflammation. The correct incorporation in the ration of vitamin A, β carotene, vitamin E and selenium plays a major role in stimulating the immune response in the cells of the mammary gland, and thereby ensuring the production of more milk with a lower SCC (Morgante *et al.*, 1995, 1999). Such incorporation is particularly useful when sheep are fed mainly conserved fodder (hay or silage), which may have undergone considerable losses in β carotene and vitamin E during storage. The sudden transition from dry fodder to young grass – which is normally rich in nitrates – is also often responsible for an increase in milk SCC and for a depression of the ovine immune system (Cuccuru *et al.*, 1994).

The microbial content of milk is a very important parameter for ovine milk quality, but nutrition has little effect on it. Bacteria and other microorganisms in the milk derive mainly from the environment, and their control depends on the efficiency of the milking practice in maintaining a clean udder.

Incorrect nutrition, e.g. excess nitrogen with inadequate fibre, may generate anomalous fermentations in the rumen and the intestines, which will produce highly contaminating faeces, due to their increased microbial contents and larger volume.

The use of poorly preserved silage, e.g. silage that has undergone anomalous fermentations, of concentrate containing contaminating components or of hay that contains soil could increase the bacterial count of milk, especially that of *Clostridium* spp. (Manfredini *et al.*, 1987; Cavani *et al.*, 1991; Deiana *et al.*, 1993; Lodi and Carini, 1993). These microorganisms have a detrimental effect on the ripening of cheese, as they are the major organisms responsible for the so-called 'late swelling' of cheese, a condition that affects cheeses with long maturation times – such as Pecorino, Romano and Fiore Sardo – and which is caused by the massive amounts of gas produced by the Clostridia (Deiana *et al.*, 1993).

8.4 Toxic Substances in Milk

Ovine milk may become contaminated by toxins such as pesticides, or by toxic metabolites of fungal origin (mycotoxins).

8.4.1 Pesticides

The generic name of pesticides includes a large number of chemicals used in the production, transport and storage of fodder. These compounds represent a danger not only because of their immediate toxicity, but also because they can be consumed by animals that produce milk, meat, etc.

Milk and dairy products are a major source of pesticide ingestion for humans (Deiana, 1991). The chemical and physical characteristics of the various compounds affect their distribution within the milk components: organochlorines are preferentially associated with the lipid fraction (Oliveri *et al.*, 1988), while organophosphates are normally associated with the proteins (Deiana, 1991). The presence of organochlorines in milk may be a direct consequence of the consumption of rations that have been contaminated, because the feedstuffs that were used to make them had previously been treated with pesticides. Their presence, however, has also been associated with sheep grazing in pastures adjacent to areas of intensive cropping where pesticides had been used extensively. The application of pesticides to extensive areas may have conveyed the pesticides or their metabolites either through wind drift on to the pasture, or by pollution of the groundwater sources used to provide water for the stock (Oliveri *et al.*, 1988).

8.4.2 Mycotoxins

Mycotoxins are secondary metabolites produced by several species of fungi, which can contaminate livestock feedstuffs at all stages of their productive cycle (cropping, harvest, transport and storage) (D'Mello and MacDonald, 1997). The mycotoxins most commonly found in livestock feedstuffs and which have been recorded in milk are:

- Ochratoxins.
- Deoxinivalenol (DON).
- Aflatoxins (AF).

Ochratoxins

These are produced mainly by the toxigenous species of *Aspergillus ochraeceus*, *Penicillium viridicatum* and *Penicillium verrucosum*, which are ubiquitous species known to cause moulding of grains and feedstuffs. Among the ochratoxins, type A is the most important because of both its hepatotoxicity and its nephrotoxicity; it is also teratogenic and carcinogenic (Kuiper-Goodman and Scott, 1989; Marquardt *et al.*, 1990).

These mycotoxins are generally not very toxic to ruminants because they are partially inactivated by the ruminal and caecal microorganisms (Kiessling *et al.*, 1984; Sreemannarayana *et al.*, 1988; Galtier, 1991). However in sheep this faculty for detoxification of mycotoxins is less well developed than in other ruminants (Xiao *et al.*, 1991; Höhler *et al.*, 1999). This is probably due to the faster rate of passage of food through the digestive tract. In dairy cattle, the transfer of ochratoxin A from the rations to the milk has been documented only recently (Skaug, 1999).

Deoxinivalenol (DON)

This is produced by fungi of the genus *Fusarium* – the causative agent of fusariosis in cereals. Contamination affects mainly the grains from wheat, barley and maize, as well as maize silage. Once again, ruminants seem to be particularly tolerant of these toxins, given the detoxifying action of the ruminal and intestinal microflora. However, the presence of DON was recorded in ovine milk, albeit in very low concentrations (Prelusky *et al.*, 1987).

Aflatoxins (AF)

These are produced by fungi of the genus *Aspergillus*. These compounds are the most dangerous to human health because of their cytotoxic, teratogenic and mutagenic properties. For this reason, we will describe in detail their passage from feedstuffs to ovine milk. The feedstuffs most at risk of contamination are maize, oilseed, groundnuts and products derived from them (Bottalico, 1999).

The International Agency for Research on Cancer (IARC) has now included aflatoxin B1 (AF-B1) and aflatoxin M1 (AF-M1) in the class of substances that are *definitely* carcinogenic for humans (Category 1). Even though AF-M1 can be synthesized by the fungal mycelium, it usually derives from the hydroxylation of AF-B1 that occurs in the liver of animals that have ingested AF-B1. AF-B1 can also be metabolized by other tissues, and this is particularly true for sheep, which have a very efficient faculty for transforming AF-B1 into AF-M1 in the nasal, olfactory and respiratory mucosae (Larsson *et al.*, 1994).

In contrast to the previously described mycotoxins, ruminants can detoxify only part of the aflatoxins ingested with feed in their digestive tract, and this is performed mainly by the protozoa living in the rumen (Kiessling *et al.*, 1984). Absorption of aflatoxins starts in the rumen and continues in the small intestine; the toxins are then transported to the liver, and its metabolites are absorbed by the blood, then excreted in the urine, bile and milk (Trucksess *et al.*, 1983). When sheep consume AF-B1 these appear in the milk within 6 h (Nabney *et al.*, 1967; Battacone *et al.*, 2000). Excretion of the toxins is affected by: (i) its rate of absorption; (ii) the gradual release of AF-B1 from the tissues into the bloodstream; (iii) the reabsorption in the small intestine via the bile (Trucksess *et al.*, 1983); (iv) the conversion rate in the liver of AF-B1 to AF-M1; and (v) the efficiency of toxin excretion in the urine and faeces (Veldman *et al.*, 1992). An important implication for human health is the quantification of the carry-over, i.e. the amount of toxin translocated into the milk in relation to the amount ingested by the animal.

In an experiment carried out in dairy ewes (Battacone, 2000) that had received different daily doses of AF-B1 (32, 64 and 128 mg), the concentration of AF-M1 in milk was related to dose level ingestion and reached a maximum by day 4. It then decreased, reaching a steady-state condition. After the ingestion of toxin had stopped, the time at which AF-M1 was no longer detectable in milk (3 days) was not related to the dietary AF-B1 level. At steady-state condition, the relationship between AF-M1 concentration in milk (mg/l) and AF-B1 intake (mg/head/day) can be expressed by the equation:

$$\text{AF-M1} = 0.00136\ \text{AF-B1} - 0.0043 \qquad R^2 = 0.77.$$

These results, compared to studies in dairy cattle, demonstrate that sheep have a lower rate of translocation of toxins to the milk (Munksgaard *et al.*, 1987; Veldman *et al.*, 1992).

8.5 Cheese Processing

The suitability of milk for cheese-making and the quality of dairy products are affected by a number of different characteristics of the milk (Bencini and Pulina, 1997). Amongst these are:

- Its chemical composition, e.g. the concentrations of fat and protein.
- The presence of flavours that may be essential for the production of typical dairy products (Rubino and Claps, 1999).
- The somatic and microbial cell counts.

The processing performance of ovine milk is normally affected positively by its fat and protein content (Ucci, 1945; Casu and Marcialis, 1966; Storry *et al.*, 1983), and negatively by its high somatic and microbial cell counts (Duranti and Casoli, 1991; Casoli *et al.*, 1992; Manfredini *et al.*, 1992; Bufano *et al.*, 1994; Serra *et al.*, 1995).

The composition and the processing performance (and hence the quality), therefore, of ovine milk results mainly from dietary effects on lactation. Particular attention has been paid to the use of silage in ovine nutrition, because in bovine milk its use has been associated with cheese defects caused by Clostridia. However, studies on the use of silage for replacing pasture and hay in dairy sheep nutrition have yielded contradictory results. Some authors (Manfredini *et al.*, 1987; Bertoni and Bernabucci, 1990; Scintu *et al.*, 1990; Cavani *et al.*, 1991) reported no effects of silage on cheese-making properties while others reported an improvement in ovine milk-clotting properties (Muscio *et al.*, 1983; Bianchi *et al.*, 1993; Leto *et al.*, 2000). Moreover, silage had no influence on the chemical and microbiological characteristics of fresh and mature cheeses (Manfredini *et al.*, 1987; Leto *et al.*, 2000), or on the flavour and aroma of cheeses matured for 120 days (Manfredini *et al.*, 1987; Cavani *et al.*, 1991). Significantly, no negative effects have been reported. The evidence above demonstrates that silage obtained from good forages, appropriately cut and stored to avoid anomalous fermentations (Cannas, 1999), can be used to replace pasture and hay for dairy sheep, without affecting the quality of the milk for cheese-making.

The use of non-structural carbohydrates (NSC = starch and sugars) – normally associated with increased amounts of concentrates in the ration and the addition of protected fat – does not seem to affect the clotting properties of ovine milk (Campus *et al.*, 1990; Chiofalo *et al.*, 1993, 1995). However, towards the end of lactation, increasing the energy content of the ration improves the processing performance of the milk (Pulina *et al.*, 1993; Serra *et al.*, 1995).

The type and physical characteristics of concentrates affect the processing performance of ovine milk. Feeding rolled, rather than crushed, grains to dairy sheep increased milk production – and fat and protein concentrations – and improved the clotting properties of the milk (Bianchi *et al.*, 1994).

Despite this, pasture seems to be the aspect of nutrition that influences the best cheese-making characteristics of the milk. Milk produced by animals grazed on pasture has better coagulation properties (Martini *et al.*, 1999), greater yields of cheese (Muscio *et al.*, 1983) and significantly lower somatic and microbial cell counts than the milk of sheep fed dry fodder (Scintu *et al.*, 1990).

The use of amino acids – cystine and methionine – is considered to be of limited value in the production of milk, whether in the form of specially rumen-protected formulations or as protein that avoids ruminal degradation (e.g. herring meal); they did not affect the clotting properties of ovine milk (Perna *et al.*, 1994; Chiofalo *et al.*, 1996; Martini *et al.*, 1999), or result in increased consistency of the curd (Cavani *et al.*, 1991). For dairy sheep fed dried forage, the administration of vitamin E affected the coagulation properties of the milk positively, probably as a consequence of the improved condition of the mammary gland (Chiofalo *et al.*, 1998).

8.6 Practical Recommendations

To improve the quality of ovine milk through nutrition, a dairy sheep farmer should consider how the ration is formulated, its distribution, of what feedstuffs it is composed and their quality. Unfortunately, some rations that could increase the fat content of milk depress the protein content, and vice versa. Therefore, they should be used as part of a 'nutritional strategy' that should aim to increase positively and simultaneously all the qualitative parameters of the milk. There are three major strategies that should be adopted for improving ovine milk quality, listed below.

8.6.1 Increasing the production of fat and protein

- Rations should have a balanced energy:protein ratio in terms of both quality and ruminal degradability, to optimize ruminal bacterial fermentation and to allow maximum production of milk and milk components.
- Rations should have a minimum fibre level of 30–32% NDF.
- Part of the fibre (at least one-third) should have adequate particle length (2–3 mm), to facilitate adequate levels of ingestion and optimal levels of ruminal pH for cellulose fermentation.
- Feeding frequency throughout the day should be increased to avoid rapid variations in ruminal pH when low fibre–high concentrate diets are used.
- Excess protein should be avoided, particularly if easily fermented in the rumen, to reduce urinary losses of nitrogen.
- Milk urea concentration should be monitored as an indicator of protein intake.
- The fatty acid composition of ovine milk fat can be modified by using rumen-protected fats.

8.6.2 Decreasing the somatic cell and bacterial counts

- Avoid sudden ration changes, which may predispose the mammary gland to infections.
- Integrate dry feed with vitamins A and E to stimulate the natural immune responses of the mammary gland.
- Harvest and preserve feedstuffs correctly to avoid contamination of milk by undesirable microorganisms such as Clostridia.

8.6.3 Minimizing the translocation of unwanted substances to the milk

- Avoid the use of feedstuffs contaminated with natural (e.g. mycotoxins) or artificial (pesticides) contaminants.
- Avoid the use of feedstuffs containing substances that can transfer off-flavours to milk.
- Avoid the use of silage that has undergone anomalous fermentation.

References

Alais C. (1988) Scienza del latte. *Principi e tecnologia del latte e dei derivati,* 2nd edition, Teniche Nuove, Milan, Italy.

Ashton W.M., Owen J.B., Ingleton J.W. (1964) A study of the composition of Clun Forest ewes' milk. *J. Agric. Sci., Camb.,* 63: 85–90.

Banni S., Martin J.C. (1998) Conjugated linoleic acid and metabolites. In: J.L. Sebedio and W.W. Christie (eds) *Trans fatty acids in human nutrition,* The Oily Press Ltd, UK, pp. 261–302.

Barillet F., Astruc J.M., Bocquier F., Jacquin M., Fraysse G., Lagriffoul G., Marie C., Pellegrini O., Remeuf F. (1998) *Influence des facteurs de production sur la composition chimique du lait valorise en fromage: le cas du lait de brebis. Basis of the quality of typical Mediterranean products,* EEAP Publication No. 90, Wageningen, Netherlands, pp. 128–144.

Battacone G. (2000) Studio della dinamica di escrezione delle aflatossine nel latte ovino. PhD thesis, Perugia, Italy.

Battacone G., Palomba M., Porcu F., Cannas A. (2000) Studio della dinamica di escrezione dell'aflatossina M1 nel latte di pecora. *Proc. XXXV Int. Symp. Zootecnia Produzioni animali di qualità ed impatto ambientale nel sistema mediterraneo,* Ragusa, Italy, pp. 151–158.

Bauman D.E., Griinari J.M. (2000) Regulation and nutritional manipulation of milk fat. Low-fat milk syndrome. Mammary gland development and prolactin receptors. In: J.A. Mol and R.A. Clegg (eds) *Biology of the mammary gland.* Kluwer Academic/Plenum Publishers, New York, pp. 209–216.

Baumgard L.H., Corl B.A., Dwyer D.A., Saebo A., Bauman D.E. (2000) Identification of the conjugated linoleic acid isomer that inhibits milk fat synthesis. *Am. J. Physiol. Regul. Integr. Comp. Physiol.,* 278: 179–184.

Bencini R., Pulina G. (1997) The quality of sheep milk: a review. *Australian Journal of Experimental Agriculture,* 37: 485–504.

Bertoni G., Bernabucci U. (1990) Possibilità di impiego degli insilati nell'alimentazione degli ovini. *Proc. Giornate di Studio, Fiera Viterbo,* Viterbo, Italy, pp. 11–22.

Bianchi M., Battaglini L.M., Fortina R. (1993) Caratteristiche del latte di pecore alimentate con unifeed. *Proc. X Congress ASPA (Italy),* pp. 333–338.

Bianchi M., Battaglini L.M., Fortina R. (1994) Effetto dell'integrazione con alimenti sotto forma fioccata oppure frantumata nella razione della pecora: indagini comparative sulle caratteristiche quanti-qualitative del latte. *Proc. XI Congress ASPA (Italy),* pp. 447–450.

Bottalico A. (1999) Muffe e micotossine delle granaglie. *Tecnica Molitoria,* 12: 195–219.

Bocquier F., Caja G. (2001) Production et composition du lait de brebis: effets de l'alimentation. *INRA Prod. Anim.,* 14: 129–140.

Bufano G., Dario C., Laudadio V. (1994) The characterizing of Leccese sheep: variations of chemical composition and lactodinamographic parameters in milk as related to somatic cell count. *Proc. Int. Symp. Somatic Cells and Milk of Small Ruminants,* Bella, Italy, pp. 368–373.

Calderon-Cortes J.F., Robinson J.J., McHattie I., Fraser C. (1977) The sensitivity of ewe milk yield to changes in dietary crude protein concentration. *Anim. Prod.,* 24: 135. (Abstr.)

Campus R.L., Papoff C.M., Cannas A., Serra A. (1990) Effetto della grassatura e della concentrazione proteica della razione sulla composizione azotata e sulla attitudine alla coagulazione del latte di pecora di razza Sarda. *Proc. VIII Congress SIPAOC (Italy),* 3.2.

Cannas A. (1995) The effects of particle size of the diet on feeding behaviour and milk production in sheep. MS thesis, Cornell University, Ithaca, New York.

Cannas A. (1999) Piatto unico anche per le pecore. *Informatore Zootecnico,* 46: 56–58.

Cannas A., Pes A., Mancuso R., Vodret B., Nudda A. (1998) Effect of dietary energy and protein concentration on the concentration of milk urea nitrogen in dairy ewes. *J. Dairy Sci.,* 81: 499–508.

Cannas A., Nudda A., Brandano P. (2001) Fibre, starch and sugar requirements in dairy sheep: effects on milk yield and its quality. *Proc. Quality of Sheep Milk*, Foggia, Italy, pp. 49–65.

Caroli A., Pagnacco G., Rognoni G., Sanna S. (1993) Il miglioramento genetico della qualità del latte ovicaprino nella pratica di allevamento. In: P. Brandano and A. Serra (eds) *Proc. III Congr. Qualità del Latte Ovino e Caprino*, Sassari, Italy, pp. 9–16.

Casals R., Caja G., Such X., Torre C., Calsamiglia S. (1999) Effects of calcium soaps and rumen undegradable protein on the milk production and composition of dairy ewes. *J. Dairy Res.*, 66: 177–191.

Casoli C., Pauselli M., Morgante M., Ranucci S., Duranti E., Mehrabi H. (1992) Comportamento reologico del latte ovino in rapporto alle caratteristiche chimico-fisiche e cellulari. *Proc. X Congr. SIPAOC (Italy)*, pp. 250–251.

Casper J.L., Wendorff W.L., Thomas D.L. (1998) Seasonal changes in protein composition of whey from commercial manufacture of caprine and ovine speciality cheeses. *J. Dairy Sci.*, 81: 3117–3122.

Casu S. (1989) In: Reda (ed.) *Gli ovini da latte. Miglioramento genetico degli animali domestici*, Rome, Italy, pp. 143–150.

Casu S., Marcialis A. (1966) Contributo alla conoscenza delle relazioni fra composizione del latte e resa in formaggio di tipo pecorino romano. *Sci. Tecn. Latt.-cas.*, 17: 204–213.

Cavani C., Bianconi L., Manfredini M., Rizzi L., Zarri M.C. (1991) Effects of a complete diet on the qualitative characteristics of ewe milk and cheese. *Small Rum. Res.*, 5: 273–284.

Chiofalo V., Savoini G., Micari P., Zumbo A., Bontempo V., Ziino M. (1993) Impiego di differenti fonti energetiche per l'alimentazione della pecora: effetti sulle caratteristiche quanti-qualitative del latte. *Proc. X Congress ASPA (Italy)*, pp. 339–344.

Chiofalo V., Savoini G., Zumbo A., Dell'Orto V. (1995) Effetto di due differenti dosaggi di somatotropina bovina ricombinante (rbST) sulla quanti-qualità di latte in pecore alimentate con diete contenenti saponi di calcio. *Proc. XI Congress ASPA (Italy)*, pp. 291–292.

Chiofalo V., Savoini G., Chiofalo L. (1996) Variation of milk yield and quality in lactation ewes by using rumen-protected amino acids. *Proc. XXXI Int. Symp. Zootecnia Food and Health. Role of Animal Products*, Milan, Italy.

Chiofalo V., Maldonato R., Latona F., Savoini G., Capogreco B. (1998) Effetto dell'integrazione nella dieta con vitamina E sulle caratteristiche quanti-qualitative e sul contenuto in cellule somatiche del latte ovino. *Proc. XIII Congress SIPAOC (Italy)*, pp. 35–38.

Chouinard P.Y., Corneau L., Kelly M.L., Griinari J.M., Bauman D.E. (1998) Effect of dietary manipulation on milk conjugated linoleic acid concentrations. *J. Dairy Sci.*, 81 (Suppl. 1): 233.

Comendador F.J., Nardo N., Bertone A., Maurizi A., Quaglia G. (1996) Effect of ultrafiltration on functional properties of sheep milk 'scotta'. *Sci. Tec. Latt.-cas.*, 47: 103–112.

Cowan R.T., Robinson J.J., Fraser C. (1979) Effect of protein content of the diet on feed intake and milk yield of ewes in early lactation. *Anim. Prod.*, 28: 453.

Cowan R.T., Robinson J.J., McHattie I., Pennie K. (1981) Effects of protein concentration in the diet on milk yield, change in body composition and the efficiency of utilization of body tissue for milk production in ewes. *Anim. Prod.*, 33: 111–120.

Cuccuru C., Zucconi A., Caria A., Ruffo G., Contini A. (1994) Analysis of risk for somatic cell count in ewes: environmental, management and microbiological factors. *Proc. Int. Symp. Somatic Cells and Milk of Small Ruminants*, Bella, Italy, pp. 174–180.

Cuesta P.A., McDowell L.R., Kunkle W.E., Wilkinson N.S., Martin F.G. (1995) Effects of high-dose prepartum injection of Se and vitamin E on milk and serum concentrations in ewes. *Small Rum. Res.*, 18: 99–103.

Deiana P. (1991) Residui di pesticidi e procesi di trasformazione del latte. *L'industria del latte*, 27: 3–32.

Deiana P., Lodi R., Malaspina P., Caredda M. (1993) Influenza dell'alimentazione sul contenuto in anaerobi sporigeni nel latte di pecora. *Il Latte* 18: 1223–1229.

Désage M., Schaal B., Soubeyrand J., Orgeur P., Brazier J.L. (1996) Gas chromatographic-mass spectrometric method to characterize the transfer of dietary odorous compounds into plasma and milk. *J. Chromatography B: Biomedical Sciences and Applications*, 678: 205–210.

Dhiman T.R., Anand G.R., Satter L.D., Pariza M.W. (1999) Conjugated linoleic acid content of milk from cows fed different diets. *J. Dairy Sci.*, 82: 2146–2156.

D'Mello J.P.F., MacDonald A.M.C. (1997) Mycotoxins. *Anim. Feed Sci. Technol.*, 69: 155–166.

Dumont J.P., Adda J. (1978) Occurrence of sesquiterpenes in mountain cheese volatiles. *J. Agric. Food Chem.*, 26: 364–367.

Duranti E., Casoli C. (1991) Variazione della composizione azotata e dei parametri lattodinamografici del latte di pecora in funzione del contenuto di cellule somatiche. *Zoot. Nutr. Anim.*, 17: 99–105.

Farid M.F.A., Hassam N.I., Zein G.N., Moussa A.A., Ghita E.I. (1979) Effect of feeding urea to lactating ewes on fatty acid composition of milk lipids. *World Rev. Anim. Prod.*, 15: 17–21.

Freetly H.C., Ferrell C.L. (1999) Relationship of portal-drained viscera and liver net flux of glucose, lactate, volatile fatty acids, and nitrogen metabolites to milk production in the ewe. *J. Dairy Sci.*, 82: 597–604.

Galtier P. (1991) Pharmacokinetics of ochratoxin A in animals. *IARC Sci. Publ.*, 115: 187–200.

Gerson T., Jihn A., King A.S.D. (1985) The effects of dietary starch and fiber on the *in vitro* rates of lipolysis and hydrogenation by sheep rumen digesta. *J. Agric. Sci. Camb.*, 105: 27. (Abstr.)

Gonzalez J.S., Robinson J.J., McHattie I. (1984) The effect of level of feeding on the response of lactating ewes to dietary supplements of fish meal. *Anim. Prod.*, 40: 39–45.

Griinari J.M. (1998) Trans fatty acids in milk fat. *Int. Work. Health benefits of old and novel compounds in milk.* Utrecht, The Netherlands.

Griinari J.M., Corl B.A., Lacy S.H., Chouinard P.Y., Nurmela K.V., Bauman D.E. (2000) Conjugated linoleic acid is synthesized endogenously in lactating dairy cows by delta(9)-desaturase. *J. Nutr.*, 130: 2285–2291.

Ha J.K., Lindsay R.C. (1991) Contributions of cows, sheep, and goat milks to characterizing branched-chain fatty acid and phenolic flavours in varietal cheeses. *J. Dairy Sci.*, 74: 3267–3274.

Ha Y.L., Grimm N.K., Pariza M.W. (1989) Newly recognized anticarcinogenic fatty acids, identification and quantification in natural and processed cheeses. *J. Agric. Food Chem.* 37: 75–81.

Höhler D., Südekum K.H., Wolffram S., Frohlich A.A., Marquardt R.R. (1999) Metabolism and excretion of ochratoxin A fed to sheep. *J. Anim. Sci.*, 77: 1217–1223.

Jiang J., Bjoerck L., Fonden R., Emanuelson M. (1996) Occurence of conjugated cis-9, trans-11-octadecadienoic acid in bovine milk; effects of feed and dietary regimen. *J. Dairy Sci.*, 79: 438–445.

Kelly L.M., Bauman D.E. (1996) Conjugated linoleic acid, a potent anticarcinogen found in milk fat. *58th Proc. of Cornell Nutrition Conference for Feed Manufacturers*, Rochester, New York, pp. 68–74.

Kiessling K.H., Pettersson H., Sandholm K., Olsen M. (1984) Metabolism of aflatoxin, ochratoxin, zearalenone, and three trichothecenes by intact rumen fluid, rumen protozoa, and rumen bacteria. *Appl. Environ. Microbiol.*, 47: 1070–1073.

Kuiper-Goodman T., Scott P.M. (1989) Risk assessement of mycotoxin ochratoxin A. *Biomed. Environ. Sci.*, 2: 179–248.

Lai P., Falchi-Delitala L. (1982) Composizione chimica di *Thymus herba barona* (L.): presumibili riflessi sugli animali in produzione zootecnica. *Il Nuovo Progresso Veterinario*, 3: 15.

Lai P., Pisanu S., Falchi-Delitala L. (1983) Ricerche sull'attività antimicrobica del latte di pecore alimentate con *Thymus herba barona* Loisel. *Il Latte*, 8: 442–445.

Larsson P., Busk L., Tjalve H. (1994) Hepatic and extrahepatic bioactivation and GSH conjugation of aflatoxin B1 in sheep. *Carcinogenesis*, 15: 947–955.

Leto G., Todaro M., Fulco A., Portolano B., Di Noto A.M., Alicata M.L. (2000) Effetti dell'insilato di sulla (*Hedisarium coronarium* L.) in rotoballe fasciate sulle caratteristiche quanti-qualitative del latte e del formaggio di pecora. *Zoot. Nutr. Anim.*, 26: 39–48.

Lin H., Boylston T.D., Chang M.J., Luedecke L.O., Shultz T.D. (1995) Survey of the conjugated linoleic acid contents of dairy products. *J. Dairy Sci.*, 78: 2358–2365.

Lodi R., Carini S. (1993) Il miglioramento della qualità microbiologica del latte per la valorizzazione dei prodotti tipici. In: P. Brandano and A. Serra (eds) *Proc. III Congr. Qualità del Latte Ovino e Caprino*, Sassari, Italy, pp. 61–66.

Luquet F.M. (1985) *Laits et produits latières. I. Les laits. Technique et Documentation*, Lavoisier, Paris.

Manfredini M., Cavani C., Chiarini R., Sanguinetti V., Zarri M.C. (1987) Effetti dell' insilato di mais sulle caratteristiche qualitative del latte e del formaggio di pecora. *Zoot. Nutr. Anim.*, 13: 21–28.

Manfredini M., Tassinari M., Zarri M.P. (1992) Caratteristiche chimico-fisiche, contenuto in cellule somatiche ed attitudine alla coagulazione di latte individuale di pecore allevate in Emilia-Romagna. *Sci. Tec. Latt.-cas.*, 43: 113–125.

Marquardt R.R., Frohlich A., Abramson D. (1990) Ochratoxin A: an important western Canadian storage mycotoxin. *Can. J. Physiol. Pharmacol.*, 68: 991–999.

Martini M., Rapaccini S., Giuliotti L. (1999) Coagulation properties of Massese sheep milk: effects of feeding management. *Proc. XIII Congress ASPA (Italy)*, pp. 487–489.

McGuire M.A., McGuire M.K., McGuire M.S., Griinari J.M. (1997) Bovinic acid, the natural CLA. *59th Proc. of Cornell Nutrition Conference for Feed Manufacturers*, Rochester, New York, pp. 217–226.

Moio L., Dekimpe J., Etievant P.X., Addeo F. (1993a) Neutral volatile compounds in the raw milks from different species. *J. Dairy Res.*, 60: 199–213.

Moio L., Langlois D., Etievant P.X., Addeo F. (1993b) Powerful odorants in bovine, ovine, caprine and water buffalo milk determined by means of gas chromatography–olfactometry. *J. Dairy Res.*, 60: 215–222.

Moio L., Rillo L., Ledda A., Addeo F. (1996) Odorous constituents of ovine milk in relationship to diet. *J. Dairy Sci.*, 79: 1322–1331.

Morgante M., Beghelli D., Ranucci S., Pauselli M., Casoli C., Duranti E. (1995) Effects of selenium and vitamin E administration on ewe mammary health: preliminary results. *XXX Symp. Int. Zootecnica Reproduction & Animal Breeding Advances and Strategy*, pp. 419–420.

Morgante M., Beghelli D., Pauselli M., Dall'Ara P., Capuccella M., Ranucci S. (1999) Effect of administration of vitamin E and selenium during the dry period on mammary health and milk cell counts in dairy ewes. *J. Dairy Sci.*, 82: 623–631.

Munksgaard L., Larsen J., Werner H., Andersen P.E., Viuf B.T. (1987) Carry-over of aflatoxin from cows' feed to milk and milk products. *Milchwissenschaft*, 42: 165–167.

Muscio A., Marsico G., Centoducati P. (1983) La qualità del latte di pecore alimentate con insilato di mais. *Ann. Fac. Agr. Univ. Bari*, 33: 71–79.

Nabney J., Burbage M.B., Lewis G. (1967) Metabolism of aflatoxin in sheep: excretion pattern in the lactating ewe. *Food Cosmet. Toxicol.*, 5: 11–17.

Naitana S., Nuvole P., Marongiu A. (1992) Lattazione. In: G. Aguggini, V. Beghelli and L.F. Giulio (eds) *Fisiologia degli animali domestici con elementi di etologia* , UTET, Turin, Italy, pp. 781–808.

Oddy V.H. (1978) Milk production in ewes fed high grain diets. *Dairy Sci. Abstr.*, 42: 4088.

Oliveri R., Corrao A., Casuccio A., Scalici M., Gullotti A. (1988) Pesticidi clororganici in campioni di latte di ovini stanziali nelle zone di Marsala utilizzate per colture protette. *Nuovi Annali di Igiene e Microbiologia*, 34: 191–201.

Palo V. (1983) Distribution of aroma in fractioned fat in Slovak ewes' cheese. *Dairy Sci. Abstr.*, 45: 4271.

Parodi P.W. (1977) Conjugated octadiecadenoic acids of milk fat. *J. Dairy Sci.*, 60: 1550–1553.

Parodi P.W. (1999) Conjugated linoleic acid and other anticarcinogenic agents of bovine milk fat. *J. Dairy Sci.*, 82: 1339–1349.

Paulson G.D., Broderick G.A., Baumann C.A., Pope A.L. (1968) Effect of feeding sheep selenium fortified with trace mineralized salt: effect of tocopherol. *J. Anim. Sci.*, 27: 195–202.

Payne E., Rattray P.V. (1980) The effect of restricted grazing on the fatty acid composition of ovine milk fat during early lactation. *British J. Nutr.*, 44: 47–52.

Pena R.N., Folch J.M., Sanchez A., Whitelaw C.B. (1998) Chromatin structures of goat and sheep beta-lactoglobulin gene differ. *Biochem. Biophys. Res. Commun.*, 252: 649–653.

Perez Alba L.M., De Souza Cavalcanti S., Perez Hernandez M., Martinez Marin A., Fernandez Marin G. (1997) Calcium soap of olive fatty acid in the diets of Manchega dairy ewes: effects on digestibility and production. *J. Dairy Sci.*, 80: 3316–3324.

Perna A., Gambacorta E., Marsico D., Cosentino E. (1994) Produttività di pecore comisane sottoposte ad integrazione con amino acidi ruminoprotetti: I. Latte. *Proc. XI Congress SIPAOC (Italy)*, pp. 451–454.

Pirisi A., Murgia A., Scintu M.F. (1994) Previsione della resa in formaggio pecorino romano e pecorino sardo in funzione del contenuto in proteine e grasso del latte di pecora. *Sci. Tec. Latt.-cas.*, 45: 476–483.

Piva G., Fusconi G. (1989) Come influire sul contenuto in grasso. *L'Inf. Agr.*, 45: 59–65.

Polidori F., Baldi A., Cheli F., Pulina G. (1991) Alimentazione e qualità del latte caprino. *III Simp Int. Zootec. Qualità del Latte Ovino–Caprino*, Varese, Italy, pp. 105–134.

Poulton S.G., Ashton W.M. (1972) Studies on ewe's milk V. The effect of high cereal diets on ewes and on the yield of milk and milk constituents. *J. Agric. Sci., Camb.*, 54: 353–359.

Prandini A., Geromin D., Conti F., Masoero F., Piva A., Piva G. (2001) Survey on the level of conjugated linoleic acid in dairy products. *Ital. J. Food Sci.*, 13: 243–253.

Prasad T., Kundu M.S. (1995) Serum IgG and IgM responses to sheep red blood cells (SRBC) in weaned calves fed milk supplemented with Zn and Cu. *Nutrition*, 11 (Suppl. 5): 712–715.

Prelusky D.B., Veira D.M., Trenholm H.L., Foster B.C. (1987) Metabolic rate and elimination in milk, urine and bile of deoxynivalenol following administration to lactating sheep. *J. Environ. Sci. Health* B., 22: 125–148.

Pulina G., Rassu S.P.G. (1991) Qualita del latte, occhio all'alimentazione. *L'Inf. Zoot.*, 38: 28–34.

Pulina G., Rossi G., Cannas A., Papoff C.M., Campus R.L. (1990a) Influenza della concentrazione proteica della razione sulla produzione quanti-qualitativa di latte in pecore di razza Sarda. *Agricoltura Ricerca*, 105: 65–70.

Pulina G., Serra A., Campus R.L., Papoff C.M. (1990b) Effetto della grassatura e della concentrazione proteica della razione sulla composizione acidica del grasso del latte di pecore di razza Sarda. *Proc. IX Congr. SIPAOC (Italy)*, 3.3.

Pulina G., Serra A., Macciotta N.P.P., Nudda A. (1993) La produzione continua di latte nella specie ovina in ambiente mediterraneo. *Proc. X Congress ASPA (Italy)*, pp. 353–356.

Pulina G., Nudda A., Battacone G., Rassu S.P.G., Cappio-Borlino A. (2000) Il problema del pagamento del latte ovino secondo la qualità. *L'Inf. Agr.*, 21: 57–61.

Rossi G., Pulina G. (1991) Il ruolo dell'alimentazione nella composizione lipidica del latte ovino. *L'inf. Agr.*, 47: 1–5.

Rossi G., Pulina G., Cannas A. (1988) Alimentazione bilanciata per avere più latte. *L'Inf. Zoot.*, 36: 41–44.

Rossi G., Serra A., Pulina G., Cannas A., Brandano P. (1991) L'impiego di un alimento unico pellettato (Unipellet) nell'alimentazione della pecora da latte: 1 Influenza della grassatura e del livello proteico sulla produzione quanti-qualitativa di latte in pecore di razza Sarda. *Zoot. Nutr. Anim.*, 17: 23–34.

Rubino R., Claps S. (1999) Sistemi alimentari e qualità del latte e dei formaggi di pecore e capre. *Proc. Seminario SIPAOC Le produzioni ovine e caprine alle soglie del terzo millennio*, Perugia, Italy, pp. 27–40.

Russel J.R., Chow J.M. (1993) Another theory for the action of ruminal buffer salts: decreased starch fermentation and propionate production. *J. Dairy Sci.*, 76: 826–830.

Scintu M.F., Pirisi A., Podda F., Sanna S., Ledda A. (1990) Alimentazione della pecora da latte con la tecnica 'unifeed: caratteristiche del latte prodotto. Primi risultati. *Proc. IX Congress SIPAOC (Italy)*, 3.4.

Serra A., Nudda A., Pulina G., Cannas A. (1995) Effects of concentrate fiber content and lambing season on summer milk quality of dairy ewes on stubble grazing. *Proc. XI Congress ASPA (Italy)*, pp. 283–284.

Sevi A., Rotunno T., Di Caterina R., Muscio A. (1998) Rumen-protected methionine or lysine supplementation of Comisana ewes' diets: effect on milk fatty acid composition. *J. Dairy Res.*, 65: 413–422.

Skaug M.A. (1999) Analysis of Norwegian milk and infant formulas for ochratoxin A. *Food Addit. Contam.*, 16: 75–78.

Sklan D. (1992) A note on production responses of lactating ewes to calcium soaps of fatty acids. *Anim. Prod.*, 55: 288–291.

Sreemannarayana O., Frohlich A.A., Vitti T.G., Marquardt R.R., Abramson D. (1988) Studies on the tolerance and disposition of ochratoxin A in young calves. *J. Anim. Sci.*, 66: 1703–1711.

Storry J.E. (1988) The effect of dietary fat on milk composition. In: *Recent developments in ruminant nutrition*. Butterworths, London, pp. 111–141.

Storry J.E., Alistair S.G., Millard D., Owen A.J., Ford G.D. (1983) Chemical composition and coagulating properties of renneted milks from different breeds and species of ruminants. *J. Dairy Res.* 50: 215–229.

Trucksess M.W., Richard J.L., Stoloff L., Mcdonald J.S., Brumley W.C. (1983) Absorption and distribution patterns of aflatoxicol and aflatoxins B1 and M1 in blood and milk of cows given aflatoxin B1. *Am. J. Vet. Res.*, 44: 1753–1756.

Ucci E. (1945) Correlazioni fra il contenuto in sostanza secca ed in lipidi del latte di pecora ed il rendimento in pecorino romano. *Ann. Istituto Sperimentale per la Zootecnia*, Rome, Italy, 3: 469–474.

Urbach G. (1990) Effect of feed on flavor in dairy foods. *J. Dairy Sci.*, 73: 3639–3650.

Van Soest P.J. (1994) *Nutritional ecology of the ruminant*, Second edition. Cornell University Press, Ithaca, New York.

Veldman A., Meijs J.A.C., Borggreve G.J., Heeres-van der Tol J.J. (1992) Carry-over of aflatoxin from cows' food to milk. *Anim. Prod.*, 55: 163–168.

Woodford J.A., Jorgensen N.A., Barrington G.P. (1986) Impact of dietary fiber and physical form on performances of lactating dairy cows. *J. Dairy Sci.*, 69: 1035–1047.

Xiao H., Marquardt R.R., Frohlich A.A., Phillips G.D., Vitti T.G. (1991) Effect of hay and a grain diet on the bioavailability of ochratoxin A in the rumen of sheep. *J. Anim. Sci.*, 69: 3715–3723.

9 Feeding Dairy Lambs

Paolo Brandano[1], Salvatore Pier Giacomo Rassu[1] and Alfio Lanza[2]
[1]*Dipartimento di Scienze Zootecniche, Università di Sassari, Italy;* [2]*Dipartimento di Scienze Agronomiche Agrochimiche e delle Produzioni Animali, Università di Catania, Italy*

There are four phases involved in raising dairy lambs: (i) newborn care and the colostrum period; (ii) suckling; (iii) weaning; and (iv) raising until replacement. Obviously the techniques adopted depend on the management system used on the farm (intensive and mainly housed, penned or kept indoors, or extensive with use of pasture).

9.1 Newborn Care and the Colostrum Period

The newborn lamb needs primary, or neonatal, care. In intensive systems the following steps must be taken: (i) clean the mouth and/or nostrils of any amniotic fluid accumulated during the final fetal phase. This should be done by hand, with the lamb being held vertically in a head-down position if possible; (ii) dry the lamb with either dry clean straw or, better still, an absorbent cloth, to remove the amniotic fluid; (iii) prepare a bed in a separate pen protected from draughts and at a temperature of 20–22°C; (iv) provide colostrum either from the mother or from another ewe, or alternatively a milk replacer; (v) separate the newborn lamb from its mother within 48 h if it is to be fed a milk replacer, in order to prevent a bond developing between the lamb and its mother, and to reduce the possibility of the transmission of infectious diseases.

It is of great importance to feed the lamb colostrum in the first hours of life, and this should continue for at least 18–36 h. Colostrum is the first secretion of the mammary gland and production normally starts before parturition, although occasionally also in the last week of pregnancy, and stops completely 4 or 5 days after lactation commences, since it is subsequently transformed into milk. It plays three roles in the feeding of the newborn lamb: as food, as a laxative and as a source of immunity against disease (Naitana *et al.*, 1992). It serves as feed due to its high content of nutrients; the quantities of these vary with the breed and the characteristics of the particular ewe. The average quantities are shown in Table 9.1.

Colostrum has a laxative function due

Table 9.1. Analysis of the average chemical composition of colostrum in Massese sheep (Casoli *et al.*, 1987).

	Time from parturition (h)		
	1	12	24
Dry matter (%)	29.6	25.3	22.6
Fat (%)	10.5	9.2	8.8
TN (%)	15.9	12.3	9.4
Casein (%)	6.0	5.4	5.2
Seroproteins (%)	9.5	6.4	3.7
Lactose (%)	2.8	3.7	4.3
Ash (%)	1.4	0.9	0.9
pH	6.37	6.42	6.50
Density	1.056	1.046	1.042

TN, total nitrogen ($N \times 6.38$).

to the presence of certain minerals (especially magnesium) that free the intestines of the newborn lamb from the meconium, which has accumulated there in the form of faeces and catabolic substances during the period inside the uterus.

Immunization is the most important function of colostrum. This occurs due to the presence of specific immunoglobulins (γ-globulins) in the maternal colostrum; these are absorbed directly by the intestine during the first hours of life, without being digested at gastric level. The absorption rate is extremely rapid in young animals during the first 6 h of life, and then progressively diminishes until it ceases altogether after 36–48 h. This is because globulin, like all the other proteins, is digested at gastric level after the first 2 days. Colostrum is indispensable in the first 2 days of the lamb's life because in the ovine, unlike in other species, the placenta is impermeable to maternal antibodies. Thus passive immunity cannot be acquired in the uterus and must be transmitted via the colostrum, preferably that of the mother, until the young lamb develops active immunity.

The immunoglobulin concentrations in the colostrum are highest in the 2 days after birth in pluriparous ewes and in ewes that have been dry for at least 2 months. The γ-globulin is absorbed in the form of protein macromolecules directly via the intestine because few proteolytic enzymes, and little hydrochloric acid, are produced in the abomasum, and the intestinal mucosa is extremely permeable in the early days of life and, above all, in the first hours after birth.

Colostrum is given *ad libitum* five or six times a day if the lamb is fed on maternal milk during the whole suckling period. It is given three times a day from a feeding bottle if the mother's milk is replaced by a milk replacer. In this case, the quantity of each feed is 4–5% of the BW of the animal (0.15–0.20 l/feed, or 0.5–0.6 l/day) (Eales and Small, 1986).

Maternal colostrum may not be available for some reason, such as the death of the ewe during birth or absence of or delay in milk secretion. In this case, it is best if colostrum from ewes from the same farm is given to the lamb. This must be taken immediately post-partum from animals with good production levels, and can be stored in frozen form for 1–2 months. This is the best alternative as the colostrum contains the specific antibodies for the most common infections on the farm. If this is not possible, artificial colostrum formulated from milk, egg yolk and antibiotics should be prepared and stored in a refrigerated form on the farm. Problems connected with the transmission of parasitic diseases in the milk (e.g. ascaridiasis) can be dealt with by a suitable treatment using either high or low temperatures.

The transition from the lamb being fed on colostrum to being fed on either natural milk or a milk replacer (transition phase) is, in natural suckling, obviously a gradual process, but this should also be the case when the lambs are fed on milk substitutes. Only after 2–3 days should the lamb be suckled on milk alone. In the transition phase, the lambs are extremely susceptible to neonatal infections such as enterotoxaemia, septicaemia and colibacillosis, despite the passive immunity provided by the colostrum. Such infections must be quickly prevented by vaccinating the mother with vaccines and/or vaccinating the newborn lamb with specific vaccines and administering massive doses of vitamins A, D and E.

9.2 Suckling

Once the initial colostral phase is finished the newborn lamb must be suckled. This phase normally lasts 5 or 6 weeks. The lamb is fed exclusively, or almost exclusively, on milk. This milk can come from the mother or from other ewes, or may be a milk replacer. The newborn lamb cannot digest other types of feed at this stage of its life. Indeed, until the lamb is completely weaned, its feeding behaviour is very similar to that of monogastric animals. Only the abomasum is functioning; the volume of the abomasum is 70% of the total gastric apparatus, with the rumen–recticulum complex and the omasum comprising the other

30%. The milk enters the abomasum directly through the closure mechanism of the reticular groove, which connects the oesophagus directly with the abomasum, bypassing the forestomachs. The lamb is no different from other young mammals in the way in which it digests milk at a gastric and intestinal level.

Casein is the principal protein in the diet and makes up 82–83% of true protein and 75–78% of TN. It is immediately coagulated in the abomasum and first digested by the rennin or chymosin. These are casein-specific proteases which can also react when the pH is very acidic during the first weeks of the lamb's life. Then – and this is of secondary importance – it is broken down by pepsin, which is a non-specific protein that reacts only when the pH is very low, and after the first few weeks of the lamb's life. The seroproteins albumin and globulin, by contrast, are not broken down by the chymosin and pepsin, and thus pass unaltered to the intestine, where they are broken down by pancreatic trypsin.

Lactose (glucose + galactose) makes up almost all of the carbohydrate content in the milk, and of the starch and fructose in milk replacers; it is almost completely digested by the intestinal lactase in the duodenum. The starch and fructose are also digested in the intestine, the former by pancreatic amylase and the latter by intestinal saccharase (sheep do not produce salivary amylase), but only after the animal reaches a certain age (1–2 months), and then only to a limited extent.

Lipids are first broken down partially by lipase in the abomasum, which principally reacts with the short-chain fatty acids (SCFA). Digestion of fats is completed in the intestine by pancreatic lipases, whose activity is assisted by the emulsifying effect of the bile.

Minerals and vitamins are absorbed only in the intestine.

If a milk replacer is used, then it must contain mainly casein, at least during the first 10 days of the lamb's life. Other proteins are much less digestible, especially if they are of vegetable origin. The milk replacer may, however, contain starch, as long as this does not make up more than 10% of the DM of the diet. If it does, it may cause intestinal disorders, which can provoke serious, and sometimes even fatal, diarrhoea.

9.2.1 Natural suckling

In the natural suckling system the lambs suckle the mother's milk directly during the ewe's early lactation period; this system is normally used on low-production/low-technology farms. Lambs suckle the mother's milk directly throughout their whole suckling period, either following the ewe to pasture for the whole day or spending the night with the mother if they are separated during the day. The suckling phase uses only the ewe's milk in the early part of her lactation period, as the lamb is then either slaughtered while still suckling or weaned once it reaches a suitable weight. The mother continues to be milked until the lactation period is over.

The growth rate of the lamb during this phase depends not only on its growth potential but also, and indeed principally, on the quantity and quality (fat and protein content) of the maternal milk and its availability (the number of feeds per day). The conversion index (kg fat–protein-corrected milk (FPCM) per kg BW) ranges from 4.5 to 6.0 and increases during the suckling period.

9.2.2 Artificial suckling

Artificial suckling may use either natural milk or a milk replacer. In the former case, the milk is given two or three times a day. This is normal practice when the lamb is orphaned or the mother is unable for some reason to feed the lamb. In the latter case, a milk replacer is given to the lamb instead of the mother's milk or milk from another ewe. This milk replacer must fulfil the lamb's feeding requirements adequately. Before describing the feeding techniques employed when using a milk replacer, the physical, chemical and nutritional characteristics of such a replacer will be described.

Milk replacers

Milk replacers are feeds that are capable of completely substituting for natural milk, whether from the mother or from another ewe, during the whole suckling period. Normally they are by-products of the dairy industry, with some components added that enable them to substitute for natural milk. They are some 40% cheaper than the milk sold for human consumption, making their use an economic proposition.

Milk replacers come in the form of powdered milk. This must be highly soluble so that it dissolves completely in the water at the moment when it is used. Normally it consists of:

- 60–75% skimmed powdered milk, a by-product of butter manufacture.
- 5–10% milk whey, a by-product of cheese-making.
- 15–25% (mainly animal) fats, usually by-products of meat processing.

Different methods are used to dry the powdered milk and milk whey described above (Piccioni, 1989). Powdered skimmed milk can be obtained by drying or dehydration, using the roller method, where the rollers are heated to 100°C, or by the spray-dry method, where it is vaporized in a heated chamber. A much finer powder is obtained with the spray-dry method, because the chamber is pre-heated to 55–60°C, the heating period is shorter and the temperature is controlled more precisely. Thus, it is more soluble, easier to coagulate and therefore easier to digest, and has greater nutritional value. This is because the proteins in general, and the sulphide amino acids methionine and cysteine in particular, are less denatured. The lysine is also more stable and the lactose is less caramelized.

The quality of powdered skimmed milk is measured by its:

- Available lysine content as a percentage of total lysine, which should not be < 80% (2.4–3% of DM).
- Casein content, which should not be less than 2.5–3% of DM.
- Coagulation time.
- Solubility.
- Acidity.

Because powdered skimmed milk is expensive it is, unfortunately, often partly replaced by alternative products (the so-called 'milk without milk'). Their casein protein content is, however, different from that of milk, and so they are less digestible. The source of the protein, which can be soybean meal, fishmeal, or milk whey, should be borne in mind when choosing a milk replacer.

If soybean meal is to be used as the protein integrator for the milk replacer it must be dehydrated, have methionine added, and its anti-pepsin and haemagglutinants deactivated by either thermal toasting or chemical treatment. The remaining oligosaccharides are removed because these are less digestible than lactose and can cause intestinal disturbances (the resulting product is called protein concentrate of soy). Finally, the anti-pepsin and haemagglutinants are treated first with alkaline solutions and then with acid solutions (resulting in soy proteinate).

Fishmeal, in the form of both fish protein concentrate and hydrolysed fish protein, has a high sulphide amino acid content, but is low in other important amino acids. It is also expensive and generally smelly.

Milk whey is a residue of coagulated bovine milk from cheese production. It is normally fed to pigs and adult cattle, but may be used to replace skimmed milk, although only partially; because its casein content is low (2 vs. 3%/l) and its lactose content is high (70 vs. 40% of DM), it should be used as de-lactosate whey, with the lactose removed (milk without milk) or as a whey protein concentrate.

Fats have to be added to the milk replacer. They are necessary in order for the fat content of the powdered skimmed milk to reach 20–30% of DM. Obviously fats which are different from, but similar in composition to, those normally found in milk are used. They are cheap, have a low melting point, are easy to emulsify and digest, and are also easy to conserve. The composition of the fats is similar to that of natural milk fats in as much as the fatty acids in the milk replacer consist of 20–30% SCFA (C4–C10), of 40–50% saturated LCFA (C14–C18) and of 30–40% unsaturated LCFA (C18:1–3).

The cost is low because animal fat by-products from the meat industry (mainly bovine tallow) and/or vegetable fats, such as coconut, palm, or soybean oils, are used. The melting point is linked both to the length of the acid chains and to the level of unsaturation or, in other words, to the number of double links. It must not be greater than the body temperature of the animal, which is 40°C. Emulsification is encouraged by the addition of emulsifying substances or surface-active agents. Soybean lecithin is excellent but expensive, while sugar glycerides are less expensive. Digestibility depends on:

- The diameter of the fat globules, which should not be greater than 5 μm.
- The structure (i.e. the chain length and non-saturated number) of the fatty acids.
- The homogenization (the type of fattening process).
- The purity of the fat (refinement, aroma and transparency).
- The age of the animals from which it came.

Conservation depends on the absence of modification processes in the fats. One modification may be acidification due to hydrolysis of the triglycerides, even if this is only partial, which frees their fatty acids. Free acidity, which is conventionally expressed as oleic acid content, must not be > 5% of fatty acid content. Another type of modification is peroxidization, or oxidation of the double links of the unsaturated fats, due to humidity and high temperatures. This destroys vitamin E, whose value should not be > 3 meq of active O_2/kg of fat. Another modification is rancidity or polymerization of the peroxides. This can greatly damage the animals' health. It is measured via the barbituric acid content, and can be prevented by adding an antioxidant, such as vitamin E, vitamin C, Se, butylidrossitoluene or BHT in doses of 1%, or by hydrogenating the polyunsaturated fatty acids (C18:2–3) to monounsaturated (C18:1). The latter are more stable but must be well emulsified, as their melting point is lower.

There are three industrial systems for fat production: (i) the mechanical system, which produces replacers that are not very water-soluble, and which and uses high temperatures (50–60°C) to reconstitute the replacer; (ii) the pneumatic system, which creates rather fine particles, then uses reconstruction temperatures of 40–45°C, similar to body temperatures; and (iii) the spray system, which creates extremely fine particles at reconstruction temperatures of 18–20°C, similar to the ambient temperature.

Lactose is the sugar which the suckling lamb utilizes most efficiently. It may come from replacer or skimmed milk, which contains 50% lactose, or from cheese whey, with 70% lactose. There are no limits to the amount of skimmed milk that can be given to the lamb, but cheese whey should not form more than 15% of traditional replacer or 50% of 'milk without milk'. The quantity of lactic acid should never exceed 2%. It is difficult to replace lactose completely because glucose, while excellent, is too expensive, and starch, although cheap, is indigestible. The products of starch hydrolysis can be used, but only gradually, and never as more than 50% of the diet and only with older lambs of 1–1.5 months. If these guidelines are not followed there may be intestinal problems such as constipation and/or diarrhoea.

The integration of the replacer is completed by the addition of fat-soluble vitamins in the following quantities:

- A, in doses of 60,000–100,000 i.u./kg.
- D, in doses of 5000–15,000 i.u./kg.
- E, in doses of 10–20 mg/kg.

Water-soluble vitamins (B_1 and B_2), minerals (Mg, Co and Cu), amino acids (lysine and methionine) and small doses of slow-acting antibiotics (virginiamycin, flavomycin, bacitramycin, and spiramycin) should be added to prevent neonatal infections such as colibacillosis.

The composition of the replacer for lambs must be such that when it is reconstituted in water at a suitable concentration it is as similar as possible to natural milk. The milk powder is generally kept in 25 kg Cellophane sacks to prevent it from becoming humid. It is composed of 3–5% water and 95–97% DM, 20–30% of the latter being fat, 20–25% protein, 35–40% sugars and 8–10% minerals and vitamins.

Feeding lambs with the replacer

Lambs are given the replacer using various methods and techniques depending on whether they are destined for breeding or being fattened for slaughter. The principal parameters to bear in mind are: concentration, quantity, feeding temperature and method of feeding.

The concentration, dilution and quantity of the replacer are always expressed in terms of kg of replacer to be diluted in a particular volume (l) or weight (kg) of water. The concentration may be expressed differently depending on whether the parameter for the water is l or kg: for example, 200 g of milk powder in 1 l or 1 kg of water would be expressed as 20% (weight of powder/volume of water) or as about 17% (weight of powder/weight of water), respectively. We suggest that, to avoid confusion, concentration should always be expressed in terms of wt/vol., i.e. kg/l.

The concentration depends on the animal's ultimate fate, but it should always satisfy its feeding requirements without exceeding its intake capacity. It is generally higher in animals which are to be fattened with milk (24%) and lower in animals which will be weaned (20%). This is the case irrespective of the final purpose of the animals: raising for breeding, or fattening for slaughter in the case of crossbreeds (Brandano and Rossi, 1971; Lanza and Galvano, 1976). Lower concentrations will not satisfy the animal's nutritional requirements and will have negative effects on its health and growth rate. Concentrations in excess of requirements may, by contrast, involve digestive problems and higher costs.

The total daily quantity of feed depends on the productive aim of the operation and on the method of feeding. The lamb should be fed a daily ration of milk equivalent to 12% BW. This requires 0.4–1.2 l/day of reconstituted replacer or 0.08–0.24 kg/day of milk powder. For the whole period the quantity of reconstituted milk required will be 40–50 l of reconstituted replacer, or 8–10 kg of milk powder, over 1.5–2 months (Brandano and Rossi, 1970; Brandano *et al.*, 1970).

The delivery temperature can be one of the following:

- The refrigerated temperature of the reconstituted replacer (3–4°C).
- The body temperature of the animal (36–38°C).
- The ambient temperature (18–20°C).

The first system requires a refrigeration plant capable of maintaining the reconstituted replacer at 3–4°C, as fermentation begins above this temperature and bacteria develop which can cause serious internal problems for the lambs. It is not used very often, because of the high conversion ratio – with resultant high replacer costs – and because a refrigeration plant is necessary. The second system means that either the replacer must be prepared and distributed immediately with water at 42–43°C (Serra *et al.*, 1988), or water heating, mixing and automatic distribution equipment must be available for the reconstituted replacer. It is, none the less, the best system physiologically, even though it is expensive and requires adequate infrastructure. The third system is cheapest (Serra *et al.*, 1988), as no particular equipment is necessary (indeed only a container and bucket are required). It is, however, the one which involves the greatest health risks, as fermentation and bacterial development can easily occur. Either it must be prepared and distributed immediately, or special replacers (acid milks) must be used which maintain their condition for at least 24–36 h. These are not very appetizing and can only be used for lambs that will be weaned and not for those being fattened for immediate slaughter, as lambs fed with this milk grow more slowly and the meat produced is poor in taste and consistency.

Feeding with the replacer can be: *ad libitum* or rationed, and continuous or interrupted. The distribution apparatus may consist of:

- A collective cylindrical container equipped with teats from which the lambs can suckle at will.
- Specifically designed individual pens.
- An automatic self-feeder fitted with a hopper which contains the replacer.
- An auger to deliver it to the mixing cups.
- A boiler to heat the water.
- A cup for mixing the replacer and water at the correct temperature and concentration.
- A tube and teat system for supplying the replacer to the animals.

9.3 Weaning

Weaning is the transition period when the animals move on from a diet of milk to one of solid feed. This may be in the form of forage such as grass, silage or hay, or of concentrates. It may also be the moment when the lamb is separated from the ewe (as in the case of natural suckling) (Lanza and Galvano, 1976), but not necessarily. In artificial suckling, the lamb is separated from the ewe at the end of the colostral phase, or, in other words, a few days after birth. Weaning may be gradual or abrupt. In the former case the transition phase from a milk-only diet to a solids-only diet of hay and concentrates, or of hay, concentrates and grass, can last 15–30 days (Rossi and Brandano, 1976). In the latter case, the adaptation period is reduced to a minimum or even eliminated.

Whichever system is used, during weaning the animal's growth rate slows down, and sometimes stops or is even reversed. As a result the animal's BW remains static or even falls. To limit this negative effect of weaning the specific guidelines and techniques described below should be used.

In a semi-extensive farming system with low levels of production, the lamb, during the suckling period, follows the ewe to pasture for the whole or part of the day. However, when it reaches a certain age (at most, 6–8 weeks), it is separated from the ewe and all her milk is milked. The lamb is weaned gradually as its intake of solid feed (almost exclusively grass) begins at a very early age (3–4 weeks). The suckling and weaning phases are normally slightly longer for female lambs destined for breeding.

In an intensive, high-producing farming system, weaning is almost always gradual as concentrates, or concentrates and hay, are almost always given to the animals during the suckling phase. The correct moment for weaning depends on the age and BW which are appropriate for that particular breed. In practice the BW is a more important criterion than the age. The correct weight for weaning is normally considered to be at least 2.5 to 3 times that at birth (9–12 kg).

During weaning the lamb changes, more or less gradually, from being monogastric, with only the true stomach (the abomasum) functional, to being polygastric – a ruminant with all four stomachs (rumen, recticulum, omasum and abomasum) functional. This change should be as gradual as possible; it involves the progressive development of the forestomachs and a relative reduction in the size of the abomasum.

The gastric apparatus is, on average, made up by volume as follows: in the first weeks of life the total volume is 0.3 l, with the abomasum forming 70–80% of this; the omasum comprises 15–20% and the rumen–reticular complex the remaining 8–10%; after weaning the rumen recticulum forms 80% of the total volume, the omasum 20%, and the abomasum only 10%.

The gastric apparatus has a capacity of 0.07–0.08 l/kg of BW in the newborn lamb, which increases to 0.2–0.25 l/kg of BW in the adult animal. Apart from this important increase in size, there is also an even more important change evident in the viscera: the relative development in size of the forestomachs and abomasum reverses as the lamb grows and the diet changes.

The development and functioning of the forestomachs principally depend on the feed used during weaning: this must be freely available to the animal and have high concentrations of energy (0.9–1 UFL/kg of DM), protein (20–22% intestinal digestible protein (PDI), of DM) and especially fibre (32–35% NDF of DM) in order to guarantee that the rumen will gradually begin to function, and to prevent intestinal disturbances.

Weaning varies according to the system of farming used. In the semi-extensive system, where natural suckling is used, it begins after 5 or 6 weeks. Until that age the lamb normally follows the ewe to pasture and takes all the milk. It continues for 2–3 weeks, during which time the lamb accompanies the ewe to pasture during the day but is separated from her at night, when the lamb is given hay and concentrates. It ends at the 8th week when the bodyweight is between 9 and 12 kg. In semi-intensive farming, where milk replacers are often used, weaning lasts about 2 weeks (the 5th and 6th). In this system either the milk replacer is gradually reduced and replaced by hay and concentrates (known as 'gradual weaning'), or supply of milk replacer is abruptly stopped and solid food supplied in the 5th week (known as 'brusque weaning'). Technically gradual weaning is preferable, and the transition from a milk-only diet to solids-only should take at least 1 week.

Only a few animals (2–3% of the males and 50% of the females) are selected for replacement. The difference in the number of males and females is due to the higher reproductive ratio of ewes with respect to rams (40:1–60:1). Those not chosen for replacement (97–98% of males and 50% of females) are slaughtered at 6 weeks as traditional suckling lambs. A certain number of lambs are fattened for at least 100 days, but more often for 6–8 months, to produce heavy lambs. In some cases these lambs are a cross between meat breed rams and dairy breed ewes (Chiofalo, 1983; Priolo *et al.*, 1997).

9.4 Fattening with Milk

Very young animals can be fattened with milk. The milk is normally replacer, because it is cheaper, but may be natural. It is nearly always given *ad libitum* to maximize the lamb's growth and to allow earlier slaughter.

Replacer is normally used for fattening lambs, because milk from dairy sheep is more valuable for cheese-making. The lamb keeps its organoleptic and commercial characteristics but can reach much greater weights than traditionally suckled lambs.

Basically, the lamb is fed *ad libitum* more (12–15% of BW/day rather than the normal 10–12%) of a replacer with a higher energy content (with a fat content of 25–30% vs. 20%) and more protein (with a protein content of 24% vs. 20%) than that of the weaner replacer. The concentrations are also higher (24% vs. 20%). The result is that intake is greater (1.5% of BW vs. 1%), as is the growth rate (200–300 g/day vs. 150–180 g/day) (Brandano and Rossi, 1971; Lucifero *et al.*, 1973; Lanza *et al.*, 1987, 1990).

However, meat must have a particular colour and taste if it is to meet customer demands and be marketable. Thus, the replacer must not alter the fundamental characteristics of the meat. The meat must be lean, and rich in proteins and unsaturated and/or SCFA. It must be very digestible, light in colour due to the lack of iron, and suitable for all types of consumer and for many culinary purposes.

Lambs fattened with milk are particularly susceptible to disease. This is because the diet is a type of force-feeding and also because lambs are kept in large numbers in confined spaces at very high stocking rates. Digestive diseases in particular are common:

- Meteorism (caused by anomalous ruminal fermentation from milk dripping from the reticular groove).
- Constipation (often caused by excessive quantities of replacer).
- Diarrhoea (caused by rapid changes in the quantity of replacer and exacerbated by sudden differences in temperature of

both the replacer and the surrounding environment).

- Viral diarrhoea (BVD).
- Enterotoxaemia and colibacillosis (both of which may be fatal).

However, there are also respiratory diseases such as bronchopneumonia (*Pasteurella* spp.), which is provoked by temperature changes, and viral influenzas, such as infectious rhinotracheitis (IBR); other diseases seen are contagious ecthyma and coccidiosis (Martin, 1986). Suitable prophylactics must be used promptly in the event of an outbreak of one of these diseases and/or adequate treatment given, if the farm is not to suffer economic damage, which may occasionally be very serious.

Fattening begins at 1–2 days after birth, and continues for 6–8 weeks until the female lambs weigh 10–12 kg and the males 14–16 kg.

The quantity of replacer necessary per head is on average 15–20 kg for 2-month-old lambs weighing 15 kg, with a replacer conversion ratio of 1.8 (Brandano and Rossi, 1971).

9.5 Replacement Ewes and Rams

Feeding young lambs is of great importance on farms which raise their own male and/or female animals to provide replacements for their flocks. The rate of replacement is the percentage of animals which must be raised in order to replace those culled because they were at the end of their productive life, diseased or for various other reasons. The replacement rate may be calculated for the whole flock, or only for adult animals. For example, in a flock of 125 sheep consisting of 100 adult and 25 young sheep, the annual replacement rate is 20% (25/125) of the whole flock or 25% (25/100) of adult females. The second method is normally used, as it is more correct.

The replacement rate varies from breed to breed and the type of farming system employed. Basically, it is inversely related to the average number of births in a productive lifespan. This can be seen by studying the relationship between the length of the productive career and the intervals between lambings. The length of the productive career is given by subtracting the age at first lambing from the age at last lactation. The lambing interval is the average time between each lambing. For example, if the productive career ends at 6 years of age, the first lambing is at 2 and the average interval between births is 1 year, the annual replacement rate is 25% ((6–2)/1= 4.0; ¼=25%)). In this hypothesis, this is the obligatory replacement rate, i.e. the number of animals which must be raised each year if the number of animals in the flock is to remain constant. To maintain flock numbers, the farmer cannot let numbers drop below this rate, but may exceed it if so desired, if the intention is to increase stocking rates for any technical or economic reason. This is termed the facultative replacement rate. The animals which have not yet given birth make up the young, or unproductive, part of the flock; the animals from first lambing to culling make up the adult, or productive, part of the flock. Obviously, the male replacement rate is much lower than the female, both because of the different reproductive ratio between the sexes (1:40 or 1:60 in the case of natural insemination) and because the respective lengths of their reproductive careers are different, with that of the ewes being longer.

Replacement techniques are very important for the successive productive and/or reproductive phases of the flock, as the success or failure of the farm depends on the length of the unproductive life. Accurate replacement feeding must guarantee that the animals are well developed, fertile, have a long lifespan, and are productive and healthy.

The most important aspects to bear in mind are diet, reproduction and the farming method employed.

The diet should contain sufficient fibre (30–32% of NDF). This is because development of the digestive system, and in particular the rumen and the cellulosolytic flora, which is essential if it is to function correctly, is closely linked to the concentration of fibre in the diet. The diet should contain enough energy (0.7–0.8 UFL/kg of DM). Lack of energy in the diet, which is often linked to excess fibre, may reduce the growth rate and balanced development of the maiden ewe. As a result, the first heat may be delayed, and consequently the first birth. In addition the body may be underdeveloped and the animal thus underweight. Excess energy in the diet, which is often the result of excessive use of concentrates, may result in the animal being too fat. This can harm the animal and have adverse effects on its reproductive life and on ovarian function, with excessive deposits of fat and difficulties at lambing time. The diet should also contain the right quantity of protein (18–20% CP of DM). Lack of protein reduces the growth rate. Young animals need more protein in the supplement if the forage consumed is of poor quality, but a reduced amount of concentrate.

In practice, the lambs should be divided into three groups, ideally homogeneous in BW and age, from weaning to first parturition:

- From 1.5–2 months to 3–4 months, fed on hay and concentrates.
- From 3–4 months to 10 months, fed on pasture grass and concentrates.
- From 10 months to first parturition, fed on hay, pasture grass and concentrates.

Concentrates should provide 60–65% of the total energy in the diet for the first group, 45–50% for the second and 20–25% for the third. Silage should be given only to the second and third groups, and never in quantities greater than 0.5–0.7 kg/head per day. Hay should be given to all the groups, as should pasture grass if possible, but above all to the third group, because it has a highly positive effect on its development.

Reproductive efficiency is principally linked to first mating and thus to first parturition. Although the first oestrus occurs quite early – at 6–8 months – it is not advisable to inseminate the animal before it has reached a certain age (10–12 months), or, better still, a certain BW (60–70% of that of a typical adult of the same breed), or a certain stature.

Very early parturition almost always results in:

- Lower milk production in the first lactation, which often does not improve in subsequent lactations.
- Greater difficulties at parturition, with the placenta often remaining in the uterus.
- Cessation of bodily development.

Delayed parturition results in a lower number of lambings in the reproductive life of the animal and a lower and more expensive replacement rate.

Correct management of the replacement involves raising the animals at pasture, with shelter being used only in the extreme seasons of winter and summer, at night and/or on very hot days. This is because spontaneous mobility (functional gymnastics), and in particular grazing, is beneficial for all the animals, and above all those still growing. Particular attention should be paid to shelter, feed and space available to the animals during the last 4–6 weeks of pregnancy.

Recent studies carried out on ewe lambs between weaning and puberty have shown the role that nutrition plays in the advance of puberty and in mammary gland development. This has notable economic importance because it correlates strongly with both unproductive life and milk production of the ewe lambs. Since puberty and the age at first mating are related to the attainment of sufficient body weight, which of course varies depending on the breed (Drymundsson, 1973; Martemucci *et al.*, 1980), we can accelerate these stages by adopting suitable feeding regimes during growth. Undernutrition during the pre-weaning period (up to the 15th week) (Rhind *et al.*, 1988) results in lower prolificacy, due probably to the lower ovulation rate and/or higher embryonic losses. In contrast, concentrate supplementation after weaning (300–500 g/day) given to ewe lambs fed on pasture reduces the age at first lambing by 1–2 months, and increases their fertility and prolificacy (Quirke, 1979; Secchiari *et al.*, 1987; Kassem *et al.*, 1989; Forcada *et al.*, 1991; Pace *et al.*, 2000).

Table 9.2. Effects of age and feeding level on mean blood concentrations of growth hormone and prolactin (Johnsson *et al.*, 1985).

Hormone	Feeding level	Age (weeks) 10	14	18
Growth hormone (μg/l)	High	1.93	1.30	1.20
	Low	3.67	1.63	1.50
Prolactin (μg/l)	High	137.0	223.6	273.1
	Low	85.6	102.5	108.9

In Mediterranean dairy sheep breeding systems, the first mating is possible at 6–7 months of age (i.e. in June) if oestrus synchronization treatments are used. These should reduce the negative effects of the long photoperiod which is typical of this season. Cannas and Rassu (unpublished) synchronized 6-month-old Sarda ewe lambs of BW 35 kg (70% of mature bodyweight) and obtained the first lambing at 12 months of age in 52% of cases.

Even if high-level feeding regimes during the pre-pubertal period are capable of advancing puberty because the animals grow more quickly, they have a negative effect on mammary gland development, and consequently on milk production during first lactation. Anderson (1975) studied mammary gland development in sheep by analysing the DNA content of the dried fat-free tissue, and observed that the growth rate was average (20%) from 3–5 months after birth, high (78%) during pregnancy, particularly in the final stage, and low (2%) during the first days of lactation. Johnsson and Hart (1985) observed that from 4 to 20 weeks after birth the parenchymal mammary growth rate is 2.4–3.7 times higher than the bodyweight growth rate, particularly in female lambs with average (100 g/day) rather than high (220 g/day) daily growth. This could be due to the positive effect of growth hormone on mammogenesis. Between 10 and 18 weeks of age plasma concentration of growth hormone is higher and the prolactin plasma level is lower (Table 9.2) in ewe lambs growing at the rate of 110 g/day and fed on a low-energy/low-protein diet than in female lambs growing at 220 g/day and fed on a high-energy/high-protein diet (Johnsson *et al.*, 1985).

Table 9.3. Feeding regime, growth rate and milk production in Sarda ewe lambs.

Parameters		LCP	MCP	HCP
Food				
Crude protein	(% kg DM)	15.16	16.58	17.96
ME	(Mcal/kg DM)	2.55	2.56	2.50
CP/ME	(g/Mcal)	59.3	64.8	71.7
Intake[a]	(% BW)	5.67	5.50	5.55
Growth	(g/day)			
4th month		157	180	163
5th month		110	123	158
6th month		46	37	57
Average		104	113	126
Milk production[b]		1062[c]	1099[c]	922[d]

LCP, low crude protein content; MCP, medium crude protein content; HCP, high crude protein content; DM, dry matter; BW, bodyweight.

[a] Average 4th, 5th and 6th months of age; [b] average first 2 months of lactation. [c,d] Significant difference between means at $P < 0.05$.

This has significant effects on the milk production of ewe lambs at first lactation. Umberger *et al.* (1985) studying Suffolk and Dorset cross ewe lambs, and Ayadi *et al.* (2003), studying Manchega ewe lambs, found that milk production was higher in female lambs with lower daily growth between weaning and puberty. Cannas and Rassu (unpublished) produced similar results when studying Sarda ewe lambs fed on isoenergetic diets with varying protein content (Table 9.3).

On the basis of results of previous studies carried out both on ewes and on dairy heifers on the interaction between growth and mammary gland development, Tolman and McKusic (2001) suggest that the optimal growth rate during pre-puberty should be between 65 and 75% of the maximum growth rate for the breed.

References

Anderson R.R. (1975) Mammary gland growth in sheep. *J. Anim. Sci.*, 41(1): 118–123.

Ayadi M., Caja G., Such X., Ghirardi J. (2003) Feeding level before puberty affects milk production during first lactation in Lacaune and Manchega ewes. *54th Annual Meeting EAAP*, Rome, Italy, p. 345.

Brandano P., Rossi G. (1970) L'allevamento artificiale degli agnelli con l'impiego di sostituti del latte e di una allevatrice meccanica. IV La produzione dell'agnellone leggero. *Alim. Anim.*, 14(1): 25–29.

Brandano P., Rossi G. (1971) La produzione dell'agnello pesante da latte. *Studi Sassaresi, Ann. Fac. Agr. Univ. Sassari*, 19: 3–11.

Brandano P., Rossi G., Lai P., Cosseddu M.C. (1970) Prova comparativa fra agnelli di razza Sarda e meticci Württemberg x Sarda, allevati con la somministrazione meccanizzata di latte ricostituito. 1. Accrescimenti, consumi alimentari e rese alla macellazione. *Alim. Anim.*, 14(4): 49–55.

Casoli C., Duranti E., Morbidini L., Panella F. (1987) Variazione della composizione chimica del colostro di pecore massesi nelle prime ore dopo il parto. *Il Latte*, 11: 1046–1050.

Chiofalo L. (1983) Prova di incrocio fra ariete Bergamasco e pecora Siciliana. Produzione dell'agnellone leggero. *Proc. 5th Congress ASPA*, Gargnano del Garda, Italy, pp. 357–364.

Dyrmundsson O.R. (1973) Puberty and early reproductive performance in sheep. I. Ewe lambs. *Anim. Breed. Abstr.*, 41: 273–289.

Eales A., Small J. (1986) *Practical lambing. A guide to veterinary care at lambing.* Longman, London.

Forcada F., Abecia J.A., Zarazaga L. (1991) A note on attainment of puberty of September-born early maturing ewe lambs in relation to level of nutrition. *Anim. Prod.*, 3: 407–409.

Johnsson I.D., Hart I.C. (1985) Pre-pubertal mammogenesis in the sheep. 1. The effects of level of nutrition on growth and mammary development in female lambs. *Anim. Prod.*, 41: 323–332.

Johnsson I.D., Hart I.C., Simmonds A.D., Morant S.V. (1985) Pre-pubertal mammogenesis in the sheep. 2. The effects of level of nutrition on the plasma concentrations of growth hormone, insulin and prolactin at various ages in female lambs and their relationship with mammary development. *Anim. Prod.*, 41: 333–340.

Kassem R., Owen J.B., Fadel I. (1989) The effect of pre-mating nutrition and exposure to the presence of rams on the onset of puberty in Awassi ewe lambs under semi-arid conditions. *Anim. Prod.*, 48: 393–397.

Lanza A., Galvano G. (1976) Svezzamento brusco a 28 e 35 giorni di agnelli di razza Comisana in allattamento naturale con miscele a diverso livello proteico. *Proc. 2nd Congress ASPA*, Bari, Italy, pp. 125–145.

Lanza A., Lanza E., Pennisi P., Aleo C., Licitra G. (1987) Produzione dell'agnello pesante con latte ricostituito a concentrazione crescente. *Zoot. Nutr. Anim.* 13: 499–509.

Lanza A., Pennisi P., Biondi L., Barresi S. (1990) Produzione dell'agnello pesante da latte alimentato con un sostitutivo a diverse concentrazioni. *Proc. 9th Congress SIPAOC*, Grado, Italy, p. 6.6.

Lucifero M., Rossi G., Brandano P., Dattilo M., Congiu F. (1973) Il momento ottimale di macellazione degli agnelli di razza Sarda allevati con la somministrazione di latte ricostituito. *Alim. Anim.*, 17(4): 19–33.

Martemucci G., Bellitti E., Todeda F., Manchisi A. (1980) Studio delle prime manifestazioni estrali delle agnelle in relazione al tipo genetico. *Ann. Fac. Agr. Univ. Bari*, 31: 185–200.

Martin W.B. (1986) In: V. Cilli (ed.) *Malattie della pecora.* Soc. Ed. Esculapio, Bologna, Italy.

Naitana S., Nuvole P., Marongiu A. (1992) Lattazione. In: C. Aguggini, V. Beghelli, L.F. Giulio (eds) *Fisiologia degli animali domestici con elementi di etologia.* UTET, Turin, Italy.

Pace V., Settineri D., Rassu S.P.G. (2000) Trifoglio sotterraneo sulla crescita e la riproduzione di agnelle di razza Sarda. Nota I. *35th Proc. Int. Symp. SIPZOO: Produzioni Animali di qualità e impatto ambientale nel sistema mediterraneo,* Ragusa, Italy, pp. 257–265.

Piccioni M. (1989) *Dizionario deglialimenti per il bestiame.* Edagricole, Bologna, Italy.

Priolo A., Lanza A., Biondi L. (1997) Effetti dell'incrocio Suffolk x Comisana. 2 Produzione dell'agnello di 100 giorni. *Proc. 12th Congress ASPA,* Pisa, Italy, pp. 227–228.

Quirke J.F. (1979) Effect of body weight on the attainment of puberty and reproductive performances of Galway and Finn x Galway female lambs. *Anim. Prod.,* 28: 297–307.

Rhind S.M., Elston D.A., Jones J.R., Rees M.E., McMillen S.R., Gunn R.G. (1988) Effects of restriction of growth and development of Brecon Cheviot ewe lambs on subsequent lifetime reproductive performance. *Small Rum. Res.,* 30: 121–126.

Rossi G., Brandano P. (1976) Prova di svezzamento di ovini di razza Sarda a diverse età. *Proc. 2nd Congress ASPA,* Bari, Italy, pp. 157–164.

Secchiari P., Trimarchi G., Ferruzzi G., Martini A., Pistoia A., Berni P., Luisi M. (1987) Effetti dell'apporto nutritivo della razione sull'inizio della carriera riproduttiva di agnelle Massesi. *Zoot. Nutr. Anim.,* 13: 223.

Serra A., Cannas A., Pulina G., Rossi G., Brandano P. (1988) Prova preliminare di impiego di un succedaneo acido del latte nell'allattamento artificiale degli agnelli. *Proc. 8th Congress SIPAOC,* Viterbo, Italy, pp. 151–161.

Tolman B., Mc KusicK B.C. (2001) The effect of growth rate on mammary gland development in ewe lambs: a review. *Proc. 7th Great Lakes Dairy Sheep Symposium,* Eau Claire, Wisconsin, pp.143–155.

Umberger S.H., Goode L., Caruolo E.V., Harvey R.W., Britt J.H., Linnerus A.C. (1985) Effects of accelerated growth during rearing on reproduction and lactation in ewes lambing at 13 to 15 months of age. *Theriogenology* 23: 555–564.

10 Digestive Disturbances and Metabolic–Nutritional Disorders

Massimo Morgante
Dipartimento di Scienze Cliniche Veterinarie, Università di Padova, Italy

10.1 Introduction

The way animals are fed in general, and the individual nutritional content of each ration, may affect both the digestive and metabolic processes, and as a result may harm the animal's health and its productive capacity. Changes in animal feeding often lead to disorders of the digestive system, which in turn influence the absorption of individual nutrients and often trigger, or predispose the animal to attack from, various other diseases. Alterations in the absorption of nutrients are evidenced by breakdown in internal homeostasis, whereby one or more critically important metabolic processes are modified. There are several examples of this, especially in sheep: e.g. hypocalcaemia, grass tetany and pregnancy toxaemia, which, although very different, are all the result of metabolic disturbance.

In the first of these, calcium (Ca) metabolism is altered, while in the case of grass tetany, magnesium (Mg) metabolism is modified, and in pregnancy toxaemia the affected metabolism is that of energy, and in particular of glucose. In ruminants, where the placenta is almost totally maternal-immunoglobulin-impermeable, even the lack of colostrum may constitute a case of nutritional disease. In fact, colostrum is not only the prime source of antibodies guaranteeing passive, circulating and local immunity against the most common infections, but also constitutes an important source of minerals and vitamins that normally exist in lower concentrations in milk (Andrews, 1990).

Before dealing with individual digestive and metabolic disturbances, we will make a number of points about how the livestock farmer, and more importantly the veterinary surgeon, should approach such problems. Firstly, in the case of sheep farming they should consider the flock as a whole, rather than dwelling on isolated cases, since the latter need to be perceived as a warning signal that may indicate that the entire flock is in the same situation, and could therefore be at risk from a given disease. This is why the detection of change is extremely important, as it enables precautionary measures to be taken with regard to the rest of the flock. In sheep farming large-scale therapeutic treatment is in fact not only extremely expensive but often impractical, and in the case of nutritional and metabolic diseases also frequently ineffective once the symptoms have appeared.

Finally, nutritional and metabolic imbalances often arise without any typical symptoms, but accompanied by a series of disturbances common to other diseases. There is a correlation here between states of deficiency and infectious and parasitic diseases. In fact, feeding represents one of the factors that leads to so-called 'conditioned diseases'. One example of the latter is clostridial infection, for which nutritional imbalance constitutes an important pre-condition. There are well-known links between an animal's nutritional status and its immune response to certain diseases caused by pathogenic agents, such as the relationship between the incidence of mastitis and the state of certain micronutrients, or that between parasitism and the protein content of the ration (Morgante and Ranucci, 1997). In the light of this, it is clear that constant surveillance of the flock is of considerable importance in establishing the degree of risk to which animals are exposed with regard to metabolic and nutritional imbalances. The present chapter focuses on the methods used to identify any nutritional imbalances or deficiencies, in order to be able to take corrective and preventative measures designed to improve the health of the livestock.

10.2 Disturbances of the Digestive System

10.2.1 General features

Disturbances of the digestive system are often the result of dietary errors resulting from continual changes in the kind of feed given to the animals, or to nutritional deficits or excesses. Such imbalances are caused mainly by the need to utilize the pasture and conserved foods according to the changing seasons, and to supplement the animal's feed with concentrated foods in relation to the productive cycle. They include problems such as bloat, ruminal acidosis and alkalosis, which in their more serious forms are characterized by clearly defined symptoms, or by lesser disturbances such as malfunction of the forestomach or of the intestinal tract, caused by alterations in ruminal fermentation or intestinal microflora, due to the presence of stimulating or inhibiting substances.

Before looking into this matter further, certain physiological particulars of the sheep's digestion, which differentiate it from that of cattle or other ruminants, will be described, since the total volume of the forestomach in relation to the animal's weight is lower, whereas the efficiency of chewing appears to be higher, when compared with the equivalent features of larger ruminants (NRC, 1985; INRA, 1988). The average time taken for solids to pass through the ovine digestive system is about 20 h, compared with 28 h in the bovine. In fact, while cattle appear better able to digest roughage, sheep are more capable of utilizing starches, and they also possess a greater intake capacity (as % DM ingested in proportion to BW). The ratio of the time the food remains in the ovine forestomach as opposed to in the intestine is inverted in the bovine (40% in the forestomach and 60% in the intestine). Therefore, intestinal disturbances of sheep could be more important than those in cattle, in that the forestomach, in certain conditions, fails to perform properly as a filter. On the other hand, the opposite could also be true: i.e. in sheep, disturbances of the forestomach could be less important.

Nevertheless, the fact remains that sheep, when compared with cattle, are less often affected by diseases of the forestomach, e.g. bloat, simple forms of indigestion, and serious forms of ruminal acidosis and alkalosis, whereas they are more often affected by enteric diseases caused by pathogens (*Clostridium*). It is not so much the full-blown, often fatal diseases that constitute the greatest threat to the productivity of the flock: the most insidious, dangerous disturbances are, on the contrary, those with no clearly defined symptoms, which thus prove difficult to diagnose, and as a result are the ones that underlie numerous other pathological conditions. Disturbances of the fermentative and digestive functions of the forestomach and intestine may lie at the root of poor utilization of nutritional components of the animal's feed (energy, proteins, vitamins, micro- and macro-elements), and of the production of toxic or inhibiting substances that may affect the functioning of other organs and systems (Bertoni, 1992).

In terms of the flock's productivity and profitability, mention ought to be made of a number of secondary diseases, e.g. mastitis, laminitis or those diseases which affect the fertility of the animal. However we still know very little about such diseases, especially those affecting sheep. As a result, in the majority of cases one can only hypothesize

Table 10.1. Disturbances of the digestive system in dairy sheep.

Disturbance	Causes	Clinical signs	Treatment and prevention
Simple indigestion	Sudden change-over to a carbohydrate-rich diet that is poor in fibre and/or high concentrations of soluble proteins and non-protein nitrogen, which may interfere with ruminal fermentation and alter the ruminal flora	•Slight degree of anorexia •Grey-brown diarrhoea with the presence of undigested grains, while in the case of animals that have been left to pasture, it is of a greenish-brown colour	•Avoid any sudden changes in feed and include adequate fibre in the diet •Grazing should be gradually increased •Use of ruminal buffers
Acute ruminal acidosis	Excessive consumption of easily fermentable carbohydrates, which cause rapid fermentation and production of lactic acid, together with a considerable reduction in ruminal pH	•Ruminal stasis and distension •Anorexia •Dehydration •Severe diarrhoea (light-coloured with strong acid smell) •Animals seem to be drunk or blind •Low ruminal fluid pH (< 5)	•Ruminal wash out and transfaunation •Correction of dehydration, electrolyte and acid–base imbalance •Avoid a sudden and large amount of highly fermentable carbohydrates
Subacute ruminal acidosis	The same causes as the acute form but less severe and more prolonged	•Slight anorexia •Intermittent diarrhoea •Lameness •Mastitis •Poor body condition	•Remove the offending food •Increase the fibre in the diet (by providing high-quality hay) •Use of ruminal buffers
Primary tympany	Grazing of pasture consisting of bloat-producing plants (lush pasture of leguminous plants)	•Distension of ruminal wall •Restlessness •Dyspnoea •Hypersalivation •Tachycardia	•Oral administration of anti-frothing substances (e.g. mineral oil)
Intestinal disorders	Dietary problems Forestomach disorders Intestinal fermentation	•Diarrhoea •Soft faecal blocks •Undigested grain in the faeces	•Reduction of forestomach disorders •Use of intestinal buffers (magnesium oxide) •Grazing should be gradually increased

about the existence of a correlation between food imbalances and the presence of certain pathological disturbances.

10.2.2 Disturbances of the forestomach

General features

The rumen is host to a continuous system of bacterial culture whereby those microorganisms best adapted to a wide range of food types are selected. There are numerous systems for controlling the rumen and its bacterial population: some are associated with the animal itself, such as salivation, re-mixing, rumination and absorption and elimination of substances, as well as voiding and eructation; others are linked to the characteristics of the feed given to the animal, such as the nutritional balance required by different forms of fermentation, the solubility of the nutrients, the size of food particles, the presence of inhibiting substances, the speed with which the rumen is filled, and finally, other characteristics linked to microorganisms, such as competition for nutrients among species, the removal of inhibitory waste products, and maintenance of redox potential (Flint, 1997; Santra *et al.*, 1998).

The only method we have of controlling ruminal fermentation is that of carefully choosing what and how we feed the animals in question. Forestomach disturbances can be basically subdivided into the following types, according to the nature (both physical and chemical) and quantity of the feed given to the animals:

- Simple indigestion.
- Acidic indigestion (ruminal acidosis).
- Primary, or frothy, bloat (tympany).

Simple indigestion

Carbohydrates and proteins are the most important factors affecting the development of fermentation. The most common cause of simple indigestion is a sudden transition to a carbohydrate-rich diet consisting of easily fermentable (concentrated) carbohydrates that have low fibre and/or high concentrations of soluble proteins and non-protein nitrogen (lush pasture), which together may interfere with fermentation and alter the ruminal flora. In the first case, where carbohydrates prevail, there is a prevalence of microbial species capable of rapid metabolism and which are tolerant of a rather low pH. This in certain situations leads to a reduction in, or even the disappearance of, certain Protozoa.

In the second case, where there is also a considerable quantity of soluble proteins or non-protein nitrogen, there may be a prevalence of microorganisms with high ureasic activity, with a resulting accumulation of ammonia (NH_3) in the rumen and the alkalizing of its contents (ruminal alkalosis), although such an event is fairly rare in sheep. Any NH_3 that escapes ruminal fermentation may enter the circulation and reach the liver, where it is transformed into urea, causing an increase in the level of the latter in the circulation – uraemia (> 0.6 mg/ml). Simple indigestion is not accompanied by any pathological variation in ruminal pH (normally 5.6–7.0).

Fermentation of carbohydrates results in the production of volatile fatty acids (VFA): the latter, if produced in large quantities, together with lactic acid (which lowers pH), can be found in an undissociated form in the rumen, where they stimulate the sensorial receptors of the ruminal epithelium, which in turn reacts by inhibiting its contractility (Smith, 1996). In such cases, this mechanism ensures a reduction in fermentation and the recovery of the organ's contractility within a few days.

The above-mentioned types of feed reduce chewing activity, and thus salivation (reducing buffer potential in the rumen), and above all prevent the stratification of the feed, which as a result travels faster. Moreover, the accumulation of osmotically active substances, first in the rumen and then in the intestine, can attract water and thus cause osmotic diarrhoea. The only symptom, apart from a slight degree of anorexia, is diarrhoea, which, in the case of concentrated feed (grain), is of a grey-brown colour and contains undigested grains, while in the case of animals that have been left at pasture, it is of a greenish-brown colour.

The prevention of these disorders, which if left for any time may develop into more serious disturbances, with adverse effects on the function of other organs and systems, must involve avoidance of any sudden changes in the type of feed; the diet must always contain adequate fibre – preferably of the long variety – with forage making up no less than 20–40% of DM. The quantity of concentrate, which should be fed twice daily, will vary according to the productive period and to the condition of the pasture, and must be given in increasing quantities starting, for example, from the second stage of the pregnancy. The concentrate must also contain a certain quantity of ruminal buffers such as sodium bicarbonate.

The optimal fermentation of carbohydrates also requires the presence of nitrogen, sulphur and essential micro- and macro-elements, while protein content also seems to play an essential part in fermentation within the rumen. Animals should be fed on suitable amounts of concentrate (low soluble protein content but rich in glucogenetic substrates and fibre) immediately prior to grazing. Even in this case, however, the animals need to be provided with forage during the time they spend in the feeding areas, and the time they spend grazing should be gradually increased.

Ruminal acidosis

Ruminal acidosis is one of the most serious fermentative disorders, and is often fatal within 24 h. This condition has been called lactic acidosis, rumen carbohydrate overload, acid indigestion, toxic indigestion, cereal overload and D-lactic acidosis. It may be sub-divided into three forms, depending on its severity and course: acute, sub-acute and chronic.

ACUTE RUMINAL ACIDOSIS. This results from the excessive consumption of easily fermentable carbohydrates, which cause rapid fermentation and the production of lactic acid, together with a considerable reduction in ruminal pH. This happens when the animal eats an excessive quantity of food containing a large quantity of starches or sugars in a short time, as the latter, in turn, may act as precursors of lactic acid if the animals have not been gradually introduced to this diet (cereals, green fodder silage, fruit and roots) beforehand. Certain characteristics of the food accelerate this phenomenon, e.g. a concurrent lack of structured fibre leads to reduced salivation, and thus to a reduction in the ingestion of salivary buffers. Some fodder silage may contain large quantities of carbohydrates and lactic acid. Wheat, barley and maize grains are considered highly toxic if ingested in large quantities, whereas oats and sorghum are less so. Cereals are more toxic if finely milled, crushed or broken up, as this exposes fermentable carbohydrates more directly to the action of the bacterial flora.

In sheep breeding, this disease is often the result of accidental circumstances or of serious management errors, such as the releasing of animals into fields of stubble after harvesting, where there are going to be considerable amounts of cereal or flattened ears on the ground.

An excess of carbohydrate encourages the growth of lactic-acid-producing bacteria, the level of which, if excessive, leads to a rise in the production of lactic acid and other products, which, in turn, increase the acidity of the rumen, killing off or inhibiting other microorganisms, thus causing forestomach dysfunction and metabolic changes in the animal.

The degree of ruminal acidosis and the gravity of the symptoms vary considerably, depending on the quantity and type of feed, and the degree to which the ruminal flora had adapted. If the ingestion of carbohydrates is excessive, the acidosis will increasingly worsen; continual production of lactate lowers the pH of ruminal fluid to 5–5.5, while ruminal osmolality rises. Both factors kill Protozoa in the rumen, together with other lactic-acid-using microorganisms, while lactobacilli constitute the largest group of lactic-acid-producing bacteria in the rumen.

These changes in ruminal fluid have numerous harmful effects on the animal. During the early stages of acid fermentation, large quantities of VFA are produced, and as these acids are weaker than lactic acid, they link up with hydrogen ions and act as buffers, with the result that they are more quickly absorbed. An excessive absorption of VFA leads to systemic acidosis. Moreover, a large quantity of non-disassociated VFA in the mucus of the rumen walls leads to the inhibition of forestomach contractility, with ruminal stasis.

The osmotic pressure of ruminal fluid increases in relation to the quantity of lactic acid: the normal osmolality of ruminal fluid is about 280 mOsm/l, but this may double in the case of acute acidosis. An increase in this osmolality draws water from the vascular system, leading to an increase and dilution of ruminal contents, with distension of the organ (hydrorumen) and a significant, clinically observable degree of dehydration.

The loss of organic liquids causes a reduction in blood volume and a decrease in peripheral perfusion, which causes hypoxic metabolism: the latter aggravates the degree of systemic acidosis caused by the absorption of lactic acid. Furthermore, the latter is highly corrosive, and when it reaches a certain concentration in the rumen it may destroy the ruminal epithelium, causing so-called 'toxic rumenitis'. This disease may be so serious as to persist in those animals which have recovered from acute acidosis, only for them to die subsequently from ruminal damage. Even if the rumen recovers, bacterial abscesses may release bacterial emboli, which can enter the circulation, leading to the formation of metastatic hepatic abscesses.

Numerous other toxic factors in addition to lactic acid are implicated in this condition. The altered metabolism of the ruminal flora may produce large quantities of histamine and other substances, which in turn may play a part in the pathogenesis of the disease. Histamine could be involved in the pathogenesis of aseptic laminitis, which occasionally accompanies ruminal acidosis (Smith, 1996). The destruction of Gram-negative ruminal bacteria has been thought to be the cause of the freeing and reabsorption of a considerable quantity of endotoxins through the damaged mucous membrane. These endotoxins could be implicated in the majority of the symptoms, such as ruminal stasis, poor tissue perfusion, heart failure, weakness and depression. The absorption of toxins may be facilitated by hepatic dysfunction (e.g. subsequent to concomitant parasitism), or to a damaged ruminal mucous membrane.

A number of sheep may be affected at the same time in diseased flocks, depending on the nature of the diet and the way they are fed. The most highly susceptible animals appear to be lactating ewes, while heavily pregnant ewes and ewe-lambs (perhaps due to their reduced ingestion capacity) seem to be less susceptible. The rate of mortality may be as high as 90% in untreated cases, and even 30–50% in those treated.

In the acute form, the first symptom is distension of the rumen and painful abdominal spasms accompanied by total anorexia; ruminal contractions decrease and the animal stops ruminating. Certain sheep seem to be drunk or partially blind; heart rate is generally high, and the majority of animals suffer from profuse diarrhoea; their faeces are light-coloured and have a strong acidic smell. If they have eaten excessive amounts of whole grain, these faeces may contain undigested grains. They gradually exhibit signs of dehydration, and in cases where there has been impairment of renal function, anuria is also evident. After 24–48 h, the animal will be forced to assume a recumbent posture: this may also be a result of the onset of acute laminitis.

Laboratory tests may be of considerable help in the diagnosis of this condition: the haemoconcentration indicates the extent of dehydration; in serious cases, the haematocrit may reach 50%; lactate levels may confirm lactic acidosis, and the urine will be acidic and concentrated. From a practical point of view, however, the simplest way to obtain diagnostic confirmation is to measure the pH of the ruminal fluid, a sample of which may be taken either using an oesophageal probe or by directly puncturing the rumen (rumenocentesis). It must be measured immediately, since pH is raised by exposure to air. A pH of < 5 is a sign of acute ruminal acidosis. Immediate microscopic examination of a few drops of fluid will show whether Protozoa are present or not (Patra *et al.*, 1996). The animal usually dies within 48 h, but those few which survive usually die shortly afterwards, from rumenitis and infection, while pregnant ewes may abort.

The main problem with treatment is that of the number of animals affected at the same time. Treatment of acute ruminal acidosis is based upon the following:

- Correction of both the ruminal and systemic acidoses.
- Prevention of production of further lactic acid.
- Replacement of electrolytic fluid losses and maintenance of blood volume.
- Restoration of forestomach and intestinal functions.

During the initial phase and in less serious cases, the food responsible for the problem must be removed and the animal fed high-quality hay instead. The more serious cases require special forms of treatment. The best measure to adopt, especially if a considerable number of animals are affected, is to introduce water into the rumen by oesophageal intubation, in order to wash it out and completely empty it. When this operation has been completed, fresh ruminal liquid taken from healthy sheep may then be introduced into the rumen. Intravenous solutions should be administered in the case of systemic metabolic acidosis and electrolytic imbalances. Buffer substances, such as magnesium oxide, may also be administered via intubation.

Prevention is based upon the dietary management of the flock – designed to prevent circumstances that may trigger the onset of the disease. When the animals are to be fed a cereal-rich diet, this must be introduced gradually, as they are first led out to pasture after a period indoors. They must always be fed hay, even at the beginning of this transitional period: the hay should be fed before the concentrate, and should always be available to the sheep. Buffers may be introduced into the feed, particularly when the latter is given as unifeed. These substances stabilize the pH, which results in the reduced growth of acid-producing microorganisms. Buffers generally include sodium bicarbonate, sodium bentonite, magnesium oxide and calcium/magnesium carbonate.

SUBACUTE AND CHRONIC RUMINAL ACIDOSIS. As in acute ruminal acidosis, these problems are caused by a carbohydrate-rich diet low in structured fibres, or may also arise when badly prepared unifeed is fed to the animals. In such cases, the problem is caused by the continuous ingestion of these foods, which in the chronic form may cause the ruminal flora to adapt to such foods, while in the subacute form it may give rise to less serious disturbances, with alterations that are not so serious as to lead to clearly evident symptoms, whereas the chronic variety often exhibits no symptoms at all.

The subacute form can be seen in animals that have just commenced lactation, where the ration contains too high a level of carbohydrates and/or the animals are put out to pasture in lush fields of grass without having been given adequate time to adapt. Symptoms seen include:

- Simple indigestion, lasting several days.
- Aseptic laminitis.
- Subclinical mastitis – detectable by an increase in the SCN (somatic cell number) of bulk milk.
- Excessive weight loss (BCS < recommended level).
- Poor milk production (relative to energy content of diet).

Other diseases with different causes may affect sheep suffering from ruminal acidosis as a result of their greater predisposition (e.g. enterotoxaemia, listeriosis, footrot). Diagnosis of the subacute form is far from easy: in fact, there are no pathognomonic signs, or any reliable tests that can be carried out. In theory, measurement of the ruminal pH of a certain number of sheep may be a valid diagnostic aid. The chronic form affects mainly weaned lambs (see Section 10.4.3).

Primary tympany (bloat)

Acute primary tympany, given the high mortality rate associated with it, can cause serious financial loss when it affects numerous animals. Tympany resulting from grazing constitutes a seasonal occurrence associated with the grazing of sheep on lush pastures. Cases of primary tympany are fairly rare in sheep, and are associated with poor husbandry or accidental situations. If sheep are suitably adapted to pasture consisting of bloat-inducing plants, they are generally unaffected by this kind of problem, although other factors do play a certain role here, such as the predisposition of the breed, the type of food supplements the animals are fed, climatic conditions, etc.

The cause of primary tympany is the formation of froth stabilizer in the rumen, which imprisons the gases that are produced by fermentation, and impedes eructation as it inhibits the fusion of air bubbles, and thus induces the accumulation of a certain quantity of free gas in the pyloric zone: the latter is an essential prerequisite of the eructation reflex. Plant lipids and salivary mucoproteins act as anti-frothing agents, causing a reduction in the superficial tension of the ruminal contents.

From the pathogenic point of view, the speed with which the frothing substances are ingested, and thus the likelihood the said mechanisms have of preventing frothing, is of considerable importance: the formation of froth is aided by diets containing frothing substances which do not stimulate salivation, which alter ruminal pH and fermentation, and which encourage the multiplication of encapsulated bacteria that produce polysaccharides. This mainly occurs when the sheep are moved to pastures of leguminous plants, which contain high levels of frothing agents – cytoplasmic proteins present in the leaves (Smith, 1996).

Symptoms of primary tympany generally emerge as little as 15 min after the bloat-inducing food has been eaten. The wall of the rumen suddenly swells, particularly around the fossa in the left flank, although the whole of the abdomen will appear enlarged. The animal becomes nervous, showing the following increasingly distressed signs:

- Dyspnoea (tachypnoea, breathing with open mouth, outstretched neck).
- Hypersalivation.
- Severe tachycardia.

The course of the disease is short (3–4 h), after which the animal dies. Ruminal probe or puncture will reveal the presence of copious frothy content.

Treatment, based on the oral administration of anti-frothing substances (e.g. mineral oil) capable of reducing the superficial tension of the ruminal contents, is often ineffective when a large number of animals are affected and symptoms are serious, given the short course of the disease. Prevention involves avoiding grazing sheep on lush pastures of leguminous plants at certain times of the year, unless they have been prepared beforehand and their diet is supplemented with fibrous forage.

10.2.3 Intestinal disorders

Diarrhoea is the chief symptom of intestinal disorders caused by dietary changes, even though it is present in every primary disease of the intestine or the forestomach (see the case of ruminal acidosis above), as well being a response in cases of sepsis, toxaemia or to certain diseases of other organs and systems. Under normal conditions, a considerable quantity of isotonic fluid enters the intestine every day, but only a small amount of this is eliminated as a component of faeces. This liquid derives from the food the animal is fed and from secretions from the proximal digestive tract. Normally, intestinal reabsorption exceeds these secretions, and therefore even slight changes in the speed of absorption can lead to diarrhoea.

In ruminants, dilution of the faeces is often the consequence of disorders of forestomach fermentation and of an increase in the speed of intestinal transit.

In sheep, the faeces normally consist of small dark-brown faecal balls with a water content of 50–60%. It is not unusual to observe the presence of faeces in soft blocks of a greenish-brown hue, especially at the beginning of grazing. The presence of watery faeces, on the other hand, may be a sign of a digestive disorder of a dietary nature. The chief causes of diarrhoea are:

- Poor absorption (villus atrophy).
- Osmotic overload.
- An increase in secretions.
- Abnormal contractility.
- An increase in the hydrostatic pressure between the enteric circulation and intestinal lumen.

In the case of disorders caused by dietary problems and by changes affecting the forestomach, the osmotic mechanism prevails as a result of an increase in the speed of passage, which leads to larger quantities of carbohydrates and liquids being taken into the intestine than can be reabsorbed. This, in turn, leads to an increase in the quantity of substances that have not been digested, or absorbed by the large intestine. As a result, there is an increase in fermentation and in the number of osmotically active particles. In ruminants, ruminal overload (e.g. an excess of carbohydrates) may lead to osmotic imbalances in the rumen and changes in the pH of the abomasum, capable of causing diarrhoea, which may be a direct consequence of disorders of the forestomach.

Dietary factors that lead to an increase in fermentation within the large intestine can also predispose the animal to infections, e.g. clostridial enterotoxaemia, infections due to *Listeria monocytogenes* (Gillespie and Timoney, 1984). The latter is mainly associated with the feeding of sheep with badly stored fodder silage (pH > 5), containing a considerable number of microorganisms.

Prevention of intestinal disorders in sheep involves the reduction of disturbances of the forestomach. However, the speed of passage in sheep is greater than that in cattle, and thus the forestomach's filter action may be more easily affected in sheep. Ovine concentrate should be supplemented with buffers capable, above all, of acting at the intestinal level, such as calcium carbonate and magnesium oxide. If they are fed grains of cereal (such as barley grain), these may get past the ruminal barrier undigested and cause abnormal fermentation in the large intestine. Thus, when this kind of feed is given to animals, the faeces should be checked for excessive amounts of undigested grain. Grain can be left to soak for 8–12 h before feeding to sheep, should it prove impossible to crush or flake it. These procedures not only make it more digestible, but also prevent the loss of grain in the faeces, thus improving productivity. Putting the animals out to graze should be done gradually, especially after the winter or after it has rained. Animals should also be fed rough forage, in order to guarantee that food remains for a longer time in the rumen, thus reducing the speed with which food passes through the alimentary tract.

10.3 Metabolic Diseases

10.3.1 General features

The animal's organs ensure the homeostasis of body fluids by means of a continuous flow of diverse metabolites into and out of these organs. Any changes in the rapidity of either input or output which cannot be promptly adjusted by an opposing factor will alter the equilibrium and may eventually cause severe upset.

There are a considerable number of mechanisms working to maintain this equilibrium: in the first place, animals have developed a complicated endocrine control system; secondly, the stability of the dynamic input/output system chiefly depends on the quantity of metabolite in circulation and on the quantity present in the reserve fluids and organs, mechanisms designed to absorb or synthesize this metabolite and to transform, utilize and excrete it.

Each of the metabolic pathways, including those of water, minerals, trace elements, electrolytes, proteins and energetic metabolites, may be affected by an imbalance between 'input' and 'output', resulting in various metabolic disorders (Payne, 1977). The majority of these disorders affect mainly animals with greater metabolic requirements, such as growing lambs and productive ewes (pregnant and lactating animals). They are the result of a compromise in the animal's capacity for dealing with the demand for high levels of production. Thus diseases of this kind can be considered to be due to human error, i.e. the wrong diet or poor animal husbandry. This alters the balance between input and output, with consequent adverse effects on the health and production of the animals.

In dairy sheep farming, however, the selection methods and type of breeding that have been adopted up until now have yet to create the serious metabolic disorders found in dairy cows, although some can be identified. In sheep farming, dietary imbalances and nutritional insufficiencies continue to play a greater part than high production levels in creating such disorders, particularly due to the close link between this kind of livestock and the importance of grazing. Thus, the nutritional insufficiencies or excesses associated with the variability of this factor over time, in both quantitative and qualitative terms, and the mistakes made in livestock management, constitute the major causes of metabolic and nutritional disorders in dairy sheep.

10.3.2 Disorders associated with the animal's energy metabolism

We have previously considered disturbances associated with excessive feeding of carbohydrates: these include digestive and systemic diseases, such as the various forms of ruminal acidosis. In the present section the consequences of energy deficit are discussed.

Pregnancy toxaemia

Pregnancy toxaemia, otherwise known as sheep ketosis or 'twin-lamb disease', is caused by a negative energy balance due to increased demand from the fetus, or fetuses, as they quickly grow during the later stages of pregnancy (the last 2–4 weeks), and to an insufficient level of energy intake from the ration (Bezille, 1996). This disease usually affects ewes from the second or third pregnancy onwards, even though cases have been reported in hoggets subject to oestrus synchronization, probably as the result of a significant increase in multiple births. An insufficient energy intake occurs when the sheep are fed poor-quality forage, or when pasture is of poor quality due to adverse weather conditions. A fairly important factor of variability is the voluntary ingestion of feed, which tends to decrease in excessively fat ewes towards the end of pregnancy. Any sudden changes in diet and/or the presence of other diseases such as footrot or intestinal parasites, as well as certain operations such as anti-parasite baths, transport or movement of the animals, may lead to a reduction in the ingestion or absorption of food, encouraging the onset of the disease. Moreover, a lack of other nutrients, such as choline or biotin, may also contribute to the onset of the disease.

Sheep, like all ruminants, cannot regulate the fetus's demand for glucose, and so must use their body reserves when dietary supplies are inadequate. The extra glucose needed to satisfy the energy requirements of the maternal and fetal tissues cannot be obtained directly from the digestion of food, so must be synthesized. More than half the necessary glucose is produced by the liver, from propionic acid derived from the fermentation of carbohydrates in the rumen; the remainder is synthesized, by means of gluconeogenesis, from glucogenetic amino acids, lactate and glycerol. When propionic acid and other precursors of glucose fail to satisfy glucose requirements, the result is hypoglycaemia, which leads to depression of the central nervous system (CNS). The fatty acids rapidly mobilized by the fatty reserves in the liver are oxidized for energy purposes inside the liver, and produce acetylCoA in the process. A deficiency of glucose precursors leads to a reduction in the intermediate metabolites in the Krebs cycle, and in particular in the oxalacetate needed to utilize acetylCoA, which as a result is partially oxidized to form ketone bodies (acetone, acetoacetic acid and β-hydroxybutyric acid), which accumulate in the bodily fluids (blood and urine), leading to metabolic acidosis, with resulting dyspnoea, depression of the CNS, dehydration, hyperazotaemia and loss of consciousness. The presence of non-esterified fatty acids (NEFA) in the liver causes an accumulation of triglycerides, which in turn leads to hepatic steatosis (Smith, 1996).

In an affected flock the onset of the disease lasts for several weeks, with a varying number of sheep showing clinical signs of the disease each day. Although the course of the disease lasts on average 1 week, the affected sheep are only identified as such during the final stages. The first sign is a reduction in their sensory capacities, when they have problems of posture, balance and walking: this is followed by recumbency, coma and death. The affected sheep find it impossible to follow the other sheep out to graze, and they seem blind, move around listlessly, wobbling and often bumping clumsily into obstacles. They are sometimes affected by convulsions, and in the final stages of the disease, they lie down in sternal or lateral recumbency in a stupor, open-mouthed, grinding their teeth and chewing on nothing. Recumbency is followed, 3 or 4 days later, by coma and death. The differential diagnosis includes other peri-partum diseases such as gangrenous mastitis, hypocalcaemia, polioencephalo-malacia, enterotoxaemia and toxicity (Ford, 1983).

During the early stages of the disease affected sheep present signs of hypoglycaemia, raised blood concentrations of NEFA, hyperketonaemia and ketonuria. In animals with advanced acidosis and renal insufficiency, the concentration of bicarbonate may be reduced by 50%. Ketonuria is usually present, and is noticeable before hyperketonaemia. Serum levels of β-hydroxybutyric acid are high (> 1 mmol/l). Anatomopathological lesions are often minimal, and are characterized by the presence of two or more fetuses, or of a single large one, which may be in an advanced state of decomposition. The liver increases in volume and becomes pale, friable and has an unctuous cut surface (due to hepatic steatosis).

Treatment is effective only in the case of sheep in the early stages of the disease, when the animal is still upright. They may be given intravenous glucose solutions, oral administration of glycerol (200 ml of a warmed 50% solution) or propylene glycol, twice a day until appetite returns. Another effective form of treatment may be the reduction of metabolic drainage by means of the removal of the fetus by Caesarean section, or by induction of delivery with administration of 10 mg dexamethasone 21-isonicotinate. However, ewes suffering from a prolonged energy deficit may show an indifferent response after glucocorticoid administration, and higher dosages may be necessary (Smith, 1996).

Prevention and control should be based on careful observation of the ewes during the last weeks of gestation. Even if cases of toxaemia in a flock are rare, the majority of the sheep may be affected by an energy deficit, with problems of growth and development of the fetus and udder, which could lead to a greater death rate among lambs, low birthweight, slow growth of surviving lambs and a reduced milk yield. Within a flock, sheep at risk may be identified with the help of a pregnancy assessment using ultrasonography: in this way, sheep that are at risk may be fed a food supplement. The concentration of β-hydroxybutyric acid may be taken as a guide to the correction of food supplements during the final weeks of pregnancy. 10% of the flock are examined, and feed is increased if the blood concentration of β-hydroxybutyric acid exceeds 0.48 mg/ml (0.8 mmol/l) (Russel, 1984).

If the sheep are fed a suitable diet during the final 3 months of pregnancy, they should put on 4 kg in weight if bearing a single lamb, and 7.5 kg if twins. If ewes are overweight at mating, their bodyweight should fall by 20% during the first 2 months of pregnancy, and then increase over the next 3 months. Those that are fat when they are about to give birth are most at risk from the disease. During the final stage of pregnancy (the last 30–40 days), the sheep should be offered a supplement of concentrate (the quantity will depend on the stage of pregnancy), in gradual increments, up to a daily dosage of 400–500 g/head till 15–20 days prior to delivery. This concentrate should contain substances with a glucogenetic action, proteins and bypass amino acids, as well as adequate amounts of minerals and vitamins (in particular cobalt and niacin). Concomitant diseases such as parasitism or infections should be treated, as they constitute secondary factors of malnutrition.

10.3.3 Disorders associated with protein deficiency

As far as protein intake is concerned, ruminants have an advantage over monogastric animals in that the ruminal microflora may synthesize new proteins from non-protein sources of nitrogen. Moreover, each molecule of urea that forms in the body as the end-product of protein metabolism may be recycled as a source of ruminal nitrogen by means of salivary enzymes. This means that ruminants may economize on the ingestion of proteins. Generally speaking, therefore, protein deficiency should not be a serious problem for sheep, except in certain circumstances: problems may arise, for example, when protein requirements increase as a result of high milk production. Furthermore, protein deficiency rarely exists in sheep, but tends to be part of a wider problem involving other nutrients such as energy sources. Reduced ingestion of proteins, in fact, depresses the activity of the ruminal microorganisms, with the result that the digestion of carbohydrates in the rumen is compromised. Moreover, some trace elements are of vital importance to the working of the rumen: these include sulphur (for the synthesis of lysine and methionine) and cobalt. In other words, it is difficult to separate the simple lack of proteins from malnutrition in general. Other diseases may also interact, causing a serious secondary protein deficiency or the exacerbation of an already existing condition. Some diseases of the liver, the intestine and the kidneys lead to an excessive loss of serum protein, e.g. fascioliasis (liver fluke), Johné's disease and ascariasis (roundworms).

Protein deficiency has a number of consequences which can affect various organs and systems. A reduction in blood proteins leads to the modification of blood osmolality and to disorders of water metabolism, with the onset of oedema, which, in the case of sheep, will be seen primarily in the intermandibular space. A reduction in protein synthesis may alter the production of protein hormones associated with reproduction, leading to problems of fertility and interference with the synthesis of the bone matrix, which in turn can lead to osteoporosis. Similar disorders can be found in the growth of horns and hooves, leading to lameness and lesions of the feet. The growth of the fleece may also be compromised.

The effects of protein deficiency can also affect the animal's energy metabolism. The latter is adversely affected by the fact that part of the blood sugar comes from glucogenetic amino acids such as alanine which, together with methionine, play an essential role in the synthesis of lipoproteins: any deficiency of these amino acids affects metabolism and the transportation of fats to such an extent that protein deficiency in sheep towards the end of pregnancy may contribute to the onset of pregnancy toxaemia. Such a deficiency may also adversely influence the immune response to the most common pathogenic agents. It is widely recognized that the nutritional state of the host influences the speed with which proper immunity against parasites and other infectious agents is acquired, and that proteins have a major role to play. Protein supplements fed to sheep are capable of improving the response of T-lymphocytes to mitogens and to the antigens of the L3 larvae of gastrointestinal *Strongyloidea*. These supplements do not appear to influence the parasitic load directly, but they reduce the loss of production from parasitic action and help eliminate parasitic eggs. The latter seems to be a result of the reduced productive capacities of the parasites due to the host's immunity (Morgante and Ranucci, 1997).

In the case of protein deficiency, non-specific symptoms include:

- Low milk production.
- Poor quality of the fleece or of the feet.
- Low fertility.
- Oedema (particularly in the inter-mandibular space).

The final diagnosis depends on an interpretation of the metabolic profile, which may show low levels of blood urea (< 2.5 mg/ml), albumin (< 0.25 mg/ml) and haemoglobin (< 0.8 mg/ml), especially in lactating sheep. A differential diagnosis should be made, excluding the effects of a secondary deficiency, such as those caused by energy or trace element deficiencies, or by other diseases such as serious parasitic infestation or paratuberculosis.

Prevention depends, first and foremost, on an accurate calculation of the actual ingestion of proteins. This may often be deceptive if the evaluation is based only on the crude protein (or nitrogen) content of the ration.

10.3.4 Disorders associated with protein excess

An excess of protein is a fairly rare occurrence in sheep farming, although certain factors may bring it about. Likewise, it is rare to come across cases of toxicity caused by the use of non-protein nitrogen sources, such as urea. Nevertheless, certain situations may lead to an excess of protein and/or non-protein nitrogen, above all when milk-producing sheep are fed large quantities of concentrates, or when spring pasture is extremely rich in non-protein nitrogen or easily degradable proteins.

In the latter case, an important role is played by even a temporary lack of an adequate quantity of easily fermentable carbohydrates capable of providing the carbon framework required for protein synthesis by the ruminal flora. In either case, the excess ammonia formed in the rumen may enter the circulation and avoid detoxifying action in the liver, especially if the latter is affected by pathological changes such as parasitism, e.g. liver fluke, and may cause problems. Nevertheless, there are no reports of this occurring in sheep, although it is easy to imagine the pathological role played by excess ammonia. However, it should be pointed out that sheep have a higher blood urea level (3.5–5.5 mg/ml) than cattle or other ruminants. The blood urea levels, as well as those present in the milk, may indicate the state of protein metabolism, even though they may have been modified as a result of other physiological situations, e.g. anorexia, or pathological conditions, e.g. renal insufficiency.

In cows, protein excess may cause reproductive disorders by means of two basic mechanisms: the first consists of a concurrent energy deficit, resulting from an imbalance between the energy and protein content in the diet; the second involves the action of the urea on the intrauterine environment by the production of prostaglandin, adversely affecting the survival of the embryo. No studies of this problem in sheep have been conducted, and it is likely that protein excess plays a smaller role in the productive performance of the ovine. This is firstly because during the mating season the pastures are unlikely to be extremely rich in protein and non-protein nitrogen; secondly, the ewes are not at the height of lactation, and therefore even if their feed was extremely rich in protein, it is unlikely that these proteins would concurrently cause an energy deficit, which at present seems to be the most serious problem in cases of low fertility. Nevertheless, an excess of protein and/or non-protein nitrogen, theoretically possible in cases of high urea levels (> 6.0 mg/ml) in the blood or in milk, may under certain conditions (the release of young animals on to protein-rich pasture) during the reproductive cycle cause ruminal alkalosis and intestinal imbalances (see above), with adverse consequences for other organs and systems, e.g. the udder and feet.

10.3.5 Mineral metabolism-related disorders

Calcium and phosphorus

Phosphorus (P) and calcium (Ca) are the most plentiful minerals in the mammalian body. Approximately 80% of all P and 99% of all Ca are in the bones and teeth. An adequate amount of Ca and P is needed for the normal growth and mineralization of bone. The remaining 20% of P is distributed throughout the soft tissues, where it has a number of different functions: it is of importance for the energy exchanges that take place as a result of the formation and utilization of high-energy compounds (ATP); it has the capacity to be excreted as both $H_2PO_4^-$ and HPO_4^{2-}, thus helping to maintain the acid–base balance. Ca also has a number of functions to perform within the soft tissues, although it is of particular importance to: normal neuromuscular excitability; permeability of capillaries and cell membranes; muscular contraction; transmission of nerve impulses; and blood coagulation (Payne, 1977).

The homeostasis of Ca, and to a degree that of P, is controlled by the parathyroid hormone (PTH) secreted by the parathyroid glands, by the calcitonin secreted by the C cells in the thyroid gland, and by 1.25 dehydrocholecalciferol (1,25–$(OH)_2D$, or vitamin D) produced by the kidneys. A reduction in the blood Ca levels stimulates the production of PTH, which in turn promotes the production of vitamin D by the kidneys, causing an increase in Ca absorption at the intestinal level (Smith, 1996). Bone tissue acts as a reserve for Ca and P, which can then be mobilized to keep blood levels normal during those periods in which there is a high demand for calcium (towards the end of pregnancy it is required for the ossification of the fetal skeleton).

In situations of deficiency, low blood levels of P stimulate the production of vitamin D, which in turn encourages the intestinal reabsorption of P. The excretion of P in ruminants involves its elimination in the saliva more than through the kidneys. PTH stimulates the elimination of P with saliva, and inhibits reabsorption by the kidney tubules. The efficacious reabsorption of P by the kidneys, together with the faculty for recovering that excreted in the saliva, makes ruminants, especially sheep, highly resistant to prolonged P deficiency (Payne, 1977)

In healthy adults, the total concentration of Ca in the blood is about 1.0 mg/ml (2.5 mmol/l). Of this, approximately 50% exists in the ionized (metabolically active) form. The concentration of inorganic P in the blood is about 0.5–0.7 mg/ml.

Hypocalcaemia

Ewes can be affected by this disorder particularly during the final 4–6 weeks of pregnancy. Compared to dairy cows, sheep require more Ca for the growth of the fetus or fetuses than they do for lactation. However, in certain circumstances, e.g. the administration of substances capable of chelating the Ca, such as tetracycline, and with high-yielding animals, hypocalcaemia may occur even during the first 6 weeks of lactation. Flaccid paralysis is the first sign of hypocalcaemia in sheep, even though the tetanic variety is more common than in cattle. Due to the fact that pregnancy toxaemia and hypocalcaemia arise during the same period, it is often difficult to distinguish between them. A useful aid to differential diagnosis may be analysis of the urine using special reactive strips, in order to check for the presence of ketone bodies in the case of toxaemia.

Lack of food, forced exercise or transportation (transport hypocalcaemia) may trigger off the disease in predisposed animals. Long-distance transportation, moreover, may be the cause of hypocalcaemia in lambs, which if associated with hypomagnesaemia, takes the form of muscular tremors and tetany rather than flaccid paralysis. In sheep, as in lambs, the course of the disease is fairly short, and if the animals are not properly treated they may die within 24–48 h. The intravenous injection of Ca gluconate solution (1 g/50 kg BW) leads to rapid recovery. Prevention of the disease is mainly dietary-based: the ewes must be carefully observed during the final 6 weeks of pregnancy, and provided with a sufficient quantity of Ca and energy in their feed (Russel, 1983). To this end, farmers are advised to use Ca-rich hay, such as that made from leguminous plants.

Magnesium

As well as catalysing the action of certain enzymes, e.g. alkaline phosphatase, magnesium (Mg) plays a role in the synthesis and lysis of acetylcholine and in the enzymatic scission of adenosine triphosphate (ATP). The excessive administration of Mg salts causes sensory and motor paralysis, together with a state of narcosis and the abolition of reflexes. Approximately 70% of the Mg present in the body is deposited in the skeleton, from where it is difficult to mobilize in deficiency; 29% is found in the soft tissues, while only 1% is present in the extracellular (interstitial and intravascular) fluid (Payne, 1977).

Forms of grass tetany arise when the blood concentration of Mg falls from the normal level of 0.2–0.3 mg/ml to < 0.11 mg/ml. The blood concentration of Mg varies within a fairly wide range of values, and depends mainly on the quantity of this element supplied in the diet, as opposed to what happens in the case of Ca, and to a certain extent that of P, where an appreciable quantity may be reabsorbed from the bones if required. Mg is eliminated through the urine, milk and faeces. Renal excretion of Mg begins when its blood concentration exceeds 0.18–0.19 mg/ml, and is linearly correlated to concentration in the blood. Thus, the homeostatic control of magnesaemia depends on dietary intake and those factors which might hinder the absorption of Mg.

Hypomagnesaemic tetany

Hypomagnesaemic tetany (otherwise known as grass tetany, lactation tetany, transport tetany, etc.) involves a deficiency of Mg ions in the blood and in the cerebrospinal fluid (CSF) that causes the loss of normal neuromuscular function, the onset of tetanic spasms and clonic convulsions. Lactation, stress, transportation and/or anorexia may trigger off the symptoms in animals fed a poor diet. Grass tetany affects mainly lactating animals when pasture consists predominantly of graminaceous plants which are low in Mg, especially if they have been fertilized using nitrogen and potassium. Adverse weather conditions, which may lead to anorexia (energy deficit), may worsen the symptoms. A ration containing less than 0.2% Mg of DM can cause the onset of the disease, which is frequently fatal and must always be suspected in cases of sudden death of grazing animals (Smith, 1996).

The administration of a Mg supplement for a long period prior to letting the animals out to graze is of little benefit in preventing the onset of this disease. Prevention should, on the other hand, consist of the suitable integration of Mg in the daily diet during grazing on graminaceous plants and during periods at risk. This supplement may be provided in various forms: for example, Mg may be added to the animal's drinking water, or they may be fed mineral supplements or concentrates, or salt may be spread over the pasture or on their feed, although the simplest, most economical method appears to be that of supplementing their diet with leguminous forage (Decante, 1996).

Sodium and potassium

Sodium (Na) constitutes the principal cation in the blood and within the extracellular system (the osmotic skeleton), while 98% of potassium (K) is present at the intracellular level, where it contributes towards maintaining membrane potential, together with normal muscular and cardiac contraction. They are absorbed by the intestine and excreted by the kidneys. The quantity of K to be found in feeds of plant origin often exceeds the animal's requirements. Both Na and K are to be found in considerable quantities in the body fluids and organs compared to their turnover: as a result, any disorder connected to their metabolism evolves rather slowly (Valercher *et al.*, 1996). In other words, the system is suitably buffered, thus enabling animals to survive relatively brief periods of deficiency without any problem.

Mammary oedema

During the final stages of pregnancy (the last few days before lambing), an excessive swelling of the mammary gland may occur in several members of the flock. The udder appears symmetrical, of a normal colour, while the animal sometimes has difficulty in walking, or even in lying down to sleep, due to the size of the udder. The teats also appear considerably larger than normal, the udder feels doughy to the touch, and when pressed with the fingertips, clear pitting remains (fovea). Both rectal temperature and mucosal colour are often normal. The most commonly affected animals are primiparous ewe-lambs and the fatter ones.

This condition is not dangerous as such, as it lasts for only a few days after lambing. The onset of the oedema, however, may cause problems both to the ewes and to the lambs: the affected udder may be subject to trauma; the increase in size of the organ and the teats may prevent feeding of the lambs or inadequate milking, which may lead to lesions of the teats and milk engorgement. Situations like this expose the organ to the risk of bacterial infection (mastitis). In general, in the more serious cases, traces of blood may be found in the milk for a period lasting a number of days. The presence of red blood cells in the milk has been linked to an increase in the quantity of somatic cells (neutrophil leucocytes) (Morgante *et al.*, 1996). This alteration gives the milk a pink colour, thus making it unsuitable for sale or for processing.

This form of mammary oedema should be distinguished from the other, commonly seen variety, which may appear in serious cases of mastitis: in the latter case, it is characterized by reddening of the skin, the presence of heat, pain on touching, the involvement of the supra-mammary lymph nodes, and in most cases by fever, anorexia and exhaustion, as well as by a considerable increase in the quantity of somatic cells in the milk. The etiopathogenesis of this disorder has still to be fully explained.

Mammary oedema occurs, above all, in animals fed large quantities of cereals or fodder, and in those with a diet containing an excessive quantity of Na and/or K. It would seem to be connected to a form of water retention of an osmotic nature, although in the case of dairy cows hormonal causes cannot be excluded. Animals affected by mammary oedema do not generally require any specific form of treatment; prevention in the form of adequate dry rations, taking care to avoid excessive quantities of Na or K, constitutes the best way of controlling this problem.

10.4 Diseases of Lambs

10.4.1 Failure of passive transfer of immunity (FPTI)

Lambs are immunologically competent from birth, and have the capacity for developing a complete immune response like an adult sheep. However, they are virgin from the immunological point of view, in that they have never been exposed to antigenic stimuli, and thus have no kind of acquired immunity and a considerable quantity of immunoglobulins in circulation. Lambs start to produce antibodies during the second week following birth, and only attain a significant level when they are in their second month. Under normal conditions, a temporary form of protection against infection is provided by the passive antibodies the mother transfers to the lamb through its colostrum. The immunoglobulins that accumulate in the colostrum are selected from those contained in the serum, and are not produced by the udder (Smith, 1996).

At birth, a lamb possesses special enterocytes in the intestinal tract, capable of absorbing large quantities of molecules such as immunoglobulins by means of a process

Table 10.2. Metabolic nutritional disorders of dairy sheep.

Disorder	Causes	Clinical signs	Treatment and prevention
Pregnancy toxaemia	Negative energy balance due to (i) increased demand from the fetuses as they quickly grow during the later stages of pregnancy; and (ii) an insufficient level of energy intake	•Reduction of sensory capacity •Depression •Recumbency •Coma and death within 4–7 days •Ketone bodies in urine	•Glucose solutions (I/V) •Glycerol or propylene glycol (orally) •Glucocorticoids •Caesarean section
Disorders associated with protein deficiency	•Primary: poor ingestion of protein with the diet •Secondary: malnutrition in general, high load of gastrointestinal and hepatic parasites	•Intermandibular oedema •Low milk production •Low fertility •Lameness •Poor quality of the fleece	•Adequate protein intake with the diet in relation to the productive period. •Control of parasitic infections.
Disorders associated with protein excess	Excess ingestion of protein and/or non-protein nitrogen	•Ruminal and intestinal imbalances •Reproductive disorders? •Blood urea > 6.0 mg/ml	•Adequate protein intake with the diet, avoiding sudden overload of easily degradable N sources
Hypocalcemia	High requirement of Ca for the growth of the fetus(es) combined with low Ca intake and/or absorption of pregnancy	•Tremors and tetany •Flaccid paralysis (less frequent) •Death within 48 h	•Ca gluconate solution (1 g Ca/50 kg BW I/V) •Adequate Ca intake in last weeks
Hypomagnesiemic tetany	•Low Mg content in the diet (pasture consisting mainly of graminaceous plants fertilized with N and K). •Predisposing factors (lactation, transport, stress, etc.)	•Tetany •Convulsions •Sudden death	•Adequate daily Mg intake with the diet supplementing the diets of sheep grazing on dangerous pasture (with < 0.2% of Mg/DM)
Mammary oedema	Excessive quantity of Na and/or K in the diet during last phase of pregnancy	•Symmetric mammary enlargement •Blood in the milk •Secondary problems for milking and sucking lambs	•Adequate amount of concentrate, Na and K in the diet before lambing

of pinocytosis. The intestinal immunoglobulin absorption window is narrow and lasts for a period of 18–24 h from birth, although it is greater during the first 6–12 h. The blocking of this intestinal function is due to the loss of these special enterocytes. A passive transit failure may occur because the colostrum does not form, or because it contains too few antibodies, or because it is lost as a result of premature lactation, with the lamb failing to ingest it, or because it is ingested but not absorbed.

The main cause of the failure of colostrum formation in sheep is either the onset of acute mastitis or the functional incapacity of the organ due to a previous, chronic mastitis. These abnormalities are often unilateral, while the other half of the udder continues to function normally. However, in the case of twin or triple lambs, the colostrum may not be enough for all. Even with one healthy udder, the weaker lambs may not get a sufficient quantity of colostrum during the first hours after birth. Another thing that may happen is that the lambs are taken away from the ewe immediately after birth, to be fed artificial milk; they may refuse to suckle the dummy with the colostrum, or the latter may be fed in too small amounts, or too late.

The inadequate intake of maternal immunoglobulins as such does not produce any particular symptoms, and thus cannot be diagnosed by clinical examination of the animal. One symptom that strongly suggests the presence of FPTI is the appearance of bacterial infection, and in particular of septicaemia, septic arthritis, pneumonia and enteritis during the first 2 weeks after birth. FTPI may be diagnosed from the low levels of immunoglobulin in the serum as early as the first 36–48 h after birth. A quick way of testing these levels is by carrying out a turbidimetric analysis using zinc sulphate, a latex agglutination test or an immunoenzymatic test. The IgG levels generally considered to be adequate appear rather arbitrary, and probably vary depending on environmental conditions. Generally, adequate values are > 80.0 mg/ml of IgG.

It is not easy to treat this condition, particularly in large flocks, and prevention is still the best method of controlling the problem. Prevention must be based on checks to ensure that the lambs have ingested a sufficient quantity of colostrum during the first 24 h after birth. Special attention must be paid to ewes giving birth to more than one lamb, and to those presenting pathological alterations in the udder. A good general rule would be to create a colostrum bank at lambing time, freezing individual batches of 500 ml of colostrum from different subjects, to be administered the following lambing time to the weaker lambs or to those with mothers at risk, even though in sheep farming it is not difficult getting colostrum from other animals since most ewes give birth around the same time. Special care must be taken of lambs that are taken away from their mothers to be artificially fed, to ensure that colostrum is administered both at the recommended times and in the recommended quantities (500–800 ml within the first 24 h of birth).

10.4.2 Neonatal hypoglycaemia

This disease may arise in those lambs, born as twins or triplets, which are immature or undersized, or where the ewe lacks milk (usually due to mastitis), generally in environmental conditions characterized by low temperatures. The animals are often found dead, but an anatomo-pathological examination shows no serious injuries: the only sign of the possible presence of the disease is an empty stomach. Diagnostic confirmation may be given by a positive response to parenteral treatment using glucose solution. During the 24 h immediately after birth, about 50% of all fats present in the newborn lambs can be found in the fatty tissue in the form of brown fat, which is utilized by the lamb for non-shivering thermogenesis (Blood *et al.*, 1989). This may range in level from 1.5–4.5% of bodyweight at birth, and smaller lambs with lower levels show a reduction in neonatal vitality. Factors affecting the onset of the disease include a low degree of interest shown by the ewe, or her complete absence, in the case of lambs which are only a few days old.

Treatment of lambs suffering from hypoglycaemia and hypothermia involves the intraperitoneal or intravenous administration of 10 ml/kg (BW) of a 20% glucose solution, before warming the body to bring it back to its normal temperature (39.5–40°C). In order to prevent the onset of this condition, lambs need 180–220 ml of colostrum/kg BW during the first 18 h from birth, in order to provide them with sufficient energy for the production of heat. During the final stages of gestation, great care must be taken over the way the ewes are fed, so that they can produce a sufficient quantity of colostrum. Colostrum may be easily obtained, however, by milking ewes with an excess yield. The chances of survival of weaker lambs may be improved by keeping ewes who have had two or more lambs in individual pens.

10.4.3 Milk indigestion

This is caused by the feeding of an excessive quantity of milk, but under normal conditions (i.e. when lambs are looked after by their mother) this problem does not arise. It is usually the result of the situation in which the lamb is fed artificially with unsuitable types or quantities of milk, or when the sheep are led out to pasture for a few hours and the lambs are left in the barn. An excessive milk intake makes it impossible for the lamb to digest all of the milk. This then reaches the colon, where it ferments to produce osmotically active sugars and acids. This attracts water into the intestine, causing diarrhoea, although this is often not very serious and can be quickly rectified by the resumption of normal feeding. In cases where the ewes are led out to graze while their lambs remain in the barn, another factor may be that the famished lambs prematurely ingest grass, forage and often faeces. They often have a greyish-yellow coloured diarrhoea without their general health being compromised in any way. Nevertheless, the persistence of this problem may predispose the lambs to complications which may sometimes be rather serious, e.g. clostridial infections. It is a good idea to examine the faeces each time the lambs have diarrhoea, in order to exclude the presence of parasitic infections, e.g. coccidiosis.

10.4.4 Chronic ruminal acidosis

Chronic ruminal acidosis affects weaned lambs fed large amounts of concentrates or cereal-rich diets. The ruminal bacterial population alters, to produce increased numbers of microorganisms that both use and produce lactic acid. There is a reduction in the quantity of cellulosolytic bacteria, and a proliferation of bacteria that degrade sugars and starch. The lactic acid fails to accumulate as it is continuously being metabolized by the bacteria. The resultant rapid fermentation produces a high concentration of VFA, leading to a slight acidosis in the ruminal liquid. A concentrate-rich ration does not help mastication, and thus salivation, and therefore the ruminal buffer system is further compromised as a result. The high levels of VFA, which stimulate the inhibitory receptors in the epithelium, and the type of feed that does not encourage the physical stimulation of the organ together inhibit ruminal contractility. Although the bacterial population is metabolically very active, the number of species present falls, and the protozoan population declines or disappears altogether, thus leading to the poor utilization of the dietary protein.

The quantity of butyric and propionic acids increases, stimulating the proliferation of the papillae in the ruminal epithelium, which in extreme cases leads to parakeratosis of the rumen, a consequent reduction in the absorption of VFA, and a greater susceptibility to trauma and inflammation of the ruminal wall's deeper tissues. Such lesions enable bacteria to enter and spread throughout the liver, and possibly other organs as well, leading to the formation of hepatic abscesses in the majority of affected animals. Lambs suffering from ruminal acidosis lose their appetite and display ruminal hypocynesia, delayed growth and, in some cases, symptoms of a chronic inflammatory reaction. Other concomitant pathological conditions which may accompany chronic ruminal acidosis include aseptic laminitis and cerebrocortical necrosis, or polioencephalomalacia (thiamine deficiency). These problems presumably arise when the acidic fermentation of the rumen leads to the production of toxic factors such as histamine and thiaminase (Smith, 1996).

10.4.5 Rickets

Young, growing animals may be affected by rickets, which is characterized by defective bone growth. The basic lesion consists of defective calcification, the persistence of hypertrophic cartilage and swelling of the osteoepiphyses. The poorly mineralized bones are prone to curvature as a result of the body's weight. This condition is most frequently caused by a lack of either vitamin D or P: in the former case, it tends to affect animals kept indoors, whereas in the latter case it tends to affect those raised outdoors. One factor that may predispose the animal towards such insufficiencies, and thus to defective bone growth, is the presence of intestinal parasites. Rickets may be a problem in certain environmental contexts, such as in some mountain areas or situations where the animals are intensively fed in order to fatten them for slaughter. Rickets may be the result of a primary P deficiency in phosphorus-poor mountain areas, or of a lack of vitamin D in lambs raised for excessively long periods in the barn. However, lambs are less likely to be affected by a primary P deficiency than are calves.

The clinically visible form of rickets is characterized by stiff movement of the limbs, joint enlargement – of the front limbs in particular – and the bead-like enlargement of the costochondral joints. The long bones present an abnormal curvature, which causes a forward displacement of the knee and a tendency for the animal to lie down; there is also anterior curvature and a tendency to fracture. Dental eruption is delayed and irregular, and the teeth appear poorly calcified, often decayed and of a brownish colour, badly aligned and prone to wear.

When the disease is suspected, increased alkaline phosphatase (ALP) levels in the blood may be detected. When the cause of the disease is a P or vitamin D deficiency, the presence of P in the blood will be lower than normal by about 0.3 mg/ml. One of the best methods of diagnosing rickets involves radiography of the bones and joints. Rickets should be distinguished from Cu deficiency and from bacterial diseases of the joints. Prevention involves the administration of an adequate Ca and P supplement, and an improvement in the hygiene and environmental conditions of the farm.

10.4.6 Urolithiasis

Urolithiasis is a disease that may affect lambs being fattened on diets rich in concentrates, and involves the formation of urinary calculi. The disease can be economically damaging when it affects a large number of animals at the same time. The formation of calculi in the urinary tract may cause the partial or total obstruction of the urethra, and the consequent death of the animal as a result of the rupture of the urethra itself or of the bladder. The symptoms are characterized by dysuria, anuria, colic, depression and preputial or abdominal oedema. The mechanism underlying the formation of these calculi is still not fully understood; however, it is generally believed that when the formation of calculogenetic crystalloids in the urine is such that it impedes their being dissolved, then they may precipitate and deposit themselves on nuclei, often formed by mucoproteins.

The degree of over-saturation of the urine depends on diet, pH, the animal's consumption of water and on the presence of crystallization inhibitors in the urine. Struvite calculi (consisting of magnesium ammonium phosphate ($MgNH_4PO_4$) salts) are the most commonly found variety in meat lambs fed on considerable quantities of concentrates. Such rations may contain large amounts of P and Mg, which increase renal excretion and the presence of crystals in the urine, particularly if the latter is noticeably alkaline. Acidification of the urine, on the other hand, is capable of dissolving these salts, thus preventing the formation of crystals.

Urolithiasis is seen mainly during the winter, when the animals drink less water, leading to the formation of less urine and an increase in the concentration of crystalloids. An increase in the concentrate:forage ratio increases the quantity of sediment in the urine; however, an increase in the Ca content of the diet protects against the formation of calculi by reducing the quantity of P absorbed. The feeding of pellets to the lambs increases both the concentration of P in the urine and the formation of urinary calculi, although the underlying mechanism in question has yet to be fully understood.

The only treatment available for animals with clinical symptoms of obstruction of the urethra is surgery, in the form of a urethrostomy. Animals at risk from struvite calculi must be fed a diet containing smaller quantities of P and Mg, but additional Ca, in the form of good-quality hay made from lucerne. A gradual increase in the salt (NaCl) content of the diet may reduce the incidence of this disease, probably as a result of its diuretic effect and the consequent dilution of the urine. Struvite crystals precipitate in an alkaline environment, and so the addition of ammonium chloride to the diet (0.5%) acidifies the urine and thus reduces the risk of the disease (Smith, 1996).

10.4.7 Nutritional muscular dystrophy

Nutritional muscular dystrophy (NMD), otherwise known as 'white muscle disease' or 'stiff lamb disease', is a hyperacute/subacute degenerative disease of cardiac and/or skeletal musculature, caused by a diet low in selenium (Se) and/or vitamin E. It affects mainly lambs born from ewes fed on a Se-deficient diet during pregnancy. The effects of Se deficiency and vitamin E deficiency are partly the result of the onset of damage to the cell membranes and to proteins, which compromises cellular physiology. Selenium, as a component of the GSH-Px, and vitamin E act as biological antioxidants by protecting the organism from damage caused by highly reactive forms of oxygen (free radicals), which are normally produced during metabolism. The exact correlation between the two elements, other metabolic factors and other trigger mechanisms present in NMD is still not clear, since many animals suffering from a deficiency of Se and/or vitamin E are not affected by muscular diseases. Poor absorption of Se often lies at the root of this deficiency in sheep: its absorption depends on the composition of the diet, on the absorption of proteins and on the presence of antagonist substances (Vrzgula, 1991). The factor that plays the most important role here is soil acidity: where the soil is rather acidic (pH 4.5–6.5), Se cannot be utilized by the plants. Judging by the close similarity between Se and sulphur (S), it is likely that soils rich in S may be lacking in Se as a result of competition between chemical elements.

There are two forms of NMD: cardiac NMD and skeletal NMD. The former is characterized by the hyperacute symptoms of cardiac decompensation, while the latter is characterized by myasthenia and difficulty in walking. In both cases, the most affected animals in the flock tend to be the younger ones in the phase of rapid growth. Evidence suggests that there may also be a form which arises prior to, or soon after, birth, whereby the animals are affected by muscle degeneration.

The onset of the cardiac form of NMD is usually rapid, with serious debilitation, and sometimes death. It involves lesions of the heart, diaphragm and intercostal muscles. The affected animals are characterized by a marked degree of depression, dyspnoea, considerable weakness, recumbency and a quick, irregular pulse, although their rectal temperature is often normal.

The skeletal form, on the other hand, usually starts more slowly, and is characterized by muscular weakness or a certain stiffness of movement. The majority of affected animals remain recumbent, and can only manage to stand for a short period at a time. The limb muscles may be swollen, hard and painful to the touch: the most commonly affected ones are the gastrocnemius, the semitendinosus, the semimembranosus, the biceps femoris and the muscles in the lumbar regions of the glutei and the neck; the tongue muscles may also be affected, causing dysphagia. Animals suffering from the skeletal variety often respond well to treatment: an improvement can be seen within 3–5 days, and the animals are able to stand and walk again. Animals suffering from an acute form of muscle degeneration often present high serum levels of enzymatic activity – CPK, AST and LDH (Smith, 1996).

Pathological examination of animals that have died, or have had to be slaughtered, shows signs of bilateral, symmetrical muscle degeneration. The degeneration of the skeletal muscles is characterized by discoloration of muscle tissue, which appears either pale pink, yellowish-grey or has a whitish hue, with a dry, opaque (wax-like) surface when cut (waxy, hyaline or Zenker's degeneration). Affected muscle is often found adjacent to healthy muscle.

The cardiac muscle is subject to degeneration similar to that affecting the skeletal muscles, often with lesions present in both ventricles.

The levels of Se present in an animal, or group of animals, can be checked by either biopsy or a blood test; vitamin E levels can be measured in blood, but blood estimation gives no information on the amounts of Se or vitamin E in the body's reserves. Tissue biopsies (especially of liver tissue), on the other hand, can provide information about the body's reserves. Whole blood is preferable to plasma or serum when measuring the level of Se. The standard concentration of Se in the liver varies from 0.9–1.75 mg/g of DM. The erythrocyte level of the Se-dependent enzyme GSH-px can also provide useful information about Se reserves; the activity of GSH-px in the erythrocytes remains constant for 4–6 days when kept at a temperature of 4°C, whereas it noticeably diminishes after this time.

Table 10.3 shows the blood concentrations of Se and GSH-px in dairy sheep reared in Italy. The critical concentration of vitamin E in the plasma ranges from 1.1–2.0 ppm in the larger animals. The vitamin E present in the plasma deteriorates rapidly: therefore samples to be used for analysis must be frozen immediately and kept at –70°C if the analysis is to be carried out later (Smith, 1996).

In the case of cardiac NMD, the damage is often widespread and life-threatening, although treatment can still be successful. The skeletal variety is easier to treat, although the prognosis is always guarded and depends on any subsequent complications that may arise. Treatment consists of the parenteral administration of Se and vitamin E solutions: these are available with concentrations of selenium ranging from 0.25–5 mg/ml in combination with 50 mg of vitamin E in the form of DLα-tocopherol acetate. The recommended dosage of Se varies from 0.55–0.67 mg/kg (2.5–3.0 mg/45 kg) BW, to be administered intramuscularly or subcutaneously. Compound medicines must not be administered in doses over and above the recommended range, as this can cause Se toxicity.

Prevention and control involve dietary supplementation with Se and vitamin E, although Se deficiency is more commonly implicated than that of vitamin E in the aetiology of NMD. Additional Se is provided in the form of food supplements: the addition of inorganic Se (0.2 ppm) may not be sufficient in high-risk areas. However, this supplementation must cover the entire year and not be limited only to periods in which concentrates are given to the animals (at the end of pregnancy and at the beginning of lactation). The administration of organic Se is more effective than that of inorganic compounds, which, once metabolized in the rumen, are transformed into compounds

Table 10.3. Selenium levels in dairy sheep reared in Italy (Morgante *et al.*, 1998).

Selenium status	Se blood level (mg/l)	GSH-px blood level level (U/g haemoglobin (Hb))
Deficient	< 0.05	< 212
Low/marginal	0.05–0.083	212–281
Marginal	0.083–0.110	281–338
Adequate	> 0.11	< 338

that are not easily absorbed. Alternatively, individual animals may be periodically given Se and vitamin E (parenterally) in order to maintain levels and/or to help the placental passage of Se to the fetus. This treatment should only be used, however, during the period in which the ewes are not lactating. A subcutaneous administration of sodium selenite (Na_2SeO_3) at the dose of 0.1 mg/kg twice before lambing can maintain normal GSH-Px levels for at least 3 months in Se-deficient dairy ewes (Morgante *et al.*, 1999).

As far as the methods of supplementing the animal's diet are concerned, samples of blood and tissue should be periodically taken from animals at risk in order to maintain adequate levels of Se. In high-risk areas, samples should be taken every 60–90 days in order to establish the Se levels in the more susceptible animals. Any adjustments to supplements must be made on the basis of these results. The feeding of animals on well-made, properly stored hay, or fresh cereals and green fodder, guarantees an adequate supply of vitamin E.

10.4.8 Copper deficiency

Copper (Cu) deficiency occurs when the amount supplied in the animal's diet is insufficient, or when the absorption or metabolism of Cu is negatively affected by other factors (secondary deficiency). A variety of factors may lead to a reduction in the absorption of Cu by the gastrointestinal tract. The recommended minimum quantity of Cu in the animal's diet (of DM) is 5 ppm for dairy sheep and 7 ppm for wool sheep. Secondary Cu deficiency is mainly associated with high levels of molybdenum, sulphates, zinc (Zn), iron (Fe) and other compounds in the diet. The excessive administration of molybdenum can lead to the formation of Cu molybdate in the rumen, which is barely soluble and is not absorbed by the intestine. Moreover, an excess of sulphides or sulphates in the animal's feed or drinking water may lead to the formation of Cu thiomolybdate (which is insoluble) in the rumen. An excess of iron (30 mg/kg of BW, or 1200 ppm) in the diet may also reduce Cu absorption.

The cause of Cu deficiency in clinically evident cases is often a combination of these factors, and may prove difficult to quantify. Cu is an essential component of numerous enzymes; the more important Cu-containing enzymes from the physiological point of view include:

- Superoxide dismutase (the cytoplasmic form, Zn- and Cu-dependent).
- Cytochrome oxidase.
- Lysyl oxidase.
- Ascorbic acid oxidase.
- Ceruloplasmin (Smith, 1996).

Moreover, a normal intake of Cu is essential for the absorption of iron and its transport from the liver to the reticuloendothelial system, and thus for the formation of haemoglobin. The exact pathophysiology of the majority of syndromes caused by Cu deficiency is still not understood. However, Cu appears to play an important role in preventing damage from cellular oxidization, as well as in the metabolism of Fe and S.

The symptoms of Cu deficiency include:

- Profuse, watery diarrhoea.
- Reduction in weight gain.
- Wasted appearance.
- Anaemia.
- Alteration in the colour or quality of the fleece (achromotrichia).
- Spontaneous fractures.
- Demyelination of the peripheral nerves (enzootic ataxia in sheep and goats).
- Lameness (physitis).

Generally, only one of these symptoms predominates in any given flock. Lambs and fetuses are more susceptible to Cu deficiency than adult sheep: the latter, on the other hand, are often subject to Cu toxicity. Lambs born from mothers suffering from Cu deficiency are more likely to show clinical signs of the syndrome than calves. Milk is a poor source of Cu, containing only 0.2–0.6 ppm in normal sheep, and 0.01–0.02 ppm in those suffering from serious deficiency.

The primary site for the accumulation of Cu is the liver, where its normal concentration ranges from 80–200 ppm of total DM. Liver concentrations of more than 350 ppm are not infrequent in sheep which receive supplements. Cu concentrations may remain normal even if those in the liver fall below 35 ppm, but if the latter fall below this figure then the former are bound to fall as well. The concentration of Cu in the plasma is generally 5% higher than that in the serum. Normal concentrations of Cu in the serum range from 0.7–1.2 mg/ml. Concentrations below 0.4 are considered distinctly low, whilst those of 0.4–0.7 are considered marginal, and hence difficult to interpret.

Treatment of Cu deficiency is normally achieved by injecting Cu-chelated compounds (1–2 mg Cu/kg BW), and the prognosis is often good, although it will depend on the severity of the symptoms (Suttle and Linklater, 1983). Lambs may be given 35 mg of Cu sulphate/head, twice a week, in order to prevent the onset of symptoms in endemic areas. Ovine Cu requirements range from 5–10 mg/kg of DM. Sheep can ingest toxic amounts if they eat supplements intended for cattle, or even lamb feed containing 20 mg/kg of DM. It is important to check the level of Cu in the liver at the slaughterhouse regularly, in order to ensure that lambs have been receiving an adequate supplement of Cu.

Table 10.4. Diseases of lambs.

Disorder	Causes	Clinical signs	Treatment and prevention
Failure of passive transfer	•Poor quality and quantity of colostrum •Colostrum ingestion failure •Absorption failure	•Major susceptibility to: septicaemia, septic arthritis, pneumonia and enteritis •Low blood IgG level	•Administration of 500–800 ml of good-quality colostrum within 24 h of birth
Neonatal hypoglycaemia	Poor milk ingestion in the first hours after birth, and low environmental temperatures	•Weakness, tremors and depression •Low body temperature and hypoglycaemia	•Adequate intake of milk •Glucose solutions (I/V or I/P) •Warming
Milk indigestion	Milk overload	•Greyish-yellow diarrhoea	•Avoid milk overload (especially artificially fed lambs)
Rickets	Vitamin D and/or P deficiency	•Stiff movements •Joint enlargement •Skeletal abnormalities	•Adequate Ca and P intake •Vitamin D administration
Urolithiasis	Concentrate-rich diet with high levels of P and Mg	•Dysuria or anuria •Preputial or abdominal oedema	•Surgery •Ca and NaCl supplementation •Urine acidification
Nutritional muscular dystrophy	Selenium and/or vitamin E deficiency Presence of antagonist elements (e.g. S)	•Dyspnoea •Skeletal muscle weakness and stiffness	•Parenteral administration of Se/vitamin E •Se supplementation of diet
Copper deficiency	•Primary: copper deficiency •Secondary: molybdenum excess	•Profuse watery diarrhoea •Anaemia •Lameness (physitis)	•Parenteral administration of copper glycinate •Adequate Cu supplementation
Iodine deficiency (hypothyroid goitre)	•Primary: iodine deficiency •Secondary: presence of goitrogenic plants	•Low growth rate •Goitre •Abortion and stillbirths, with goitre and alopecia	•Adequate iodine supplementation

10.4.9 Iodine deficiency (hypothyroid goitre)

The clearest symptom of iodine (I) deficiency is the swelling of the thyroid gland, or goitre, especially in newborn animals. Sheep and goats are more prone to this deficiency than cattle. However, adult animals may also be affected by insidious, subclinical forms, which lead to a reduction in milk yield, embryonic mortality/losses and a high rate of mortality among newborn animals. Iodine deficiency is quite common in sheep.

The majority of forage contains low levels of I as regards the animals' requirements. Inland areas tend to become increasingly poor in I as a result of their constant impoverishment. This situation may be exacerbated by other factors, such as a high Ca intake, which limits the absorption of I, or by the feeding of *Brassicae*, which produce thiocyanate in the rumen; this can restrict I uptake.

Almost all the I present in the body is stored in the thyroid in the form of thyroglobulin and thyroxine (T4). Iodine is the base ingredient of the thyroid hormones – triiodothyronine, or thyronine (T3), and tetraiodothyronine (T4), only the first of which represents the free, physiologically active form. These have important functions in the control of energy exchanges, metabolism and tissue growth.

Iodine deficiency leads to the onset of goitre because the paucity of thyroxine causes stimulation of the thyroid by thyroid-stimulating hormone (TSH) produced by the pituitary gland, inducing hyperplasia of the thyroid.

A reduction in basal metabolism leads to the onset of a series of non-specific clinical symptoms, including insufficient development in growing animals, low milk yields and the general debilitation of sheep, accompanied by a gradual decline in their sexual functions. A more specific diagnostic symptom is the occurrence of abortions and stillbirths that may exhibit goitre and alopecia.

A precise diagnosis can be made by measuring I level in the blood, in the urine and even in the milk, although it is not easy to interpret the results obtained, in as much as this deficiency is a long-term problem, and the thyroid gland accumulates sufficient I for use during periods of reduced intake. In sheep, less than 80 mg/ml of I in the milk may be a sign of insufficient I in the feed. Once it has been established that there is a deficiency, simple correction of the diet should by itself resolve the problem. Many complex foods contain a sufficient quantity of I, such as iodized salt blocks, and should be made available to the animals. The recommended daily dose is 0.8 mg/kg DM of feed in the case of lactating ewes, and 0.12 mg/kg DM for lambs and non-lactating ewes. An excessive quantity of I supplement may lead to clinical problems of excess, or iodism.

References

Andrews A.H. (1990) Colostrum. Part of nature's survival kit. *Proc. Alltech's Sixth Annual Symp.*, pp. 227–293.

Bertoni G. (1992) Ruolo dell'alimentazione nel modificare i parametri qualitativi del latte ovino. *Proc. Seminario SIPAOC,* Perugia, Italy, pp. 11–27.

Bezille P. (1996) Tossiemia da gestazione e ipocalcemia nella pecora. *Summa,* 9: 87–91.

Blood D.C., Radostitis O.M., Henderson J.A. (1989) *Veterinary medicine* 9th edn., Baillière Tindall, London.

Decante F. (1996) Tetania da erba: fisiopatologia e prevenzione. *Summa,* 9: 67–75.

Flint H.J. (1997) The rumen microbial ecosystem – some recent developments. *Trends in Microbiology,* 5: 12, 483.

Ford E.J. (1983) Pregnancy toxaemia. In: W.B. Martin (ed.) *Diseases of sheep.* Blackwell Scientific Publications, Oxford, UK.

Gillespie J.H., Timoney J.F. (1984) In: Hagan and Bruner (eds) *Malatttie infettive degli animali domestici,* Grasso pubbl., Bologna, Italy.

INRA (1988) *Alimentation des bovins, ovins et caprins.* INRA, Paris.

Morgante M., Ranucci S. (1997) Fattori nutrizionali predisponenti e condizionanti l'insorgenza di malattie diffusive. *Proc. seminario 'Patologie diffusive dei piccoli ruminanti: scenari, strategie e definizione dei ruoli'*, Perugia , Italy, pp. 37–59.

Morgante M., Capuccella M., Ranucci S., Beghelli D., Tesei B. (1996) Le cellule nel latte di pecora: variazioni quali-quantitative durante i primi mesi di lattazione. *Proc. XII Congress SIPAOC (Italy)*, pp. 57–62.

Morgante M., McCoy M.A., Pauselli M., Beghelli D., Ranucci S. (1998) Selenium status in Sardinian dairy ewes. *Proc. VI Congress of Fe.Me.S.P.Rum.*, Postojna, Romania, pp. 111–114.

Morgante M., Beghelli D., Pauselli M., Dall'Ara P., Capuccella M., Ranucci S. (1999) Effect of administration of Vitamin E and selenium during the dry period on mammary health and milk cell counts in dairy ewes. *J. Dairy Sci.*, 82: 623–631.

National Research Council (NRC) (1985) *Nutrient requirements of sheep*, 6th revised edn. National Academic Press, Washington, DC.

Patra R.C., Lal S.B., Swarup D. (1996) Biochemical profile of rumen liquor, blood and urine in experimental acidosis in sheep. *Small Ruminant Research*, 19: 177–180.

Payne J.M. (1977) *Metabolic diseases in farm animals*. Williams Heinemann Medical Books Ltd, London.

Russel A.J.F. (1983) Deficiencies of macro-elements in mineral metabolism. In: W.B. Martin (ed.) *Diseases of sheep*. Blackwell Scientific Publications, Oxford, UK.

Russel A.J.F. (1984) Means of assessing the adequacy of nutrition of pregnant ewes. *Livestock Prod. Sci.*, 11: 429–435.

Santra A., Karim S.A., Mishra A.S., Chaturvedi O.H., Prasad R. (1998) Rumen ciliate protozoa and time utilisation in sheep and goat. *Small Ruminant Research*, 30: 13–18.

Smith B.P. (1996) *Large animal internal medicine*. 2nd edn. The C.V. Mosby Company.

Suttle N.F., Linklater K.A. (1983) Disorders related to trace element deficiencies. In: W.B. Martin (ed.) *Diseases of sheep*. Blackwell Scientific Publications, Oxford, UK.

Valercher J.F., Schelcher F., Foucras G., Espinasse J. (1996) Equilibrio idro-elettrolitico: meccanismi regolatori e patologia. *Summa*, 9: 57–65.

Vrzgula L. (1991) *Metabolic disorders and their prevention in farm animals*. Elsevier, Amsterdam.

11 Grazing Management and Stocking Rate with Particular Reference to the Mediterranean Environment

Giovanni Molle[1], Mauro Decandia[1], Sebastiano Ligios[1], Nicola Fois[1], Timothy T. Treacher[2] and Maria Sitzia[1]
[1]*Istituto Zootecnico e Caseario per la Sardegna, Olmedo, Italy;* [2]*51 Western Road, Oxford, UK*

11.1 Pasture Eco-physiology

Grazing is the interaction between animals using the pasture and the pasture itself. Indeed the grazing behaviour of herbivores is strongly influenced by the characteristics of the pasture, while its growth habit, growth rate and eco-physiological development are in turn affected by the grazing intensity.

Forage plants can be perennial species, whose life cycle develops over a period of years, or annual species, which germinate, flower, and die during a single growing season (autumn–spring or spring–summer). In both cases, the growth rate is strictly linked to the climatic conditions and thus to the seasons, which govern the transition from the vegetative phase of herbage growth to the reproductive phase, during which the plant flowers and seeds. The autumn–spring cycle is the most common for annual forage in Mediterranean climates. During the winter vegetative phase the biomass principally consists of leaves of high nutritive value. As the photoperiod lengthens and the temperature starts to rise, the reproductive phase begins. In grasses this phase is marked by a fall in quality of the forage due both to the increase in the fibrous content of the plant as the reproductive stem develops and hardens, and to the fact that the plant does not produce new leaves. In addition, all of the plant's reserves are concentrated in the seeds. In legumes the fibrous proportion also increases as flowers form and the pods develop, but the quality of herbage available for grazing is generally higher than that for grasses during the pasture growing cycle (Table 11.1) as well as standing hay in summer (Sitzia and Fois, 1999; Fois *et al.*, 2000c).

Apart from climatic factors such as humidity and temperature, herbage growth is influenced by the type and fertility of the soil. In the Mediterranean environment growth usually slows down during the colder months and may actually stop at high altitudes. Basically growth depends on the ability of the plant to intercept sunlight and transform non-organic carbon into carbohydrates. In this respect the quantity of green leaves is of fundamental importance, and this is expressed by the leaf area index (LAI), which measures the green area per unit of ground surface area. The plant has an efficient LAI if green tissue is continuously produced and replaced.

After the leaves have been grazed the plants replace them in a time period which

Table 11.1. Average dry matter availability (DMA, t/ha), crude protein (CP, %DM) and NDF (%DM) of some Mediterranean forage species (Fois *et al.*, 2000a,b).

	Burr medic (*Medicago polymorpha*)		Sulla (*Hedysarum coronarium*)		WFC[a]	Annual ryegrass (*Lolium rigidum*)		Natural pasture[b]	
	Winter	Spring	Winter	Spring	Winter	Winter	Spring	Winter	Spring[c]
DMA	1.7	4.6	1.9	6.0	2.0	1.7	6.1	1.0	5.1
CP	26.6	22.1	22.0	16.9	18.7	17.5	11.7	16.3	5.9
NDF	33.6	42.1	33.7	43.6	42.4	38.2	54.8	48.3	69.0

[a] Winter forage crops: oats (*Avena sativa*), hairy vetch (*Vicia villosa*), Persian clover (*Trifolium resupinatum*); [b] 60% grasses; [c] late spring.

depends on the species (leaf emergence rate, average life span of the leaves) and the intensity of use. In general, intensive use, which results in removal of most of the leaves, lengthens the period necessary for regrowth and thus the length of time before the plant can be grazed again.

The production of biomass depends on the number of plants per unit area as well as on the intrinsic characteristics of the forage species used. In grasses, tillering is thus of fundamental importance as a guarantee of good levels of production. This takes place during the vegetative growth of the plant and is favoured by a good light level at the base of the tiller of the mother plant. If water and light are limited, few secondary tillers are produced and this can eventually result in excessive thinning-out of the pasture and the possible death of the plants. Apart from seeding density, one of the factors which principally favours tillering is grazing. Indeed, in a pasture of Italian ryegrass, continuously grazed at different heights using variable stocking rates, the tiller number was inversely related to height of the pasture and thus to grazing intensity (Sitzia *et al.*, 1997). In a pasture of annual ryegrass under rotational grazing, the number of tillers per hectare was higher than under continuous grazing at the same stocking rate (6 ewes/ha), due to the higher instant stocking rate (Table 11.2). Grazing can also affect the number of pods, or the production of stolons or stems in clover or medics.

The available biomass, expressed in tons of dry matter per hectare (t DM/ha), consists of all the dead and green matter present per unit of surface area. It can be estimated by measuring the height of the pasture, as there is a strong correlation between the parameters of pasture height and total mass. The height of the pasture may be undisturbed pasture height, which is measured by simple ruler (Filigheddu and Pulina, 1986) or, with more precision, by a sward-stick (Barthram, 1985). Another method, which takes into consideration the density as well as the height of the pasture,

Table 11.2. Variation in the pasture density (no. of tillers/m^2) of annual ryegrass and of Italian ryegrass when different grazing methods and stocking rates are used (Sitzia *et al.*, 1996; Porqueddu *et al.*, 1997).

Grazing method	Species	Stocking rate	Pasture density
Rotational	Annual ryegrass	Fixed	2750
Continuous	Annual ryegrass	Fixed	2250
Continuous	Italian ryegrass	Variable, high	4850
Continuous	Italian ryegrass	Variable, medium	3250
Continuous	Italian ryegrass	Variable, low	2900

Table 11.3. Linear regression equations (y = a + bx) between the pasture height (cm) measured with the herbometer (EH = x) and with the sward-stick (SSH = y) (Carta, 1993; Sitzia, personal communication).

Species	Grazing method	Season	Range		a	b	R^2
			SSH	EH			
Italian ryegrass	Continuous	S	1.2–17.2	0.3–14.5	1.21	1.05	0.89
Italian ryegrass	Not grazed	S	3.6–41.4	1.5–25.5	0.54	1.42	0.91
Annual ryegrass	Rotational	W–S	9.3–30.6	4.2–20.2	3.59	1.29	0.97
Meadow[a]	Rotational	W	0.9–49.9	1.6–41.1	1.33	1.23	0.95

S: spring; W: winter.
[a] Annual ryegrass or pure sulla.

uses a herbometer consisting of a weighted disc or square plate mounted on a graduated pole (Holmes, 1984). These two measurements are closely correlated (Table 11.3).

After measuring pasture height with a herbometer, the available biomass can be calculated using the relationships shown in Table 11.4.

11.2 Behaviour of Grazing Sheep

Grazing is a process which involves the following behaviour: (i) movement of the animal associated with the choice of the feeding site and, within it, the patch or feeding station and finally the biting point (Bailey *et al.*, 1996); and (ii) eating, which includes biting and rapid chewing of the herbage, and ends when the bolus is swallowed.

When grazing, sheep tend to select a diet of a quality that is above the average of the herbage on offer. They select the green components, and in particular the leaves of the plants and, when grazing heterogeneous pastures, they tend to prefer the legumes (Hodgson, 1982; Penning *et al.*, 1995, 1997; Molle *et al.*, 2000, 2003). By this behaviour, sheep enhance their intake rate (IR, intake of dry matter per time unit of grazing). This is the most efficient feeding strategy for small ruminants as a high IR results in a high passage rate of nutrients, which helps to overcome the limitation of ruminal volume on intake in small ruminants (Prache *et al.*, 1998).

The selection of the diet of a grazing sheep is determined by: (i) information gathered by the senses as well as subconscious knowledge of the post-ingestive effects of feeds; and (ii) learning from the mother while young, both pre- and post-weaning, and from other members of the flock (Prache *et al.*, 1998).

When grazing on a monoculture with a homogeneous pasture structure, diet selection is obviously limited, and feeding behaviour depends principally on the accessibility of the herbage (see Section 11.4).

Table 11.4. Linear regression analyses (y = a + bx) between the pasture height (x, mm) measured using the herbometer and biomass availability (y, t DM/ha) (Sitzia *et al.*, 1997, 1998, 2000).

Species	Grazing method	Season	Range (mm)	a	b	R^2
Annual ryegrass	Rotational	Winter–spring	56–417	0.116	0.013	0.84
Italian ryegrass	Rotational	Winter–spring	37–290	0.016	0.01	0.75
Italian ryegrass	Continuous	Winter–spring	30–90	0.32	0.04	0.61
Italian ryegrass	Continuous	Late spring	30–90	0.22	0.07	0.85
Sulla	Rotational	Winter–spring	58–678	0.793	0.01	0.75
Burr medic	Rotational	Winter–spring	12–330	−0.026	0.016	0.57

When grazing on a heterogeneous pasture of a monoculture, by contrast, there will be more selection due to the presence of patches of vegetation at different stages of maturity. Most selection is evident on pastures consisting of several species at different stages of maturity.

Recent studies have found 'opportunist' selective behaviour in sheep. Sheep have a certain selection threshold, above which they accept a diet of lower quality (Roguet *et al.*, 1998). For example, where legumes and grasses are found intimately mixed there is less selection than when they are offered separately in large patches in blocks or strips (Baumont *et al.*, 2000). In addition, selection is also affected by the presence of chemical compounds in the herbage, which may either increase, e.g. soluble carbohydrates (Ciavarella *et al.*, 2000), or decrease herbage intake, e.g. anti-nutritional factors, such as tannins (Silanikove *et al.*, 1997). Generally, the selection of feed tends towards optimization of conditions in the rumen (Cooper *et al.*, 1996).

The intake per metre of walking is an important index of selection activity. During the day there are phases of intense feeding per distance walked, and phases of selective feeding in which the ruminant consumes less feed for every metre walked (Meuret, 1997).

11.3 Grazing Methods

11.3.1 Objectives

Managing the complex grazing ecosystem is principally aimed at optimizing animal production while maintaining the resource (pasture). With this aim in mind, intake of nutrients from the grazed herbage must be maximized, as pasture is generally the cheapest source of feed. It is of particular importance to increase the length of the intensive grazing phases, during which the greatest quantity of nutritive material is consumed, at the expense of the selective phases. This reduces the energy expended by the animal during grazing relative to the amount of nutrients ingested.

A long-term grazing strategy designed to maximize animal productivity should also incorporate the following objectives:

- The pasture must be capable of adequate recovery after grazing.
- The pasture must be kept in a 'leafy' state for as long as possible, in order to delay the loss in nutritional value which takes place when it enters the reproductive phase.
- The persistence of the pasture should be maximized by management to reduce the spread of weeds, and to facilitate the germination of self-seeding annual species and the survival of perennial forage species.
- The parasite burden of the animals should be reduced as much as possible.
- The environmental impact on the whole grazing ecosystem must be minimized.

11.3.2 Description of methods

The principal grazing methods that can be used to achieve these objectives are shown in Fig. 11.1.

Continuous stocking

This method involves the uninterrupted use of a particular grazing area by the animals for a long period, normally the whole of the grazing season. Continuous grazing does not mean that there are no intervals in the use of each plant (tiller or grazing point), since, after defoliation, the plants usually become less accessible to the animal. The plants can be grazed after regrowth to a minimum height (ideally 20–30 mm), at which they are again accessible to the animals.

Continuous grazing can be *set* stocking, in which there are a constant number of animals on the grazing area, or *variable* stocking. In the latter, the stock number is adjusted to pasture growth so that a targeted level of herbage allowance (g DM/animal/day) is provided in order to match livestock requirements. Continuous variable stocking may also be achieved by varying

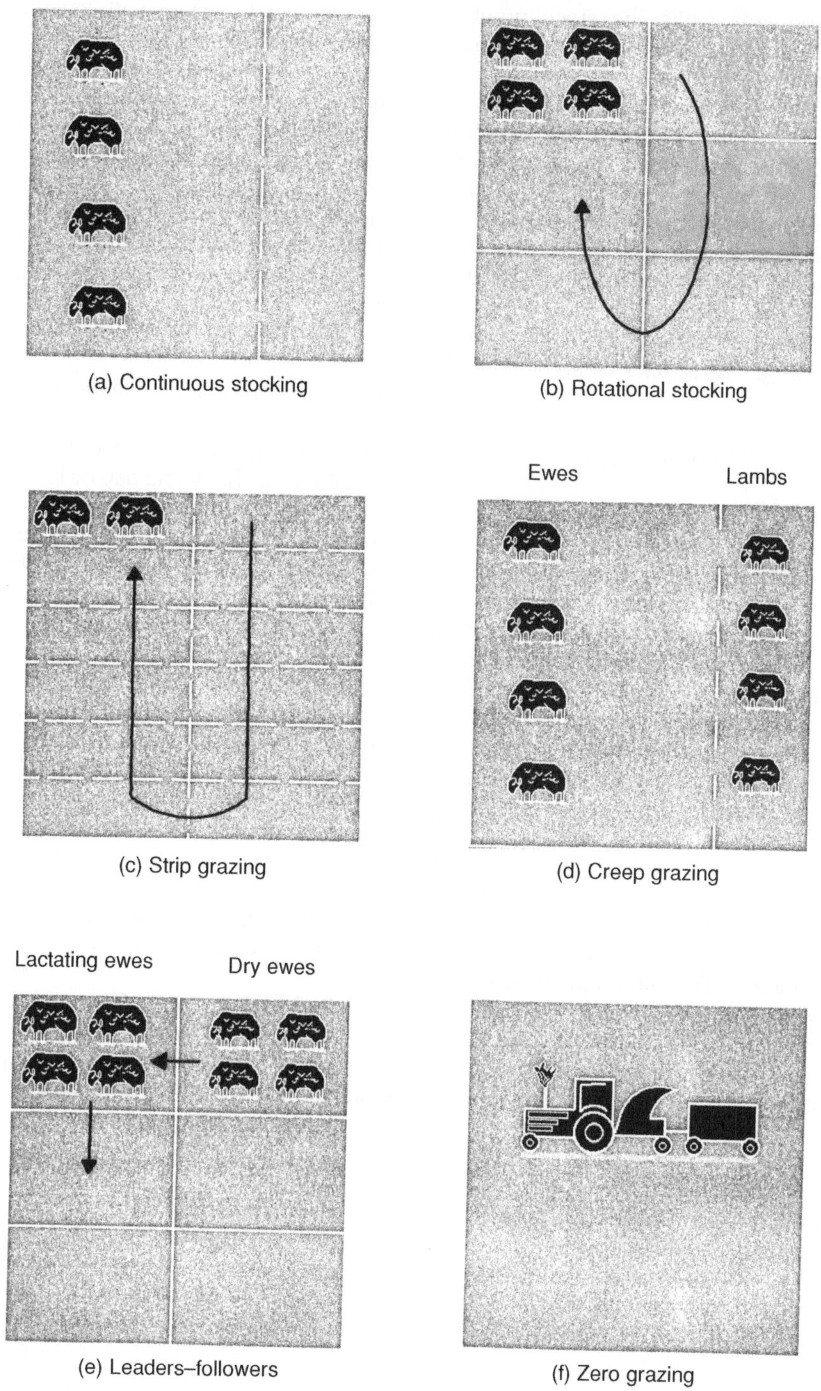

Fig. 11.1. Grazing methods. Dotted lines indicate moveable fences. The creep grazing system shows a fence with selective gates that permit the lambs, but not their mothers, to pass through to the area reserved for grazing by the lambs (Hodgson, 1990).

the grazing area while the stock number remains the same; this is often the best option to meet animal requirements. In addition, if the stocking rate is adjusted to the growth of the pasture, the negative effects of animal selectivity are reduced, in particular the spread of ungrazed patches. Selective behaviour can put the persistence of the most palatable species at risk, and thus increase the proportion of unpalatable weeds. These effects are modulated by stocking rate, herbage availability and by the resilience of grazed species when defoliated.

Rotational grazing

In this method the pasture is grazed for certain periods, interspersed with breaks which permit undisturbed growth of the pasture. The pasture is subdivided into paddocks which are used in succession, and the period of time for which each paddock is used is called the grazing period. This is followed by the rest period, and the sum of the two periods is called the grazing cycle.

In rotational grazing the stocking density i.e. the number of animals actually being grazed at any point in time, and not just the stocking rate, is important. One type of rotational grazing is strip-grazing, in which the area is adjusted on the basis of the daily feed requirements of the animals. This is usually done daily, although it may be increased to a maximum of 2 or 3 days.

Other types of rotational grazing, albeit little used in the Mediterranean environment, are:

- Creep grazing: one category of animals, usually suckling lambs, has access to a separate area of pasture, generally of higher quality. This allows the lambs to select a diet of higher quality without competition with their mothers.
- Leaders–followers: in this case a flock is divided into two groups with different nutritional requirements (e.g. lactating and non-lactating sheep), and the grazing period is divided into two sub-periods. The pasture is grazed first by the group which has greater requirements (leaders), and then by the group with lower requirements (followers). Thus, in certain conditions, the feeding requirements of the different animals can be satisfied in a more balanced way. For instance, milking sheep will consume a larger quantity of leaves, which are the most nutritional part of the pasture, while non-lactating sheep are forced to consume the stems, which have lower nutritional value.

Rationed grazing and complementary grazing

Both the techniques described above can be used either for the whole day or for a limited number of hours per day in what is termed rationed grazing. If access to the pasture is less than 12 h daily, intake is likely to be restricted (Hodgson, 1982). Rationed grazing is of particular importance in the management of dairy sheep, whose access to pasture is limited both by the need to remove them during milking and by the fact that they are often housed at night, to protect them from cold or from predators. Rationed grazing is often used at the beginning of the grazing season when there is not enough forage available to satisfy the nutritional needs of the animals, and the animals require high-value supplements, which are expensive and sometimes not easily available.

When sheep are allocated nutritionally high pasture for a few hours daily in order to complement a nutritionally low pasture grazed for the remaining daytime, the method is termed complementary grazing.

Zero grazing and deferred grazing.

Zero grazing is the term for the method in which herbage is cut and carted, usually with machinery, and trough-fed to the flock. It is used where only very small areas of grassland are available or where the areas of grassland are too far from the sheep house for the flock to walk to each day in a reasonable time.

If one refrains from grazing a pasture for a limited period until the available biomass is sufficient, this is termed deferred grazing. Deferred grazing is also used when the trampling associated with grazing could damage the soil structure (Ligios *et al.*, 1997), which often happens in wet conditions during autumn and winter.

11.3.3 Criteria for choosing grazing methods

In normal conditions, the use of one of the grazing methods discussed above throughout the grazing season will probably not give optimum results. A number of comparisons between continuous stocking and rotational grazing of cattle in temperate climates have shown that using rotational grazing, rather than continuous stocking, has little advantage: a 6–7% increase in meat production and an even smaller increase in milk production. This is particularly true when stocking rate is adequate for efficient utilization of pasture production (Hodgson, 1990).

In lactating sheep, grazing improved pasture of annual ryegrass, subterranean clover and burr medic for the entire grazing season at a stocking rate of 6 ewes/ha; no significant improvement in production was observed when using rotational grazing as opposed to continuous stocking (Sitzia *et al.*, 1996). The available biomass, however, tended to be greater when rotational grazing was used.

Short-term studies (Avondo *et al.*, 1994) in spring and summer on sheep whose access to pastures of oats or clover was limited to 6 h a day, with a high stocking rate of 15 ewes/ha, indicated that mean daily intake and milk yield were higher with rotational grazing (1295 g DM and 615 ml, respectively) than with continuous stocking (1125 g DM and 460 ml). These differences, while small in absolute terms, were statistically significant and may be associated with the greater availability of biomass in rotational grazing (4.8 t DM/ha) than with continuous stocking (3.7 t DM/ha).

While there is relatively little information on Mediterranean pastures grazed by sheep, general criteria for grazing management can be established (Table 11.5). These criteria provide only broad guidelines because, as previously stated, changing the stocking rate or the grazing pressure could result in a different method giving better production levels.

Table 11.5 does not preclude the use of multiple methods on the same grazing area during the course of one grazing season. For example, Italian ryegrass could be

Table 11.5. Guidelines for the choice of grazing methods.

	Conditions of use			
Type of pasture	Forage habit of growth	Forage production	Stocking rate[a]	Grazing method
Natural or semi-natural pastures, cereal stubble	Prostrate or slightly upright	Low–medium (1–5 t DM/ha)	Low–medium (1–5 head/ha)	Continuous stocking
Forage crops of grass and legumes	Upright	Medium–high (5–10 t DM/ha)	Medium–high (5–10 head/ha)	Rotational grazing
High-quality legume forage crops, forage crops containing anti-nutritional factors	Various	Various	Various, generally high (10–15 head/ha)	Rationed grazing
Sorghum spp.[b], Maize[b]	Upright (unsuitable for grazing)	Very high (> 10 t DM/ha)	Very high (equivalent to > 15 head/ha)	Zero grazing

[a] These are guidelines for annual stocking rates based on long-term experiments carried out by the Istituto Zootecnico e Caseario under semi-arid Mediterranean conditions; [b] only under irrigation.

grazed initially using the rationed grazing method during late autumn, when quantity is limited, but it is the best quality of feed available nutritionally for the ewes in late pregnancy and the suckling period. A rotational grazing system can then be used, with a move towards shorter grazing cycles and higher stocking densities during the transition from winter to spring. This can be implemented, for example, by excluding some paddocks from grazing in spring and using them for hay or silage production. Finally, in summer, when only very mature herbage is available, continuous stocking would be the most appropriate choice.

In self-reseeding species (e.g. subterranean clover and annual medics), during the first grazing season, the pasture must not be grazed during, or immediately after, the flowering period, especially if the pasture has been grazed rotationally at a high stocking density during the rest of the grazing season. This deferred grazing is aimed at favouring the establishment of a large seed bank in the soil. After the first reseeding, these species can be either rotationally grazed or continuously stocked at a low stocking rate during the flowering–seeding phase.

More sophisticated grazing methods (e.g. leaders–followers), aimed at satisfying the part of the flock with higher feeding requirements, are applicable when the grazed herbage represents the main component of the diet.

On particular farms the choice of grazing system will be affected by other factors, such as the cost of pasture fencing and drinking points, the time and cost involved in leading the flocks to the pasture, the mechanization of forage cutting, conservation and feeding.

It is not possible to generalize about the impact of grazing methods on pasture persistence and other ecological factors relevant to farm management, e.g. spread of unpalatable weeds, parasite burden etc. Generally, the grazing method itself seems to have less impact than the actual stocking rate. However, further studies are needed to evaluate the implications of grazing methods.

11.4 Effect of Grazing Methods on Feeding Behaviour of Grazing Sheep

In grazing animals intake is regulated by physical, metabolic and behavioural mechanisms. The activities of selection and walking have specific energy costs. These energy costs are minimized when sheep are fed from troughs and the diet is easily accessible.

Herbage intake depends on three main factors:

- Intake capacity, which is related to the characteristics of the animals, such as size, physiological state and productive level.
- Forage ingestibility, which depends on chemical characteristics which affect the rate of particle breakdown in the rumen, e.g. fibre content and presence of anti-nutritional factors.
- Forage accessibility.

The Australian Grazfeed pasture management system (CSIRO, 1993) defines this as the ease of harvesting of the herbage from the grazed pasture.

The following equations explain this schematically:

$$I = PI \times A$$

where: I is the intake of herbage (DM); PI is the potential DM intake, which depends upon both the animal's intake capacity and the forage ingestibility; and A is the accessibility expressed as a decimal, varying from 0 to 1 as the herbage varies from totally inaccessible to fully accessible.

The accessibility of the pasture modifies the mass of each bite, bite frequency and total daily grazing time. These parameters are correlated to intake according to the following equation:

$$I = BM \times BF \times GT$$

where: I is the intake of herbage (DM); BM is the bite mass (g DM); BF is the bite frequency (number of bites per minute of grazing); and GT is the grazing time (min/24 h).

The product of BM × BF is intake rate (IR) in g DM/min of grazing.

The biomass per unit of area, which can be estimated by pasture height measurements, and pasture structural composi-

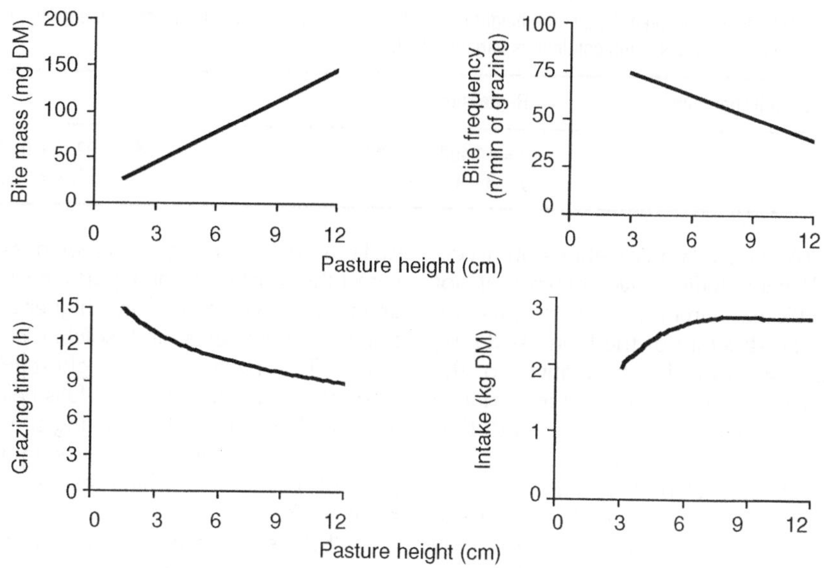

Fig. 11.2. Responses by feeding behaviour variables to pasture height in mutton sheep continuously stocked on perennial ryegrass (Penning, 1986; Penning *et al.*, 1991).

tion (percentage of leaves, stems, flowers, seed heads and dead matter) are the most important factors which affect herbage accessibility and thus herbage intake. For example, it has been found that when the pasture height is reduced, BM is also decreased, and below a certain threshold this reduction is only partially compensated by an increase in BF and GT (Penning, 1986; Penning *et al.*, 1991) (Fig. 11.2). While there are general relationships between the characteristics of pasture and feeding behaviour, this behaviour varies depending on the grazing methods, the forage species and other factors. Given the limitations of the available data on Mediterranean pastures, the purpose of the following sections is to define guidelines for a set of grazing conditions with particular reference to Mediterranean grass and legume-based pastures. The last section provides guidelines for grazing temperate-climate pastures.

11.4.1 Grass – continuous stocking

The height of the pasture is an effective indicator of herbage mass, in particular during the vegetative phase. In pastures of Italian ryegrass (*Lolium multiflorum* Lam.) continuously stocked during the period of vegetative growth there is a positive linear relationship between height (measured by a weighted scale grass-meter) and herbage mass. There is a difference between winter–early spring and late spring, which is associated with variations in the structure of the pasture (Fig. 11.3).

Pasture height was also found to be a good predictor of herbage intake by sheep (Table 11.6).

Pasture height also affected feeding behaviour: sheep grazing on pasture heights that were proved to limit the intake showed a higher GT than their counterparts (791, 647 and 578 min/24 h, respectively), for groups that grazed on pasture with heights of 30, 60 and 90 mm in spring (Casu *et al.*, 1995).

Average accessibility was calculated at 0.65–0.70 when the height greatly limited intake (20 mm in winter, 30 mm in spring), and was 0.75–0.87 for the highest pasture (80 mm in winter and 90 mm in spring (Bocquier *et al.*, 1993). This indicates that pasture height and hence herbage mass do

Table 11.6. Regressions between the pasture height (x, mm) and intake (y, g DM), calculated as the group mean, in sheep grazing Italian ryegrass using continuous variable stocking.

Period (range of height, mm)	Regressions	
Winter (20–80)	$y = 14.452\,x + 1268$	$R^2 = 0.92$, $P < 0.01$
Spring (30–90)	$y = 38.567\,x - 0.244\,x^2 + 333$	$R^2 = 0.92$, $P < 0.05$

not completely explain limitations on accessibility. Recent studies have shown that not only the biomass and height of the pasture, but also the structure of the biomass during the passage from the vegetative to the reproductive phase, are of great importance in modifying feeding behaviour and thus intake by grazing ruminants.

Indeed, ease of prehension on mature pasture depends not only on the height and density of the pasture but also on the ease of severance of herbage and on the presence of barriers to defoliation (Prache and Peyraud, 1997), such as leaf sheath and dead herbage. In these conditions the sheep make a choice, trying wherever possible to ingest the most nutritional parts, particularly green parts, which also have higher concentrations of proteins and soluble carbohydrates. This results in the highest IR; only when the IR for this material falls below the average IR for the entire feeding site do the sheep move to graze the less nutritional parts of the pasture, which results in a lowering of IR.

Obviously, the more leafy the pasture the less the limitation on accessibility, and thus the higher the intake rate. This explains why there is no difference in intake between 60 mm and 90 mm pastures (a significant

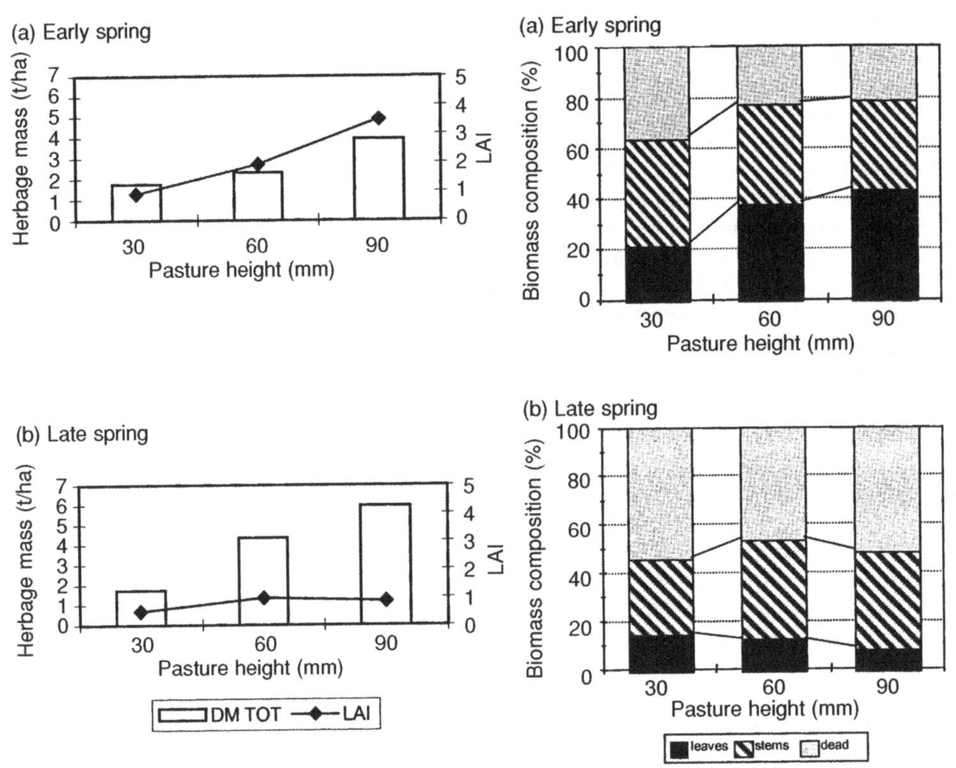

Fig. 11.3. Relationship between pasture height and herbage mass (DM), leaf area index (LAI) and herbage composition of Italian ryegrass continuously stocked by sheep (Sitzia *et al.*, 1997).

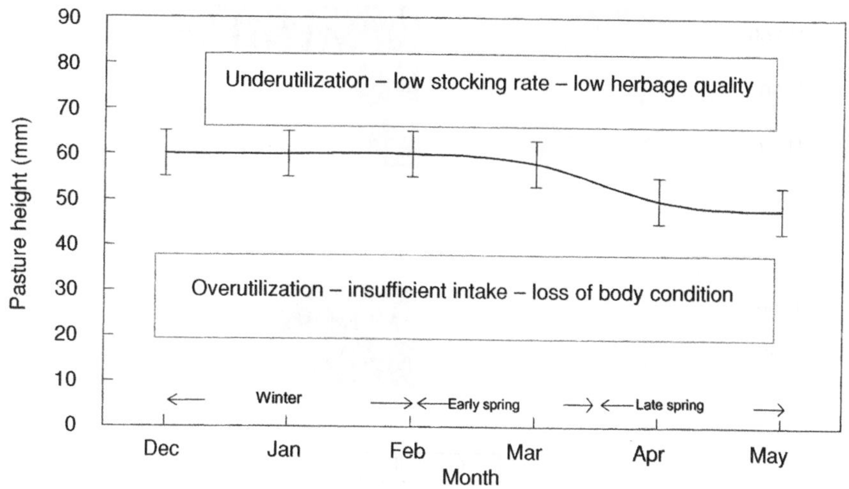

Fig. 11.4. Optimal profile of the height of the pasture for Italian ryegrass grazed by dairy sheep using the continuous stocking method.

effect in the quadratic equation, see Table 11.6). Indeed, in the plot grazed at 90 mm pasture height, the fibrous part of the biomass increases and the nutritional proportion of the diet becomes less accessible (Fig. 11.3).

On the basis of the above results, continuously stocked Italian ryegrass should be grazed in winter at a height of 60 mm (2 t DM/ha), which permits the pasture to capture as much light energy as possible. In spring it should be used at progressively lower heights in order to maintain the quality of the pasture (Fig. 11.4). It should not, however, be used at heights of less than 30 mm (1.5–2 t DM/ha).

When the pasture is mature, after ear emergence, the available biomass is not a significant indicator of herbage accessibility. In these conditions, even with continuous stocking, the available green biomass per unit of area plays a predominant role in determining grazing behaviour and intake at pasture. When the pasture is heterogeneous, and *Dactylis glomerata* L. is present in both its vegetative and reproductive form, the availability of green leaves in quantities of < 0.5 t DM/ha greatly reduces the IR and the sheep begin to consume reproductive stems, which they had initially rejected (Prache *et al.*, 1998).

11.4.2 Grass – rotational and rationed grazing

In rotational grazing there are two modifications in the relationship between available biomass and feeding behaviour: (i) the rapid variation of the herbage on offer during the grazing period causes an extremely dynamic interaction between the sheep and the pasture; and (ii) the structural composition of the pasture is notably different when rotational grazing is used, and defoliation is at much greater intervals than under continuous stocking. This causes important variations in the behaviour of the animals, whose intake rate falls considerably during the grazing period, but, in contrast with continuous stocking, their grazing time does not markedly increase, and in some cases, does not increase at all (Penning *et al.*, 1994; Molle *et al.*, 2004). Thus, given a certain pasture height (herbage mass), less herbage is usually ingested in rotational grazing than in continuous stocking (Fig. 11.5).

While the pasture height is used in equations for the above-mentioned parameters, it has recently been established that the green biomass per unit of surface area, or the allowance of green material per head per day, is a better variable for determining the grazing behaviour of the animals,

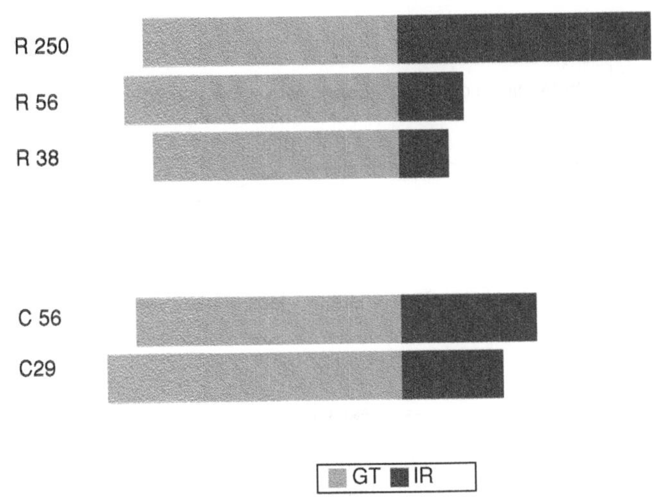

Fig. 11.5. Grazing time (GT) and intake rate (IR) in mutton sheep, continuously stocked (C) on perennial ryegrass pasture at either 30 or 60 mm high, or rotationally grazed. The figures which follow the letters indicate the pasture height in mm, measured by pasture stick (Penning *et al.*, 1994).

because it takes into account the effect of barriers of stems and dead parts of the plants (Delagarde *et al.*, 2001).

In rationed grazing where the time of grazing is restricted each day, IR is affected. In these conditions it may be of interest to calculate the hourly consumption (g DM/h), which is related not only to the available biomass and its composition but also to the total time that the animals are allowed to graze the pasture. For example, lactating sheep receiving a complete diet indoors followed by periods of grazing (Table 11.7) had mean intakes of Italian ryegrass ranging from 250 to 200 g DM/h at the beginning of the grazing period to 180–120 g DM/h at the end of the grazing period, when the pasture height was 30–40 mm. Thus the highest hourly consumption levels occurred when the time the sheep had been on the pasture was lowest (Fois *et al.*, 1995). A series of studies, under similar grazing conditions, found that average consumption varied from a minimum of 147 g DM/h in winter to a maximum of 275 g DM/h in spring (Fois *et al.*, 1997; Sitzia *et al.*, 1998). Similar results have been found with Latxa lactating dairy sheep in Spain: comparison between two groups spending 4 and 7 h daily on a grass-based pasture demonstrated that the proportion of the time devoted to grazing was 53% (7 h) and 77%

Table 11.7. Effect of total hours on pasture on feeding behaviour and intake in Sarda sheep grazed on Italian ryegrass, as a complement to a complete diet (0.85 UFL and 160g CP per kg DM).

Lactation month	Hours on pasture (h/day)	Pasture height (mm)	Grazing time (% total time)	Intake rate (g DM/min grazed)	Hourly intake (g DM/ h)
2–3	2	80	84	5.2	253
2–3	2	45	74	3.1	143
4–5	3	130	63	5.2	197
4–5	3	30	87	3.5	177
6	4	40	66	3.0	119

(4 h), and the hourly intake ranged on average between 166 (7 h) and 244 g DM/h (4 h) (Perojo *et al.*, 2003).

11.4.3 Legume-based pastures

Ingestion of legumes usually results in a higher rate of passage and digestion than grasses. The behaviour of sheep grazing legumes differs from those on grass, with both bite mass and intake rate being greater, and rumination time lower. In general this results in greater daily intake and higher performance (Table 11.8).

Studies on white clover in northern Europe using continuous stocking have shown that the herbage mass influences sheep behaviour in a similar way to perennial ryegrass. Sheep grazing monocultures of white clover and perennial ryegrass ranging in height from 30 to 60 mm have greater daily intakes on clover than on ryegrass (+ 556 g DM/head) (Penning *et al.*, 1995). This is because clover is more accessible than ryegrass, with leaves more easily ingested as they are in the surface horizon of the pasture. The greater rate of passage of clover results from differences in chemical composition, with clover generally having a lower fibre content than grasses. The fibre has a lower hemicellulose content and is broken down more quickly into small particles, which exit more quickly from the rumen.

Although, in general, information on grazing of white clover may be extrapolated to other legumes, intake may be affected by differences in morphology which affect the accessibility of different leguminous species, and in some cases by the presence of antinutritional factors that may limit consumption. In particular, the presence of condensed tannins may affect the feeding behaviour of the ruminants, reducing intake rate and total hourly consumption even when feed is fully accessible (Silanikove *et al.*, 1994; Decandia *et al.*, 2000; Landau *et al.*, 2000).

However, moderate levels of tannins (3–4% DM), in legumes such as sulla, despite their tendency to reduce intake rate, allow the nitrogen in the pasture to be used more efficiently, and thus increase the absorption of amino acids from the gut (Barry and McNabb, 1999).

11.4.4 Grazing on mixed pastures of grasses and legumes

Feeding behaviour on mixed pasture basically depends on the available biomass and its quality, in the same way as in monocultures. Indeed, in sheep grazing for about 5 h per day on natural pastures, intakes correlate well with the available biomass and the CP content of the pasture, which are good indicators of pasture quality (Avondo *et al.*, 1996) (see Chapter 5).

In mixed pastures, both the number of species and their spatial distribution play an important role in determining feeding behaviour.

Table 11.8. Feeding behaviour of lactating sheep suckling twins grazing on monocultures of either perennial ryegrass or white clover continuously stocked to maintain a pasture height of 60 mm (Gibb and Orr, 1997).

	Perennial ryegrass	White clover
Bite mass (fresh matter) (mg)	303	547
Bite mass (mg DM)	83	93
Mastications per bite (n)	1.7	1.3
Intake rate (g DM/min grazing)	4.5	5.9
Grazing time (n hours/24 hours)	9.7	8.8
Rumination time (n hours/24 hours)	6.0	4.9
Total intake (kg DM/day)	2.7	3.3
Lamb growth (g/head/day)	310	370

Comparisons of mixed with pure legume diets do not always show the superiority of the latter in terms of consumption and performance (Molle *et al.*, 1998, 2000, 2003). This is due to the sheep's partial preference for legumes as compared to grasses (Newman *et al.*, 1992). A mixed diet compared with a pure legume diet may result in lower fermentation in the rumen and thus lower concentrations of metabolites (e.g. volatile fatty acids) in the bloodstream; these control the satiety centres, thus reducing the sheep's appetite.

11.4.5 Grazing pastures in temperate areas

In temperate regions, where pastures are based on perennial grasses, particularly perennial ryegrass (*Lolium perenne*), herbage growth is strongly seasonal due to variations in temperature and light throughout the year. Studies in the 1980s showed that the single criterion of pasture height, in general, provided a link between pasture growth and long-term productivity on one hand, and sheep behaviour and performance on the other (Treacher, 1990). Although these studies related to ewes suckling lambs and producing high milk yields in relatively short lactations, they should form a satisfactory basis for the grazing management of dairy ewes, whose lactations are generally longer, but with lower peak yields, than those of suckling ewes.

Detailed studies of the balances between photosynthesis, gross tissue production and the amounts of herbage, either consumed by the grazing sheep or dying, in pastures maintained at different heights, indicated that the greatest harvest of herbage per unit area by sheep occurred on pastures maintained at surface heights of 40–60 mm (Parsons and Chapman, 2000). Also, tiller densities of pastures maintained at 40–60 mm in spring and early summer are high, and the elongation of reproductive tillers, with its associated deterioration of digestibility, is almost completely eliminated. These short pastures maintain a fairly constant leaf area index throughout the season, which results in similar growth rates in autumn to those of taller pastures in which the leaf area, although higher in spring, declines through the summer and autumn as the pasture structure deteriorates.

Guidelines for management of sheep pastures in the temperate northern hemisphere have been designed and tested (Treacher, 1990). These entail the use of continuous variable stocking, together with supplementation with concentrate feeds and conservation cutting at various stages, in order to maintain the annual profile of pasture height. The optimum profile consists of reaching a height of 40 mm as soon as possible in spring, preventing the height from exceeding 60 mm during the peak of pasture growth in May and June, and then maintaining a height of 60 mm into late summer. Milk production of suckling ewes is close to their full potential on pastures in this range of heights.

In autumn, pasture height may be allowed to increase a little, to approximately 80 mm, thus increasing green leaf mass to maintain pasture intake by the ewes. The pasture should then be grazed down to 30 mm in order to minimize cold damage in winter. If the pasture is very short in spring, supplementation with concentrate feed will be needed to prevent reductions in milk production. The period when supplementation is necessary is, however, generally short and no response is likely once herbage growth is sufficient to maintain a pasture height of 40 mm. In mid-lactation it is vital to maintain a height < 60 mm, to achieve high utilization of herbage and prevent deterioration of the pasture structure and reductions in nutritive value. Generally it is not possible to increase stocking rates and, therefore, areas of land are left ungrazed to increase stocking density on the grazed areas. Herbage on these ungrazed areas is then conserved as hay or, preferably, silage, as this releases the land for regrazing sooner.

11.5 Determination of Sustainable Stocking Rate

The stocking rate (SR) is the variable which has the greatest impact on the ecology of

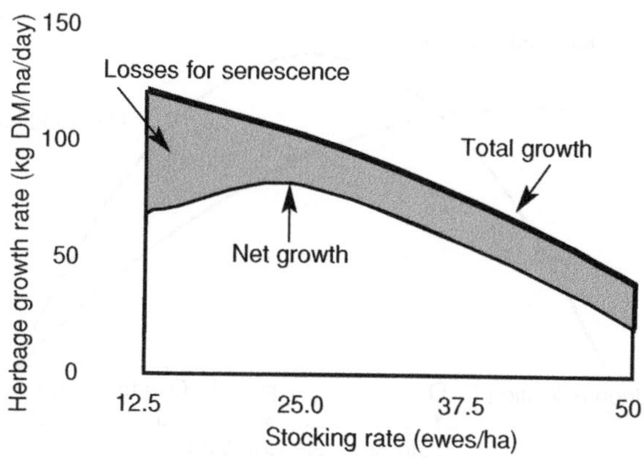

Fig. 11.6. Pattern of total and net herbage production in relation to the stocking rate (Hodgson, 1990).

grazing systems. It is the ratio between the total number of grazing animals and the total surface area grazed, expressed as number per ha (n/ha). The impact of a particular SR on an ecosytem may be affected by the size or shape of the area being grazed (Vallentine, 1990). It is also important to define the length of time that the SR is used during the year. Grazing pressure (number of animals per unit weight of herbage on offer) may be a more appropriate expression of grazing intensity. It is, however, difficult to estimate on individual farms.

As the stocking rate is increased the total growth rate of the pasture declines, but this is partially compensated for by the lower rate of loss of biomass due to senescence, at least until a certain threshold is reached. Above this level, overgrazing results in a lower net biomass growth (Fig. 11.6). In addition the stocking rate, and thus the intensity of grazing, affects the regrowth

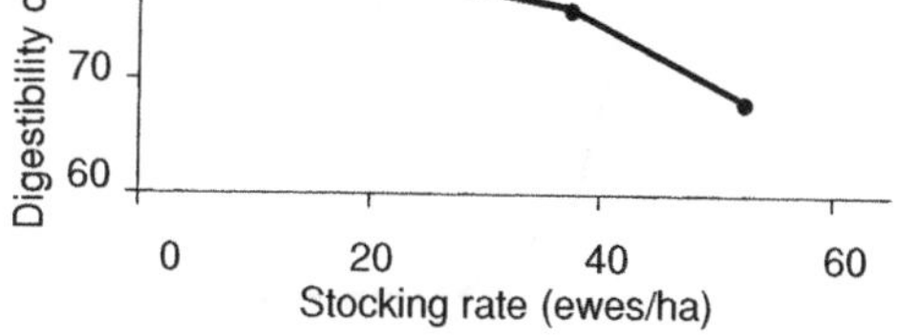

Fig. 11.7. Influence of stocking rate on digestibility of the herbage selected by grazing sheep (Hodgson, 1990).

pattern. Very intensive grazing, with the removal of most of the leaf mass, giving low values of residual leaf area index (LAI), prolongs the time necessary for the pasture to grow again and thus increases the interval between grazings.

Increasing the stocking rate generally results in the pasture being leafier (Fig. 11.3, Section 11.4) but with very high stocking rates the digestibility of the herbage consumed may be reduced because the sheep cannot select a higher-quality diet and must consume parts of the plants which they normally reject (Fig. 11.7). Thus, when the available biomass and the pasture height fall below a certain level, the herbage dry matter intake is limited, and with very high stocking rates the intake of digestible matter is further reduced.

The combined effects of the factors discussed above result in the relationship between stocking rate and production per animal, and per unit of surface area (Fig. 11.8). Normally when the stocking rate rises above the threshold of minimum stocking (point D), individual performance is reduced but there is a corresponding increase in production per hectare. However, above a certain stocking rate (point E), both the individual production and the production per hectare fall due to the effects of overstocking.

Stocking rate not only affects the quantity of animal output but may also affect the quality of milk and meat produced. For

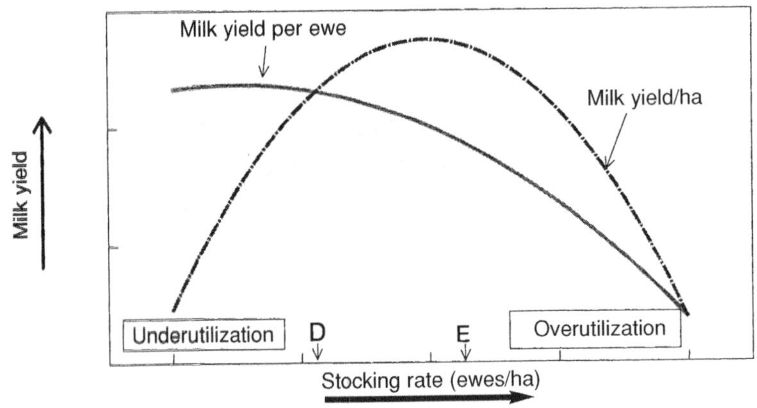

Fig. 11.8. The relationship between stocking rates and milk production (Van Soest, 1983).

example, when dairy sheep are continuously stocked at high stocking rates there is a significant reduction in the protein content of the milk (Molle *et al.*, 1997).

To establish the pasture carrying capacity for a given farm or region, a point of equilibrium must be found between the stocking rate which gives the highest production per individual animal (D) and that which gives the highest production per hectare (E), while also taking into account the quality of the product.

Within this range, one criterion for determining the correct stocking rate is the amount of conserved forage needed to meet flock requirements during the period of insufficient herbage growth. This forage can be harvested during the period of maximum production of the pasture and fed to the animals during periods when pasture production is low. The exclusive use of grazing for feeding sheep is normally limited to certain periods of the year when the biomass availability of pasture is high, and thus able to satisfy the nutritive requirements of the animals.

Figure 11.9 shows a model for establishing the stocking rate in relation to pas-

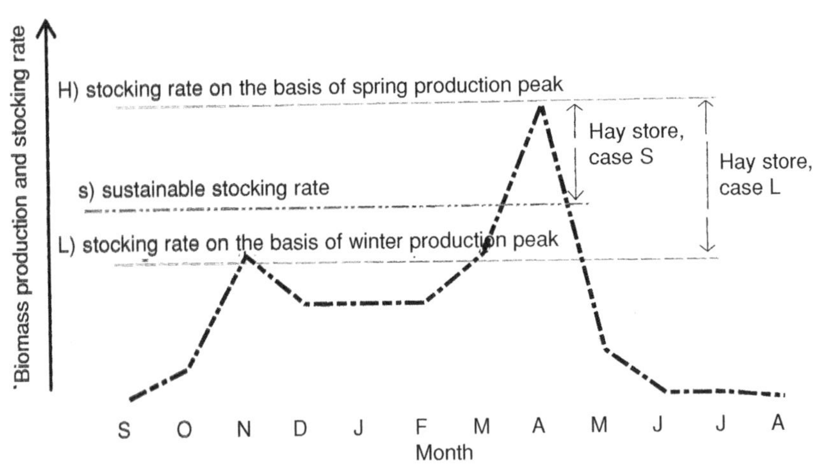

Fig. 11.9. Relationship between pasture production, stocking rates and supplementary forage production.

ture production and the need for supplementary forage. Two examples are shown, an extremely low (case L) and an extremely high (case H) stocking rate, which correspond, respectively, to the winter and spring peaks of herbage production. In case L, supplementary forage production would be excessive with the risk of underutilization of grazed pasture in spring; in case H no supplementary forage would be produced. The intermediate case (S) is an example of a sustainable stocking rate that allows the production of sufficient supplementary forage to meet the winter feed requirements of the flock.

These general guidelines are mainly focused on the feed budget and are not by themselves enough to make an accurate evaluation of the stocking rate that is sustainable in the long-term on a specific farm. Indicators of the environmental impact of grazing, in particular the persistence of the pasture, must be taken into account in selecting a stocking rate that will be sustainable in the longer term. Animal welfare and the consistency of animal performance from year to year, all evaluated over a period of years, must also be taken into consideration.

Because a farm is principally a business, an economic evaluation is indispensable when finalizing the analysis and implementing it. Recent papers (Ligios *et al*., 1998; Pulina and Zucca, 1999; Pulina *et al*., 1999) have discussed the economic and environmental criteria for establishing sustainable stocking rates.

In the following section we examine three possible scenarios, with particular reference to the feed budget.

Continuous stocking without supplementation

In this case, the stocking density and the average seasonal stocking rate can be based on the growth curves of the constituents of the pasture. More precisely the average seasonal stocking rate (SR) can be calculated from the following equation:

$$SR = \frac{GR}{PI} \times CU$$

where: SR is the stocking rate (number of sheep/ha); GR is the growth rate of the pasture per unit area (kg DM/ha/day). For example, in Australia under Mediterranean climate conditions, pasture growth rates range from 5–20 kg DM/ha/day in winter to 40–120 kg DM/ha/day in spring, depending on the forage species and on the weather pattern (CSIRO, 1993); PI is the potential herbage intake (kg DM/day); and CU is the coefficient of utilization of the pasture. The latter is conventionally set at around 0.65 (CSIRO, 1993), which appears reasonable, taking in to account losses due to trampling and dung deposition.

Rotational grazing without supplementation

In this case, a correct assessment of stocking rate for the period involves an estimation of the biomass on offer at the beginning of each grazing period and of the growth rate of the pasture. The equation for this estimation is as follows:

$$SR = \frac{\left(\frac{HM}{D} + GR\right)}{PI} \times CU$$

where: SR is the stocking rate (no. of sheep/ha); HM is the herbage mass on offer at the beginning of each grazing period (kg DM/ha); D is the duration of each grazing period (days); GR is the growth rate per unit area (kg DM/ha/day); PI is the potential herbage intake (kg DM/day); and CU is the coefficient of pasture utilization.

Available biomass in a paddock at the start of grazing can be estimated from measurements of the pasture height (Tables 11.3 and 11.4). Table 11.9 shows the ranges of the herbage mass in relation to pasture height, measured with a weighted square grassmeter, for some species of autumn–winter forage crops rotationally grazed by dairy sheep in Mediterranean environments (Sitzia *et al*., 1998, 2000). However, there is a shortage of data on the growth rates and coefficients of utilization of pasture rotationally grazed in Mediterranean environments.

Table 11.9. Pasture height (cm) and corresponding quantities of available biomass (t DM/ha).

Pasture height (cm)	Annual ryegrass[a]	Italian ryegrass[b]	Burr medic[c]	Sulla[d]
5–10	0.8–1.4	0.5–1.0	0.8–1.6	1.3–1.8
11–20	1.6–2.8	1.1–2.0	1.8–3.2	1.9–2.7
21–30	2.9–4.1	2.1–3.0	3.4–4.8	2.8–3.7

[a] *Lolium rigidum*; [b] *Lolium multiflorum*; [c] *Medicago polymorpha*; [d] *Hedysarum coronarium*.

Grazing with supplements

When using supplements, the potential herbage intake (PI) in the equations above must be corrected for the effect of substitution, which is the reduction in the voluntary intake of herbage (g) resulting from the intake of 100g of the supplementary feed expressed as a proportion. For concentrates, the reduction (Q) is given by the equation:

$$Q = CI \times S$$

where: CI is the intake of concentrates (kg DM/head/day); and S is the substitution rate of the concentrates.

When the pasture is in a vegetative phase, before the start of the reproductive phase, S is 0.4–0.6 where intake is limited by herbage accessibility, and 0.80–1.20 where pasture conditions do not limit intake (D'Urso *et al.*, 1993; Avondo *et al.*, 1995; Molle *et al.*, 1997). In general, substitution rates are greater when herbage quality is high. For example, the substitution rate is zero, or even negative, when supplements with a high protein content are used on pastures of very mature herbage, provided herbage mass is not restricted.

Where conserved forage (hay or silage) is offered as a supplement, it is assumed that the substitution rate is 1, and that the reduction in intake of grazed herbage is equal to the intake of hay or silage.

References

Avondo M., Foti F., Bognanno M., Scerra V. (1994) Effetti del sistema di pascolamento e del livello di integrazione alimentare sull'ingestione di erba di pecore in lattazione e sulle caratteristiche quali-quantitative del pascolo. *Atti XI Congresso Nazionale S.I.P.A.O.C., Perugia (Italy), 1–4 June*, pp. 409–412.

Avondo M., Licitra G., Bognanno M., Kestkaran A.N., Marletta D., D'Urso G. (1995) Effects of the type and level of supplementation on grazing behaviour of lactating ewes in a Mediterranean natural pasture. *Livestock Production Science* 44: 237–244.

Avondo M., Marletta D., Bordonaro S., Guastella A.M., D'Urso G. (1996) Sheep grazing behaviour in Mediterranean semi-extensive systems. In: M.G. Keane and E.G. O'Riordan (eds) *Pasture ecology and animal intake*, pp. 191–196.

Bailey D.W., Gross J.E., Laca E.A., Rittenhouse L.R., Coughenor M.B., Swift D.M., Sims P.L. (1996) Mechanisms that result in large herbivore grazing distribution pattern. *J. Range Manage.*, 49: 387–397.

Barry T.N., McNabb W.C. (1999) The implications of condensed tannins on the nutritive value of temperate forages fed to ruminants. *British Journal of Nutrition*, 81: 263–272.

Barthram G.T. (1985) Experimental techniques: the HFRO sward-stick. *The Hill Farming Research Organisation. Biennal Report 1984–1985*, pp. 29–30.

Baumont R., Prache S., Meuret M., Morand-Fehr P. (2000) How forage characteristics influence behaviour and intake in small ruminants: a review. *Livestock Production Science* 64, 15–28.

Bocquier F., Guillouet Ph., Barillet F., Ligios S., Molle G., Sanna A., Casu S., Caja G., Such X., Gasa J., Ferret A., Oregui L., Urarte E., Agabriel J., Chapciaux P., Espinasse C. (1993) A computer program for diet formulation in dairy sheep: evaluation of food intake predictions. *Proc. of the 5th International Symposium on Machine Milking of Small Ruminants*, Budapest, Hungary, pp. 608–621.

Carta W. (1993) Prime osservazioni sull'uso di metodi non distruttivi per la stima della produzione dei pascoli mediterranei. Laurea thesis, Ist.di Agronomia e Coltivazioni erbacee, Università di Sassari, Italy.

Casu S., Ligios. S., Molle G., Fois N., Decandia M., Sitzia M., Roggero P.P. (1995) Influence de la hauteur de l'herbe offerte sur les caractéristiques de la prairie et les performances de brebis laitières avec ou sans complémentation. *Seminaire International: Nutrition et Alimentation des Brebis Laitières.* Saint-Affrique, Aveyron, France.

Ciavarella T.A., Dove H., Leury B.J., Simpson R.J. (2000) Diet selection by sheep grazing *Phalaris aquatica* L. pastures of differing water-soluble carbohydrate content. *Aust. J. Agric. Res.*, 51: 757–764.

Commonwealth Scientific and Industrial Research Organisation (CSIRO) (1993) Grazfeed: a nutritional management system for grazing animals. Horizon Technology PTY Limited, Roseville, NSW, Australia, p. 157

Cooper S.D.B., Kryazakis I., Oldham J.D. (1996) The effects of physical forms of feed, carbohydrate source, and inclusion of sodium bicarbonate on diet selections of sheep. *J. Anim. Science*, 74: 1240–1251.

Decandia M., Sitzia M., Cabiddu A., Kabaya D., Molle G. (2000) The use of polyethylene glycol to reduce the anti-nutritional effects of tannins in goats fed woody species. *Small Ruminant Research*, 38: 157–164.

Delagarde R., Prache S., D'Hour P., Petit M. (2001) Ingestion de l'herbe par les ruminants au pâturage. *Fourrages* 166: 189–212.

D'Urso G., Avondo M., Biondi L. (1993) Effect of supplementary feeding on grazing behaviour of Comisana ewes in a Mediterranean semi-extensive production system. *Animal Feed Science and Technology*, 42: 259–272.

Filigheddu S., Pulina G. (1986) Stima della produzione quantitativa e qualitativa dei prati-pascoli avvicendati della Gallura in base alla rilevezione di alcuni parametri tecnici. In: *Foraggicoltura e Zootecnia Gallurese in regime asciutto*, ERSAT Publishing, pp. 129–143.

Fois N., Molle G., Ligios S., Casu S. (1995) Effetto del pascolo razionato sull'ingestione e le prestazioni produttive di pecore alimentate con diete unifeed. *Proc. XLIX Convegno SISVET* , pp. 893–894.

Fois N., Molle G., Sitzia M., Pani R., Moniello G., Pinna W. (1997) Alimentazione con unifeed ed unifeed + pascolo in pecore di razza Sarda: aspetti quanti–qualitativi della produzione lattea. *Proc. XII Congresso Nazionale ASPA*, pp. 255–256.

Fois N., Ligios S., Sitzia M. (2000a) Organic and conventional dairy sheep farming systems in the Mediterranean environment. First year results. In: K. Søegaard, C. Ohlsson, J. Sehested, N.J. Hutchings, T. Kristensen (eds) *Proc. of EGF Grassland Science in Europe.* Vol. 5, pp. 517–519.

Fois N., Ligios S., Sitzia M. (2000b) Preliminary results of an organic dairy sheep farming system in Sardinian lowland. *Proc. of the 5th International Livestock Farming Systems' Symposium*, EAAP publication No. 97, pp. 120–122.

Fois N., Ligios S., Sitzia M. (2000c) Stubble management of *Medicago polymorpha* L. and pod consumption by grazing ewes during summer. *Options Méditerranéennes*, 45: 365–368.

Gibb M., Orr R.J. (1997) Grazing behaviour of ruminants. *IGER Innovations*, pp. 54–57.

Hodgson J. (1982) Ingestive behaviour. In: J.D. Lever (ed.) *Herbage intake handbook*, British Grassland Society, pp. 113–138.

Hodgson J. (1990) *Grazing management, Science into Practice.* Longman, Harlow, UK, p. 203.

Holmes C.W. (1984) The Massey grass meter. *Dairy farming annual.* Massey University, pp. 26–30.

Landau S., Silanikove N., Nitsan Z., Barkai D., Baram H., Provenza P.D., Prevolotsky A. (2000) Short-term changes in eating patterns explain the effects of condensed tannins on feed intake in heifers. *Applied Animal Behaviour Science*, 69: 199–213.

Ligios S., Sulas L., Molle G., Fois N. (1997) Utilizzazione e gestione di colture foraggere in sistemi asciutti per ovini da latte. *Rivista di Agronomia*, 31 (suppl.): 326–331.

Ligios S., Natale M.A., Oppia P., Bucarelli G., Casu S. (1998) Influence de quelques facteurs de production sur les resultats technico-économiques des exploitations ovins lait en Sardaigne. *Atti della riunione del Reseau Ovins et Caprins, sub-network Sistemi di produzione*, Olmedo, Italy.

Meuret M. (1997) Préhensibilité des aliments chez les petits ruminants sur parcours en landes et sous-bois. *INRA Prod. Anim.*, 10 (5): 391–401.

Molle G., Ligios S., Fois N., Decandia M., Casu S., Bomboi G. (1997) Responses by dairy ewes to different pasture heights under continuous stocking either unsupplemented or supplemented with corn grain. *Options Méditerranéennes, Serie A*, 34: 65–70.

Molle G., Sitzia M., Decandia M., Fois N., Ligios S. (1998) Feeding value of Mediterranean forages assessed by the n-alkane method in grazing dairy ewes. In: G. Nagy and K. Petö (eds) *Proc. of 17th General Meeting of EGF*, Debrecen, Hungary, pp. 365–368.

Molle G., Sitzia M., Decandia M., Fois N., Scanu G., Ligios S. (2000) Intake and performance of dairy ewes grazing Mediterranean forages either as pure or mixed pastures. *Options Méditerranéennes*, 52: 187–192.

Molle G., Decandia M., Fois N., Ligios S., Cabiddu A., Sitzia M. (2003) The performance of Mediterranean dairy sheep given access to sulla (*Hedysarum coronarium* L.) and annual ryegrass (*Lolium rigidum* Gaudin) pastures in different time proportions. *Small Ruminant Research*, 49: 319–328.

Molle G., Decandia M., Sitzia M., Cabiddu A., Fois N., Ligios S., Giovanetti V., Rutter S.M. (2004) Foraging behaviour of sheep rotationally grazing annual ryegrass (*Lolium rigidum* Gaudin). In: A Lüscher, B. Jeangros, W. Kessler *et al.* (eds) *Proceedings of the 20th General Meeting of EGF, Grassland Science in Europe*, vol. 9, 575–577.

Newman J.A., Parsons A.J., Harvey A. (1992) Not all sheep prefer clover: diet selection revisited. *J. Agric. Sci. Camb.*, 119: 275–283.

Parsons, A. J., Chapman, D.F. (2000) The principles of pasture growth and utilization. In: A. Hopkins (ed.) *Grass: its production and utilization.* 3rd edition. Blackwell Science Ltd., Oxford, UK, pp. 31–89.

Penning P.D. (1986) Some effects of pasture conditions on grazing behaviour and intake by sheep. In: O. Gudmundsson (ed.) *Grazing research at the northern latitude*, Plenum Publishing Corporation, pp. 219–226.

Penning P.D., Parsons A.J., Orr R.J., Treacher T.T. (1991) Intake and behaviour responses by sheep to changes in sward characteristics under continuous grazing. *Grass and Forage Science*, 46: 15–28.

Penning P.D., Parsons A.J., Orr R.J., Hooper G.E. (1994) Intake and behaviour responses by sheep to changes in sward characteristics under rotational grazing. *Grass and Forage Science*, 49: 476–486.

Penning P.D., Parsons A.J., Orr R.J., Harvey A., Champion R.A. (1995) Intake and behaviour responses by sheep, in different physiological states, when grazing monocultures of grass and white clover. *Applied Animal Behaviour Science*, 45: 63–78.

Penning P.D., Newman J.A., Parsons A.J., Harvey A., Orr R.J. (1997) Diet preferences of adult sheep and goats grazing ryegrass and white clover. *Small Ruminant Research*, 24:175–184.

Perojo A., Garcia-Rodriguez A., Arranz J., Oregui L.M. (2003) Effect of time spent on pasture on milk yield, body reserves, herbage intake and grazing behaviour. *Proceedings of the First Joint Seminar of the FAO-CIHEAM Sheep and Goat Nutrition and Mountain and Mediterranean Pastures Sub-Networks*, Granada, Spain, Abstracts, p. 65.

Porqueddu C., Fois N., Sitzia M., Sulas L. (1997) Forage production and persistence of a sowed self-reseeding pasture and related animal performances under rotational and continuous grazing. *Improving forage crops for semi-arid areas workshop.* Mallorca, Spain.

Prache S., Peyraud J.-L. (1997) Préhensibilité de l'herbe pâturée chez les bovins et les ovins. *INRA Prod. Anim.*, 19 (5): 377–390.

Prache S., Gordon I.J., Rook A.J. (1998) Foraging behaviour and diet selection in domestic herbivores. *Ann. Zootech.*, 47: 335–345.

Pulina G., Zucca C. (1999) Un nuovo indicatore territoriale per la valutazione dell'impatto ambientale del pascolamento. *L'Informatore Agrario*, 42: 105–109.

Pulina G., Salimei E., Masala G., Sikosana J.L.N. (1999) A spreadsheet model for the assessment of sustainable stocking rates in semi-arid and sub-humid regions of South Africa.

Roggero P.P., Sitzia M., Molle G., Fois N. (1998) Spring grazing management of Italian ryegrass with dairy sheep in a rainfed Mediterranean environment. In: W.P. Papanastasis and D. Peter (eds) *Proc. of the International Workshop*, 'Ecological basis of livestock grazing in Mediterranean ecosystems' Thessaloniki, Greece, 268–271.

Roguet C., Dumont B., Prache S. (1998) Selection and use of feeding sites and feeding stations by herbivores: A review. *Ann. Zootech.*, 47: 225–244.

Silanikove N., Nitsan Z., Perevolotsky A. (1994) Effect of daily supplementation of polyethylene glycol on intake and digestion of tannin-containing leaves (*Ceratonia siliqua*) by sheep. *Journal of Agricultural and Food Chemistry*, 42: 2844–2847.

Silanikove N., Gilboa N., Perevolotsky A. (1997) Effect of foliage-tannins on feeding behaviour in goats. *Options Méditerranéennes, Serie A*, 34: 43–46.

Sitzia M., Fois N. (1999) Semi e legumi di *Medicago polymorpha* L. come risorsa alimentare estiva per gli ovini in ambiente mediterraneo. *Rivista di Agronomia*, 33:185–188.

Sitzia M., Sulas L., Porqueddu C., Fois N. (1996) Dairy sheep farming system in Sardinia: comparison between rotational grazing and continuous stocking. In: G. Parente, J. Frame and S. Orsi (eds) *Proc. of EGF Grassland Science in Europe* Vol. 1, 611–614.

Sitzia M., Roggero P.P., Fois N., Molle G. (1997) Grazing management of an Italian ryegrass sward with dairy sheep in the Mediterranean environment. *Proc. of XVIII International Grassland Congress (Canada)*, 29: 165–166.

Sitzia M., Fois N., Ligios S., Molle G., Decandia M. (1998) Grazing as a complement to complete diet for dairy ewes. In: G. Nagy and K. Petö (eds) *Proc. of EGF Grassland Science in Europe*, Vol. 3, 191–194.

Sitzia M., Ligios S., Fois N. (2000) *Medicago polymorpha* L. forage production and its quality when grazed by ewes. *Options Méditerranéennes*, 45: 191–194.

Treacher, T.T. (1990) Grazing management and supplementation for the lowland sheep flock. In: C.F.R. Slade and T.L.J. Lawrence (eds) *New developments in sheep production.* Occasional Publication No.14. British Society of Animal Production, pp. 45–54.

Vallentine, J.F. (1990) *Grazing management.* Academic Press, San Diego, California, p. 659.

Van Soest P.J. (1983) *Nutritional ecology of ruminants.* O & B Books Inc., Corvallis, Oregon, p. 374.

Index

Printed and bound by CPI Group (UK) Ltd, Croydon, CR0 4YY
16/09/2026
14963436-0001